# Advanced Mathematics

*Precalculus with Discrete Mathematics and Data Analysis*

# Student Resource Guide

## for Study and Review

**McDougal Littell/Houghton Mifflin**
Evanston, Illinois
Boston    Dallas    Phoenix

# About the *Student Resource Guide for Study and Review*

The *Student Resource Guide for Study and Review* is a set of lessons designed to help students study and review effectively. This guide is intended to be used in conjunction with *Advanced Mathematics, Precalculus with Discrete Mathematics and Data Analysis*, © 1994 or © 1992.

Each lesson in the *Student Resource Guide* covers a group of related sections in *Advanced Mathematics*. The components of the lessons are described below.

- An **Overview** gives a brief summary of the sections.
- An illustrated list of **Key Terms** provides vocabulary review.
- **Understanding the Main Ideas** summarizes, illustrates, and connects concepts and results presented in the sections.
- **Checking the Main Ideas** uses basic exercises to let students confirm their understanding of important facts and ideas.
- **Using the Main Ideas** presents detailed worked-out **Examples** with **cautions** about common errors.
- **Exercises**, mainly at "A" level, let students practice and apply material covered in the sections.

A two-page **Chapter Review** is provided for each chapter in *Advanced Mathematics*. Each Chapter Review includes the sections described below.

- A **Quick Check** allows students to identify topics to review from the current chapter.
- A **Practice Test** helps students prepare for a chapter test.
- A **Mixed Review** offers ongoing cumulative review of all chapters.

Answers are provided in a separate **Answer Section**.

A **Spanish Glossary** gives Spanish equivalents of important terms used in *Advanced Mathematics*. A definition of each term is given in Spanish.

## Acknowledgments

The author wishes to thank **Lillian Preston Seguin**, Mathematics Specialist, Grades 6–12, Brownsville Independent School District, Brownsville, Texas, for her contribution to the *Student Resource Guide*.

## Credits
Cover: Russell Brough
Illustrations: Cameron Gerlach

ISBN: 0-395-42170-5

56789–BW–00 99 98

# Table of Contents

# Linear Functions

## Sections 1-1, 1-2, 1-3, and 1-4

**OVERVIEW**   Lessons 1-1, 1-2, and 1-3 review the essentials of coordinate geometry: distance, midpoint, and slope formulas, and equations and graphs of lines. In Section 1-4, these concepts are applied to mathematical models that are linear functions.

## KEY TERMS

| KEY TERMS | EXAMPLE/ILLUSTRATION |
|---|---|
| **Linear equation** (p. 1) <br> an equation that can be written in the *general form* $Ax + By = C$, where $A$ and $B$ are not both 0 | $y = \frac{3}{2}x - 3$ <br> $y + 7 = -3(x + 5)$ |
| **Solution of an equation** (p. 1) <br> an ordered pair of numbers that makes a linear equation true | $(4, 3)$ is a solution of $y = \frac{3}{2}x - 3$, because $3 = \frac{3}{2}(4) - 3$. |
| **Graph of an equation** (p. 1) <br> all the points in the coordinate plane corresponding to solutions of an equation | 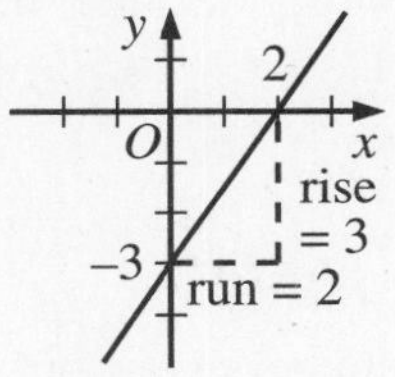 Graph of $y = \frac{3}{2}x - 3$ |
| **x-intercept** (p. 1) <br> the $x$-coordinate of the point on a graph with $y$-coordinate 0 | The graph above has $x$-intercept 2. |
| **y-intercept** (p. 1) <br> the $y$-coordinate of the point with $x$-coordinate 0 | The graph above has $y$-intercept $-3$. |
| **Slope of a nonvertical line** (p. 7) <br> the ratio of rise to run | The slope of the line above is $\frac{3}{2}$. |
| **Linear function** (p. 19) <br> a relationship of the form $f(x) = mx + k$ | $C(x) = 0.25x + 30$ |
| **Zero of a function** (p. 19) <br> a number that when substituted for the variable gives a function value of 0 | $C(-120) = 0.25(-120) + 30 = 0$, so $-120$ is a zero of $C$. |
| **Mathematical model** (p. 20) <br> one or more functions, graphs, tables, equations, or inequalities that describe a real-world situation | 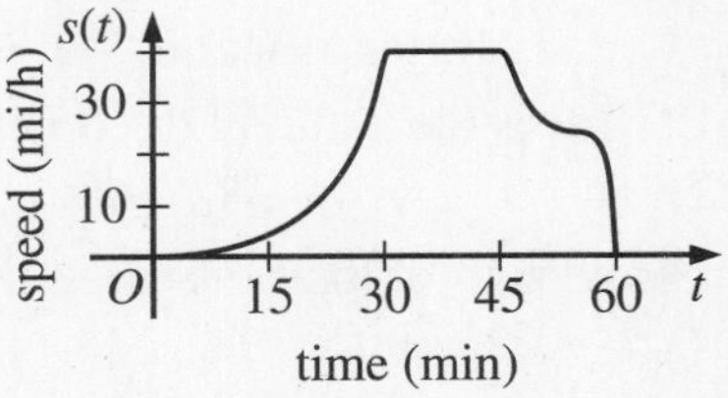 |

# UNDERSTANDING THE MAIN IDEAS

## Coordinate geometry

For any two points $P(a, b)$ and $Q(c, d)$:

For $P(-3, 0)$ and $Q(4, -1)$:

- the length of $\overline{PQ}$ is $PQ = \sqrt{(a - c)^2 + (b - d)^2}$. $\longrightarrow$ $PQ = \sqrt{(-3 - 4)^2 + (0 - (-1))^2}$
$$= \sqrt{50} = 5\sqrt{2}$$

- the midpoint $M$ of $\overline{PQ}$ is $\left(\dfrac{a + c}{2}, \dfrac{b + d}{2}\right)$. $\longrightarrow$ $M = \left(\dfrac{-3 + 4}{2}, \dfrac{0 + (-1)}{2}\right)$
$$= \left(\dfrac{1}{2}, -\dfrac{1}{2}\right)$$

- if $a \neq c$, the slope of line $PQ$ is a constant, equal to $\dfrac{b - d}{a - c}$. $\longrightarrow$ slope $= \dfrac{0 - (-1)}{-3 - 4} = -\dfrac{1}{7}$

- If $a = c$, then line $PQ$ is a vertical line and has no slope. $\longrightarrow$ The line through $(-2, 4)$ and $(-2, -1)$ has no slope.

## Linear equations

- The graph of a linear equation in $x$ and $y$ is a line. To graph a linear equation, find the coordinates of several points on the line, plot the points, and draw the line through them.

$3x - 2y = -6$

| $x$ | $y$ |
|-----|-----|
| 0 | 3 |
| $-2$ | 0 |
| 2 | 6 |

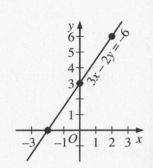

- Forms of a linear equation

*general form:* $Ax + By = C$, $A$ and $B$ not both zero $\longrightarrow$ $3x - 2y = -6$
$3x + 6 = 2y$

*slope-intercept form:* $y = mx + k$ where $m$ is the slope and $k$ is the $y$-intercept $\longrightarrow$ $\dfrac{3}{2}x + 3 = y$

To find the slope-intercept form, solve for $y$. slope $\longleftarrow$ $y$-intercept $\longrightarrow$

*point-slope form:* $\dfrac{y - b}{x - a} = m$ where $m$ is the slope and $(a, b)$ is on the line $\longrightarrow$ $\dfrac{y - 6}{x - 2} = \dfrac{3}{2}$

*intercept form:* $\dfrac{x}{a} + \dfrac{y}{b} = 1$ where the $x$-intercept is $a$ and the $y$-intercept is $b$ $\longrightarrow$ $\dfrac{x}{-2} + \dfrac{y}{3} = 1$

- Two different nonvertical lines are parallel if and only if their slopes are equal. $\longrightarrow$ The graphs of $y = -5x$ and $y = -5x + 4$ are parallel lines.

- Two nonvertical lines are perpendicular if and only if the product of their slopes is $-1$. $\longrightarrow$ The graphs of $y = 5x$ and $y = -\dfrac{1}{5}x - 5$ are perpendicular lines, since $5\left(-\dfrac{1}{5}\right) = -1$.

## Linear functions

For any linear function $f(x) = mx + k$: $\longrightarrow$ $f(x) = 3x - 12$

- $f(a)$ is the value of the function when $a$ is substituted for $x$. $\longrightarrow$ $f(-1) = 3(-1) - 12 = -15$
- if $f(a) = b$, then $(a, b)$ is on the graph of $f$. $\longrightarrow$ $(-1, -15)$ is on the graph of $f$.
- the zero of the function is the $x$-intercept of the graph; $k$ is the $y$-intercept. $\longrightarrow$ Since $f(4) = 3(4) - 12 = 0$, the $x$-intercept is 4; the $y$-intercept is $-12$.

## CHECKING THE MAIN IDEAS

1. Find the length, the coordinates of the midpoint, and the slope of the line segment with endpoints $A(4, 6)$ and $B(-4, 0)$.

2. Sketch the graph of $2x + 5y = -20$. Label three points on the line.

3. Find the slope and intercepts of the line with equation (a) $2x - 3y = 9$ and (b) $y = 5$.

4. Classify each pair of lines as parallel, perpendicular, or neither. (See Example 2 on text page 10.)

   **a.** $y = -3x + 3, 3x - y = 6$    **b.** $y = -3x + 3, x - 3y = 6$

5. Find an equation of the line described. Choose the most convenient form.

   **a.** The line with no slope and $x$-intercept $-4$

   **b.** The line through $(3, -1)$ and $(-1, 1)$ (See Example 2 on text page 14.)

   **c.** The perpendicular bisector of the segment joining $(0, -6)$ and $(4, 4)$ (See Example 4 on page 15.)

6. Let $f(x) = \frac{1}{2}x - 4$. Find $f(6)$ and the zero of $f$.

7. *Critical Thinking* Describe the graph of $Ax + By = C$ when (a) $A = 0$, (b) $B = 0$, and (c) $C = 0$.

## USING THE MAIN IDEAS

**Example 1** Solve the equations $4x - y = -2$ and $5x + 4y = -20$ simultaneously.

**Solution** Make a quick sketch to estimate the number of solutions and the approximate coordinates of an intersection point. The graph shows that the intersection point is located near $(-1, -3)$.

 **Caution:** If the graphs are parallel lines, there is no common solution. If the graph of both equations is the same line, every point on the line is a common solution.

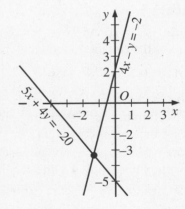

**Method 1: Substitution**

Solve the first equation for $y$.
$$4x - y = -2 \rightarrow 4x + 2 = y$$
Substitute in the second equation.
$$5x + 4y = -20 \rightarrow 5x + 4(4x + 2) = -20$$
Solve for $x$.     $21x + 8 = -20$
$$x = -\frac{4}{3}$$
Solve for $y$.     $y = 4\left(-\frac{4}{3}\right) + 2 = -\frac{10}{3}$

The solution is $\left(-\frac{4}{3}, -\frac{10}{3}\right)$.

**Method 2: Multiplication with Addition or Subtraction**

Multiply the first equation by 4 so the coefficient of $y$ are opposites.

$$4(4x - y) = 4(-2) \rightarrow 16x - 4y = -8$$

$$5x + 4y = -20 \qquad \underline{5x + 4y = -20}$$

$$21x \qquad\quad = -28$$

$$x = -\frac{4}{3}$$

Substitute $-\frac{4}{3}$ for $x$. $\qquad 4\left(-\frac{4}{3}\right) - y = -2 \longrightarrow y = -\frac{10}{3}$

The solution is $\left(-\frac{4}{3}, -\frac{10}{3}\right)$. Notice that $\left(-\frac{4}{3}, -\frac{10}{3}\right)$ or $\left(-1\frac{1}{3}, -3\frac{1}{3}\right)$ agrees with the estimated solution $(-1, -3)$.

**Example 2** Let $f$ be a linear function such that $f(4) = 360$ and $f(6) = 720$.

   **a.** Find an equation for $f(x)$.

   **b.** Give a geometric interpretation of the function $f$. Describe the graph of this geometric interpretation of $f$.

**Solution**   **a.** Let $f(x) = mx + k$.

     $720 = m \cdot 6 + k \qquad \leftarrow$ Since $f(6) = 720$

     $360 = m \cdot 4 + k \qquad \leftarrow$ Since $f(4) = 360$

     $360 = 2m \qquad\qquad\quad \leftarrow$ Subtract the second equation from the first.

     $180 = m$

     $360 = 180 \cdot 4 + k \quad \leftarrow$ Substitute 180 for $m$.

     $-360 = k$

     Thus, $f(x) = 180x - 360$.

   **b.** If $x$ represents an integer greater than 2, then $f(x)$ is the sum of the measures of the angles of a polygon with $x$ sides. For this interpretation, the graph is the set of points with coordinates $(3, 180)$, $(4, 360)$, $(5, 540)$, and so on.

## Exercises

**8.** Solve $x - y = 1$ and $9x + y = 6$ simultaneously. Then sketch the graphs of the equations and label the intersection point.

**9.** Given $A(-6, -1)$, $B(0, 1)$, $C(5, -2)$, and $D(-4, -5)$, show that quadrilateral $ABCD$ is a trapezoid.

**10.** Refer to the points in Exercise 9. Find an equation of the line through $B$ which is parallel to the line through $A$ and $D$.

**11.** *Writing* Write a paragraph in which you discuss at least four different ways that a line can be specified uniquely, that is, exactly one line is determined.

**12.** *Application* A group of climbers leaves the 11,200 ft peak of Mt. Massasoit at 2 P.M., descending at 1500 ft/h. The base of the mountain is at an elevation of 4300 ft.

   **a.** Express the elevation, $E$, of the climbers as a function of the time, $t$, in hours after 2 P.M.

   **b.** At what time should the climbers arrive at the base of the mountain?

## Sections 1-5, 1-6, 1-7, and 1-8

| OVERVIEW | Section 1-5 introduces techniques for simplifying complex expressions. Section 1-6 reviews various methods for solving quadratic equations. Sections 1-7 and 1-8 explore quadratic functions and models algebraically and graphically. |

### KEY TERMS

| KEY TERMS | EXAMPLE/ILLUSTRATION |
|---|---|
| **Imaginary unit $i$** (p. 26)<br>$i = \sqrt{-1}$ and $i^2 = -1$ | $\sqrt{-2} = i\sqrt{2}$ |
| **Complex number** (p. 27)<br>any number of the form $a + bi$, where $a$ and $b$ are real numbers and $i$ is the imaginary unit ($a$ is the *real part* and $b$ is the *imaginary part* of $a + bi$.) | $\underbrace{-\sqrt{5}}$<br>real part, imaginary part 0<br><br>$\underbrace{0.25i}$<br>imaginary part, real part 0<br><br>$\underbrace{\frac{1}{3}} - \underbrace{8i}$<br>real part   imaginary part |
| **Imaginary number** (p. 27)<br>a complex number $a + bi$ with $b \neq 0$ | $2i\sqrt{2}, 9 + 4i$ |
| **Pure imaginary number** (p. 27)<br>a complex number with $a = 0$ and $b \neq 0$ | $-i\sqrt{5}, 2.6i$ |
| **Complex conjugates** (p. 27)<br>two complex numbers $z = a + bi$ and $\bar{z} = a - bi$ | $1 + i\sqrt{2}, 1 - i\sqrt{2}$ |
| **Quadratic equation** (p. 30)<br>any equation that can be written in the *standard form* $ax^2 + bx + c = 0$, where $a \neq 0$ (A *root* or *solution* of a quadratic equation is a value of the variable that satisfies the equation.) | $x^2 + x - 2 = 0$ has roots 1 and $-2$ since $1^2 + 1 - 2 = 0$ and $(-2)^2 + (-2) - 2 = 0$. |
| **Completing the square** (p. 30)<br>the method of transforming a quadratic equation so that one side is a perfect square trinomial | $x^2 + 4x + 1 = 0$<br>$x^2 + 4x = -1$<br>$x^2 + 4x + \left(\frac{4}{2}\right)^2 =$<br>$-1 + \left(\frac{4}{2}\right)^2$<br>$(x + 2)^2 = 3$ |

| | |
|---|---|
| **Discriminant** (p. 31)<br><br>   the value $b^2 - 4ac$ for a quadratic equation $ax^2 + bx + c = 0$ | $3x^2 + x - 2 = 0$<br>$ax^2 + bx + c = 0$<br>$a = 3, b = 1, c = -2$<br>$b^2 - 4ac = 1^2 - 4(3)(-2)$<br>$\qquad\qquad\quad = 25$ |
| **Quadratic function** (p. 37)<br><br>   a function of the form $f(x) = ax^2 + bx + c, a \neq 0$, whose graph is a parabola | 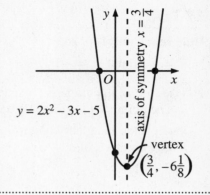 |
| **Quadratic model** (p. 44)<br><br>   a quadratic function that models a real-world situation | If a certain type of carpeting costs \$40/yd$^2$, then the cost of carpeting a square room $y$ yards on a side is $C(y) = 40y^2$. |

## UNDERSTANDING THE MAIN IDEAS

### Complex numbers

- To simplify an expression that contains the square root of a negative number, begin by writing the expression in terms of $i$.

  $$\sqrt{-27} - \sqrt{-3} = i\sqrt{27} - i\sqrt{3}$$
  $$= (3\sqrt{3} - \sqrt{3})i = 2i\sqrt{3}$$

- To add or subtract complex numbers, group the real parts and the imaginary parts.

  $$(5 - 7i) - (-2 + i) = 5 - 7i + 2 - i$$
  $$= 7 - 8i$$

- To multiply two complex numbers use FOIL (first, outer, inner, last) and substitute $-1$ for $i^2$.

  $$(2 + 3i)(5 - i) = 10 - 2i + 15i - 3i^2$$
  $$= 10 + 13i - 3(-1) = 13 + 13i$$

- To simplify an expression with a complex number in the denominator, multiply the numerator and denominator by the complex conjugate of the denominator.

  $$\frac{1}{1 + i\sqrt{3}} = \frac{1}{1 + i\sqrt{3}} \cdot \frac{1 - i\sqrt{3}}{1 - i\sqrt{3}} = \frac{1 - i\sqrt{3}}{1 - 3i^2}$$
  $$= \frac{1 - i\sqrt{3}}{1 - 3(-1)} = \frac{1}{4} - \frac{\sqrt{3}}{4}i$$

### Solving a quadratic equation: $ax^2 + bx + c = 0$

- Use factoring if $a$, $b$, and $c$ are integers and $b^2 - 4ac$ is a perfect square.

  $$2x^2 - 7x - 4 = 0$$
  $$(2x + 1)(x - 4) = 0$$
  $$2x + 1 = 0 \text{ or } x - 4 = 0$$
  $$x = -\frac{1}{2} \text{ or } x = 4$$

- Solve by completing the square if $a = 1$ and $b$ is even.

$$x^2 + 4x + 1 = 0$$
$$(x + 2)^2 = 3 \quad \text{(See the example in Key Terms.)}$$
$$x + 2 = \pm\sqrt{3} \quad \text{(Take the square root of both sides.)}$$
$$x = -2 + \sqrt{3} \text{ or } x = -2 - \sqrt{3}$$

- Use the *quadratic formula* $x = \dfrac{-b \pm \sqrt{b^2 - 4ac}}{2a}$ otherwise.

(See Example 3 on text page 31.)

### Graphing a quadratic function

- $y = ax^2 + bx + c$ (See Example 1 on text page 38.)
  If $a > 0$, the graph is $\cup$-shaped.
  If $a < 0$, the graph is $\cap$-shaped.
  The point $(0, c)$ is on the graph.

$$\left(\frac{-b + \sqrt{b^2 - 4ac}}{2a}, 0\right) \text{ and } \left(\frac{-b - \sqrt{b^2 - 4ac}}{2a}, 0\right) \text{ are on the graph.}$$

The equation of the axis of symmetry is $x = -\dfrac{b}{2a}$.

The vertex has $x$-coordinate $-\dfrac{b}{2a}$.

- $y = a(x - h)^2 + k$
  If $a > 0$, the graph is $\cup$-shaped.
  If $a < 0$, the graph is $\cap$-shaped.
  The vertex is $(h, k)$ and the axis is $x = h$.

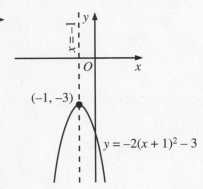

$$y = -2(x + 1)^2 - 3$$

### Quadratic models

If you have a quadratic model $f(x) = ax^2 + bx + c$, you can use the model to predict data values (see Example 2 on text page 45) or to maximize or minimize the function $\left(\text{evaluate } f \text{ when } x = -\dfrac{b}{2a}\right)$.

## CHECKING THE MAIN IDEAS

1. $(a + bi)^2 = \underline{\quad ? \quad}$
   A. $a^2 - b^2$        B. $a^2 + b^2$
   C. $a^2 + 2abi - b^2$        D. $a^2 + 2abi + b^2$

2. Simplify (a) $\dfrac{\sqrt{5}}{\sqrt{-20}}$ and (b) $\dfrac{4}{1 - i} + i$.

3. Solve $x^2 - 2x + 5 = 0$ by using two different methods. Show your work.

4. Sketch the graph of $y = -\dfrac{1}{2}x^2 - 2x + 6$. Label the vertex, axis of symmetry, and intercepts of the parabola.

---

**5.** The table below shows the area $A$ of a sector of
a circle with perimeter 12 and radius $r$.

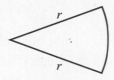

| $r$ | 0 | 2 | 5 |
|-----|---|---|---|
| $A$ | 0 | 8 | 5 |

    **a.** Find a quadratic function that models this data. (See Example 1 on
text page 44.)

    **b.** Find the maximum possible area.

    **c.** *Critical Thinking*   Find the domain and the range of the function.

## USING THE MAIN IDEAS

**Example 1** Solve $\sqrt{x + 3} = 3x - 1$.

**Solution**

$\sqrt{x + 3} = 3x - 1$

$(\sqrt{x + 3})^2 = (3x - 1)^2$     ← Square both sides.

$x + 3 = 9x^2 - 6x + 1$

$0 = 9x^2 - 7x - 2$     ← Write in standard form.

$b^2 - 4ac = (-7)^2 - 4(9)(-2) = 121 = 11^2$

Since $b^2 - 4ac > 0$, the equation has two different
real roots. Since $b^2 - 4ac$ is a perfect square, the
equation can be solved by factoring.

◀▰▰▰▰ **Caution:** If $b^2 - 4ac < 0$, the
roots are complex conjugates; if
$b^2 - 4ac = 0$, there is one real
double root.

$9x^2 - 7x - 2 = 0$

$(9x + 2)(x - 1) = 0$

$9x + 2 = 0 \text{ or } x - 1 = 0$

$x = -\dfrac{2}{9} \text{ or } x = 1$

*Check:*    $\sqrt{x + 3} = 3x - 1$        $\sqrt{x + 3} = 3x - 1$

$\sqrt{-\dfrac{2}{9} + 3} \stackrel{?}{=} 3\left(-\dfrac{2}{9}\right) - 1$    $\sqrt{1 + 3} \stackrel{?}{=} 3(1) - 1$

◀▰▰▰▰ **Caution:** Squaring both
sides of an equation does
not give an equation with
the same solutions as the
original equation. Always
check each possible root.

$\dfrac{5}{3} \neq -\dfrac{5}{3}$            $\sqrt{4} = 2$

$-\dfrac{2}{9}$ is *not* a root.        1 is a root.

The only solution is 1.

**Example 2** Sketch the graphs of the line $4x - y = 9$ and the parabola
$y = 2(x - 1)^2 - 3$. Find the coordinates of any points of
intersection. Use algebra to justify your answer.

**Solution** • Graph the line using the intercepts and one other point.

     • To graph the parabola, begin by identifying the vertex:

$y = 2(x - 1)^2 - 3 \Rightarrow y = 2(x - 1)^2 + (-3)$

$y = a(x - h)^2 + k$

Thus, $h = 1$ and $k = -3$.

    

The vertex is $(1, -3)$ and the axis of symmetry is $x = 1$. Since the value of $a$ is 2, which is positive, the parabola is $\cup$-shaped. Calculate a few points on one side of the $y$-axis and use symmetry to plot the corresponding points on the other side of the $y$-axis. Sketch the parabola by connecting the points. The sketch does not allow you to tell for certain if there are 0, 1, or 2 solutions.

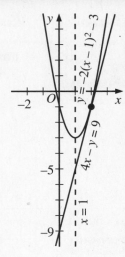

- $4x - y = 9 \rightarrow y = 4x - 9$     $\leftarrow$ Solve the linear equation for $y$.

$$y = 2(x - 1)^2 - 3$$
$$4x - 9 = 2(x^2 - 2x + 1) - 3 \leftarrow \text{Substitute for } y.$$
$$0 = 2x^2 - 8x + 8$$
$$0 = 2(x^2 - 4x + 4)$$
$$0 = (x - 2)^2 \qquad \leftarrow \text{Divide both sides by 2; factor.}$$
$$0 = x - 2 \qquad \leftarrow \text{Take the square root of each side.}$$
$$2 = x$$
$$y = 4x - 9$$
$$\phantom{y} = 4(2) - 9 = -1$$

There is one point of intersection, $(2, -1)$.

## Exercises

**6.** Simplify $\dfrac{1 - 2i}{3 - i}$.

**7.** Solve $\sqrt{6 - 2x} = x - 3$.

**8.** Sketch the graph of $y = (x + 1)^2 + 3$. Label the vertex, axis of symmetry, and intercepts of the parabola.

**9.** *Writing* Explain how the discriminant can be used to show that the line $4x + y = 4$ and the parabola $y = 2x^2 - 5x + 6$ do not intersect.

**10.** *Application* The manager of a new restaurant needs to set the price for the daily dinner special. She has estimated the revenue, $R$, that the restaurant could expect for each of three different prices, $p$.

| $p$ | \$11 | \$14 | \$18 |
|-----|------|------|------|
| $R$ | \$2640 | \$2856 | \$2808 |

  **a.** Explain why the data is or is not reasonable to you.

  **b.** Would a linear or quadratic model fit the data values better? Why?

  **c.** Find a function that models how the revenue, $R$, varies according to the price, $p$, charged.

  **d.** What price will maximize the revenue?

**11.** Is the statement $\dfrac{\sqrt{a}}{\sqrt{b}} = \sqrt{\dfrac{a}{b}}$ *always*, *sometimes*, or *never* true when $a$ and $b$ are real numbers and $b \neq 0$? Explain.

# CHAPTER REVIEW

## Chapter 1: Linear and Quadratic Functions

**Complete these exercises before trying the Practice Test for Chapter 1. If you have difficulty with a particular problem, review the indicated section.**

1. If $A = (-5, 0)$ and $B = (2, 1)$, find the length of $\overline{AB}$ and the coordinates of the midpoint of $\overline{AB}$. *(Section 1-1)*

2. Solve the equations $x + y = 2$ and $3x - y = 2$ simutaneously. Then graph the equations and label the point of intersection with its coordinates. *(Section 1-1)*

3. Find the slope of the line through $(3, -2)$ and $(-3, -2)$. *(Section 1-2)*

4. Tell if the graph of the given equation is *parallel to* the graph of $2x + 3y = 6$, *perpendicular to* the graph of $2x + 3y = 6$, or neither. *(Section 1-2)*

   **a.** $3x - 2y = -6$      **b.** $y = -\frac{2}{3}x$      **c.** $y = -\frac{3}{2}x - 1$

5. Write an equation of the line through $(-3, -8)$ and $(-5, 2)$. *(Section 1-3)*

6. Find an equation of the line perpendicular to the line $4x - y = 8$ and with $y$-intercept $-3$. *(Section 1-3)*

7. The graph of the linear function $f$ has slope 0.5. The zero of $f$ is 4. Find an equation for $f(x)$. *(Section 1-4)*

8. Simplify each expression. *(Section 1-5)*

   **a.** $\sqrt{-6} \cdot \sqrt{-15}$      **b.** $(7 - i) - (-2 + 4i)$      **c.** $\dfrac{1 + i}{1 - i}$

9. Solve. *(Section 1-6)*

   **a.** $3x^2 - 2x + 2 = 0$      **b.** $3x^2 - 7x + 2 = 0$      **c.** $x^2 + 8x = 1$

10. Sketch each parabola. *(Section 1-7)*

    **a.** $y = x^2 - 6x + 8$      **b.** $y = -\frac{1}{2}(x - 2)^2 + 3$

11. Use the values $f(1) = 3, f(2) = 5,$ and $f(3) = 9$ to find an equation of the form $f(x) = ax^2 + bx + c$. *(Section 1-8)*

If $A = (-5, 4)$ and $B = (1, -6)$, find:

1. the length of $\overline{AB}$.      2. the slope of $\overline{AB}$.

3. the coordinates of the midpoint of $\overline{AB}$.

**Write an equation of the line described.**

4. The line through $(4, 7),$ and $(-2, 1)$.

5. The line parallel to the line $4x - 6y = -1$ and with $y$-intercept $-5$

6. Find the coordinates of the point of intersection of the lines $2x + y = -2$ and $4x + 3y = -7$.

**In addition to a basic monthly charge of \$20, a cable TV company charges \$3.76 for each pay-per-view movie.**

7. Express the total monthly charge, $C$, as a function of the number of pay-per-view movies, $m$, watched.

8. Mia's bill for May was \$72.64. How many pay-per-view movies did she watch?

**Simplify.**

9. $\sqrt{-12}\sqrt{-6}$      10. $\dfrac{2-3i}{-3-i}$      11. $i^8 - i^3 + i^6$

12. Solve $2x^2 + 7 = 3x$.

13. Sketch the graph of the parabola $y = -2(x - 3)^2 + 2$. Label the vertex, the axis of symmetry, and the intercepts.

14. Sketch the graphs of $y = x^2 - 4x + 3$ and $x + y = 7$. Find the coordinates of any points of intersection.

**A fireworks rocket is fired vertically. After $t$ seconds, its height above the ground in feet is modeled by the function $h(t) = -16t^2 + 160t$.**

15. How many seconds is the rocket in the air?

16. Find the maximum height reached by the rocket.

---

**MIXED REVIEW**

*Chapter 1*

**Consider the points $A(-2, -5)$ and $B(4, -1)$.**

1. Find the length of $\overline{AB}$.

2. Find an equation of the line through $A$ and $B$.

3. Find an equation of the perpendicular bisector of $\overline{AB}$.

4. Find an equation of the parabola with vertex at $A$ and passing through $B$.

5. Estimate the solution of the system $x + 2y = -3$ and $4x + 5y = -5$ by graphing. Then find the exact solution and compare it to your estimate.

6. Find an equation of the line with $y$-intercept 2.5 and parallel to the line $3x - 2y = -6$.

**Simplify.**

7. $\sqrt{-3} \cdot \sqrt{-27} + \sqrt{-18}$

8. Find the value of $k$ if the line through $(2, k)$ and $(-2, 5)$ is perpendicular to the line $y = \dfrac{1}{3}x - 2$.

9. Sketch the graphs of $y = x^2 + x - 6$ and $x + y = 2$. Label the coordinates of any points of intersection.

10. Let $f(x) = 2x - 7$. Find $f(2.6)$ and the zero of $f$.

Denzel wants to use 100 ft of fencing to fence off a rectangular garden. The table below shows the area in square feet of the garden, $A(x)$, for a garden having one side of the specified length.

| $x$ | 5 | 10 | 15 | 20 |
|------|-----|-----|-----|-----|
| $A(x)$ | 225 | 400 | 525 | 600 |

11. Is a linear model or a quadratic model more appropriate for this relationship? Explain.

12. Find a function that models the information in the table.

13. What garden dimensions result in the maximum possible area?

14. Solve $5x^2 + 3x = 1$.

---

    

# Zeros and Factors of Polynomial Functions

## Sections 2-1 and 2-2

| OVERVIEW | These sections present some techniques for identifying zeros and factors of polynomial functions, and for evaluating and dividing polynomials. |
|---|---|

### KEY TERMS

| KEY TERMS | EXAMPLE/ILLUSTRATION |
|---|---|
| **Polynomial in $x$** (p. 53)<br>an expression that can be written in the form<br>$a_nx^n + a_{n-1}x^{n-1} + \ldots + a_2x^2 + a_1x + a_0$ where $n$ is a nonnegative integer<br>*Terms:* $a_nx^n, a_{n-1}x^{n-1}, \ldots, a_1x, a_0$<br>*Coefficients:* the numbers $a_n, a_{n-1}, \ldots, a_1, a_0$<br>*Degree:* the value of $n$ (the highest power of $x$) | $x^3 + 2x^5 - x^2 + 4 - 6x$<br><br>$x^3, 2x^5, -x^2, 4, -6x$<br>$1, 2, -1, 4, -6$<br>The highest power is 5, so the degree is 5. |
| **Constant** (p. 53)<br>a polynomial of degree 0 | $-0.5$ |
| **Linear term or polynomial** (p. 53)<br>a term or polynomial of degree 1 | Linear term: $-6x$<br>Linear polynomial: $-1 + 7x$ |
| **Quadratic term or polynomial** (p. 53)<br>a term or polynomial of degree 2 | Quadratic term: $-x^2$<br>Quadratic polynomial: $3x - x^2$ |
| **Cubic term or polynomial** (p. 53)<br>a term or polynomial of degree 3 | Cubic term: $x^3$<br>Cubic polynomial: $-x^3 + x - 1$ |
| **Quartic term or polynomial** (p. 53)<br>a term or polynomial of degree 4 | Quartic term: $3x^4$<br>Quartic polynomial: $x^4 - x^2 + 5$ |
| **Zero of a polynomial function** (p. 53)<br>a value of $x$ such that $P(x) = 0$, or a *root* of the equation $P(x) = 0$ | For $P(x) = x^3 - 2x^2 + 4x + 7$,<br>$P(-1) = (-1)^3 - 2(-1)^2 + 4(-1) + 7 = 0$, so $-1$ is a zero of $P$. |
| **Synthetic substitution** (p. 55)<br>a short-cut method for evaluating a polynomial $P(x)$ for any value of $x$ | See Example 3 on text page 55. |
| **Synthetic division** (p. 59)<br>a short-cut method for dividing a polynomial by a divisor of the form $x - a$ | See Example 1 on text page 59. |

## UNDERSTANDING THE MAIN IDEAS

**Evaluating a polynomial** ────────────────→ $P(x) = 3x^2 - 5x + 1$

- Method 1: Substitute the number ──────→ $P(-1) = 3(-1)^2 - 5(-1) + 1 = 9$
  or expression for the variable and simplify. → $P(2x) = 3(2x)^2 - 5(2x) + 1 = 12x^2 - 10x + 1$

- Method 2: Use synthetic substitution. ──────→

$$\begin{array}{r|rrr} -1 & 3 & -5 & 1 \\ \text{Add.} & & -3 & 8 \\ \hline & 3 & -8 & 9 \end{array} = P(-1)$$

- The Factor Theorem states that if $a$ is a ────→ $P(-1) = 9$, so $x - (-1)$, or $x + 1$ is *not* a
  constant, then $x - a$ is a factor of a      factor of $P(x)$.
  polynomial $P(x)$ if and only if $P(a) = 0$.

### Synthetic division

- To divide a polynomial $P(x)$ synthetically by $x - a$ (where $a$ is a
  constant), use the synthetic substitution process for evaluating $P(a)$.

*Long division*

$$\begin{array}{r} 3x - 8 \qquad \leftarrow \text{quotient} \\ x + 1 \overline{)\, 3x^2 - 5x + 1} \quad \leftarrow \text{dividend} \\ \underline{3x^2 + 3x} \qquad\quad \\ -8x + 1 \quad \\ \underline{-8x - 8} \quad \\ 9 \quad \leftarrow \text{remainder} \end{array}$$

$x + 1 = x - (-1) = x - a$; so $a = -1$

*Synthetic division*

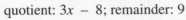

$$\begin{array}{r|rrr} a & \multicolumn{3}{l}{\text{coefficients of dividend}} \\ \downarrow & \multicolumn{3}{l}{\text{(polynomial inside division bar)}} \\ \hline -1 & 3 & -5 & 1 \\ & & -3 & 8 \\ \hline & 3 & -8 & \boxed{9} \quad \leftarrow \text{remainder} \end{array}$$

coefficients of quotient

quotient: $3x - 8$; remainder: 9

Both forms of the division show that $\dfrac{3x^2 - 5x + 1}{x + 1} = 3x - 8 + \dfrac{9}{x + 1}$,

or equivalently, $3x^2 - 5x + 1 = (3x - 8)(x + 1) + 9$.

- The remainder theorem states that when a polynomial $P(x)$ is
  divided by $x - a$, the remainder is the constant $P(a)$. Also, if $P(x)$
  has degree $n$, then the quotient will have degree $n - 1$.

     *ADVANCED MATHEMATICS Student Resource Guide* **13**

## CHECKING THE MAIN IDEAS

1. State whether the function is a polynomial function. Give the zeros of each function, if they exist. (See Example 1 on text page 54.)

   **a.** $f(x) = 9 - \dfrac{1}{x^2}$     **b.** $g(x) = \dfrac{x^2 - 4x}{4}$     **c.** $h(x) = (x + 3)(x^2 - 5)$

2. Let $P(x) = x^4 - 3x^2 + 1$. Replace each __?__ with >, <, or = to make a true statement.

   **a.** $P(-4)$ __?__ $P(-3)$        **b.** $P(\sqrt{2})$ __?__ $P(-\sqrt{2})$

3. Which one of the following is a factor of $P(x) = x^3 - 3x^2 - 5x + 4$?

   **A.** $x - 4$      **B.** $x - 3$      **C.** $x + 4$      **D.** $x + 3$

4. Find the quotient and the remainder when $3x^3 + 5x^2 - x + 4$ is divided by $x + 2$.

5. *Critical Thinking* Discuss which method, direct substitution or synthetic substitution, would be the easier way to evaluate:
   (a) $P(1 + \sqrt{3})$ where $P(x) = x^4 - x^3 + 3x^2 - 2x - 1$, or
   (b) $P(-2)$ where $P(x) = x^9 - x^4 + 5$. Explain your answers.

## USING THE MAIN IDEAS

**Example 1** If $P(x) = -x + 2x^3 + 4$, find $P(1 - i)$ and $P\left(-\dfrac{1}{2}\right)$.

**Solution** **Method 1:** Direct Substitution

$$P(1 - i) = -(1 - i) + 2(1 - i)^3 + 4$$
$$= -1 + i + 2(-2 - 2i) + 4$$
$$= -1 - 3i$$

◀━━━ **Caution:** $(1 - i)^3 = (1 - i)^2(1 - i)$
$$= [1 - 2i + (-1)](1 - i)$$
$$= -2i(1 - i)$$
$$= -2i + 2i^2$$
$$= -2i - 2$$

$$P\left(-\dfrac{1}{2}\right) = -\left(-\dfrac{1}{2}\right) + 2\left(-\dfrac{1}{2}\right)^3 + 4$$
$$= \dfrac{1}{2} + 2\left(-\dfrac{1}{8}\right) + 4$$
$$= 4\dfrac{1}{4}$$

**Method 2:** Synthetic Substitution

$$P(x) = 2x^3 + 0x^2 - 1x + 4$$

◀━━━ **Caution:** Write the terms of $P(x)$ in descending powers of x. Insert zeros as placeholders for missing terms. Insert implied coefficients of 1 or −1.

```
1 - i │  2      0       -1        4
      │       2 - 2i   0 - 4i   -5 - 3i
      ───────────────────────────────────
         2    2 - 2i   -1 - 4i  │-1 - 3i   ← P(1 - i)
```

```
 -1/2 │  2      0       -1        4
      │        -1       1/2      1/4
      ───────────────────────────────────
         2     -1      -1/2    │4 1/4    ← P(-1/2)
```

*Note:* Since $P(1 - i) \neq 0$ and $P\left(-\dfrac{1}{2}\right) \neq 0$, neither $1 - i$ nor $-\dfrac{1}{2}$ is a zero of the function $P(x)$.

**Example 2** If $x = -1$ is a root of $x^3 - x^2 + 2 = 0$, find the remaining roots.

**Solution** If $-1$ is a root of the polynomial equation, then $x - (-1)$, or $x + 1$, is a factor of $P(x) = x^3 - x^2 + 0x + 2$ and $P(-1) = 0$.

$$
\begin{array}{r|rrrr}
-1 & 1 & -1 & 0 & 2 \\
   &   & -1 & 2 & -2 \\
\hline
   & 1 & -2 & 2 & \boxed{0} = P(-1)
\end{array}
$$

$$
\begin{array}{ccc}
\downarrow & \downarrow & \downarrow \\
1x^2 & -2x & +2 \quad \leftarrow \text{Quotient}
\end{array}
$$

Thus, $x^3 - x^2 + 2 = (x + 1)(x^2 - 2x + 2) = 0$.

To find the other two roots, solve $x^2 - 2x + 2 = 0$.

$a = 1, b = -2, c = 2$

$$
x = \frac{-b \pm \sqrt{b^2 - 4ac}}{2a} = \frac{-(-2) \pm \sqrt{(-2)^2 - 4(1)(2)}}{2 \cdot 1}
$$

$$
= \frac{2 \pm \sqrt{-4}}{2}
$$

$$
= \frac{2 \pm 2i}{2}
$$

$$
= 1 \pm i
$$

The remaining roots are $1 + i$ and $1 - i$.

## Exercises

6. If $f(x) = 4x^4 - 3x^2 - 1$, find (**a**) $f(1 - \sqrt{3})$ and (**b**) $f\left(-\dfrac{1}{x^2}\right)$.

7. If $x = 2$ is a root of $4x^3 - 15x - 2 = 0$, find the remaining roots.

8. **Writing** Write a paragraph in which you discuss whether $x + 1$ is always a factor of $P(x) = x^n - 1$, where $n$ is a positive integer.

9. If $-\dfrac{1}{2}$ is a zero of $f(x) = 4x^3 + kx^2 - 3$, find the value of $k$.

10. **Application** The height of a box is 2 ft longer than each edge of the square base.
    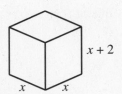
    **a.** Express the volume of the box, $V(x)$, as a polynomial function.
    **b.** Show that if the volume of the box is 16 ft$^3$, then $x$ can represent a length of 2 ft. (*Hint:* Show that $x = 2$ is a root of $V(x) = 16$, or $V(x) - 16 = 0$.
    **c.** Show that there is no other possible length that $x$ can represent for which the volume of the box is 16 ft$^3$.

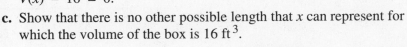

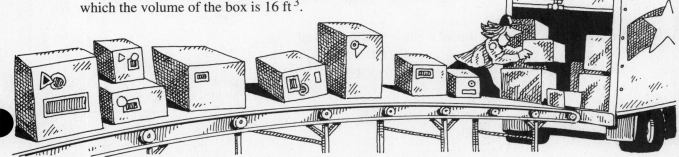

# Graphs: Maximums and Minimums

## Sections 2-3 and 2-4

| OVERVIEW | Section 2-3 is concerned with the equations and graphs of polynomial functions. Section 2-4 presents techniques for finding maximums and minimums of quadratic and cubic functions. |
|---|---|

| KEY TERMS | EXAMPLE/ILLUSTRATION |
|---|---|
| **Cubic function** (p. 62)<br> a polynomial function that can be written in the form $f(x) = ax^3 + bx^2 + cx + d, a \neq 0$ | $f(x) = 2x^3 - 3x^2 - 4x + 1$ |
| **Quartic function** (p. 64)<br> a polynomial function that can be written in the form $f(x) = ax^4 + bx^3 + cx^2 + dx + e, a \neq 0$ | $f(x) = x^4 - 3x^2 + x - 5$ |
| **Double root** (p. 64)<br> $x = c$ is a double root of $P(x) = 0$ if $(x - c)^2$ is a factor of $P(x)$. | 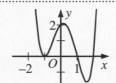<br>$P(x) = (x + 1)^2(x - 1)(x - 2)$<br>$-1$ is a *double root* of $P(x) = 0$. |
| **Triple root** (p. 65)<br> $x = c$ is a triple root of $P(x) = 0$ if $(x - c)^3$ is a factor of $P(x)$. | 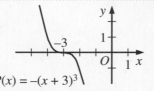<br>$P(x) = -(x + 3)^3$<br>$-3$ is a triple root of $P(x) = 0$. |
| **Maximum or minimum value of a quadratic function** (p. 68)<br> the value of $f(x) = ax^2 + bx + c$ when $x = -\dfrac{b}{2a}$ | 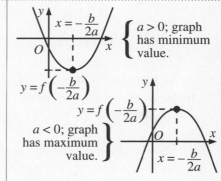 |
| **Local maximum/minimum value of a cubic function** (p. 69)<br> the values at the highest point of the peak (maximum) and at the lowest point of the valley (minimum) of the graph of a cubic function | 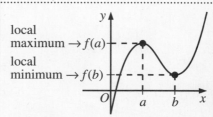 |

## UNDERSTANDING THE MAIN IDEAS

### Graphing a polynomial function $P(x)$

- If $P(x)$ is not in factored form, use factoring or synthetic division to write $P(x)$ in factored form.
- Set each factor equal to zero to find the zeros of $P(x)$. The zeros are the $x$-intercepts of the graph.
- Plot the zeros on a number line and use a sign analysis to check whether the function is positive or negative in the intervals determined by the zeros. (Choose a convenient test value for $x$ in each interval.)
- Graph the function as a smooth curve using a table of values. Keep in mind the basic shapes of a cubic function and a quartic function (see text pages 62 and 64). Check that the graph meets or crosses the $x$-axis at each zero, is above the $x$-axis in each interval where the function is positive, and is below the $x$-axis in each interval where the function is negative. If $x = c$ is a double root of $P(x) = 0$, then the graph is tangent to the $x$-axis at $x = c$. If $x = c$ is a single or triple root, then the graph crosses the $x$-axis at $x = c$. (See Example 1 on text page 63.)

### Finding the maximum or minimum value of a quadratic function $f(x)$

- Write $f(x)$ in the form $f(x) = ax^2 + bx + c$.
- Let $x = -\dfrac{b}{2a}$ and evaluate $f\left(-\dfrac{b}{2a}\right)$, which is the required maximum (if $a < 0$) or minimum (if $a > 0$).

### Finding the local maximum or minimum of a cubic function $f(x)$

- Use a graphing calculator or computer to graph $y = f(x)$. Then use the trace or zoom feature to approximate the required value. Alternatively, use a computer program like the one on text page 70 to evaluate $f(x)$ in a suitable interval to approximate the maximum or minimum value in that interval.

## CHECKING THE MAIN IDEAS

**Match each equation with its graph.**

1. $y = -x^3 + 3x^2$

2. $y = \frac{1}{2}x^3 - 4x$

3. $y = (x + 2)(x - 1)^3$

4. $y = 2x^3 - x^4$

A.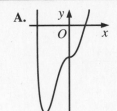

B.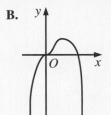

C.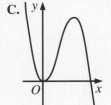

D.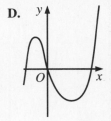

**Sketch the graph of each polynomial function.**

5. $y = (x + 1)^2(x - 2)^2$

6. $y = x^3 + x^2 - 4x - 4$

7. A ball is thrown vertically upward with an initial speed of 60 ft/s. Its height in feet $t$ seconds later is given by $h(t) = 60t - 16t^2$. What is the maximum height reached by the ball?

8. ***Critical Thinking*** Explain how the sign of the coefficient of the leading term (the term with greatest degree) can help you decide if the graph "points up" or "points down" as the value of $x$ gets greater and greater.

## USING THE MAIN IDEAS

**Example 1** Give an equation for the quartic function shown at the right. Since the $y$-axis has no indicated scale, more than one answer is possible.

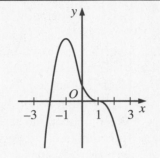

**Solution** Notice that the zeros of the function are $-2$ and $1$. Therefore, $x - (-2)$, or $x + 2$, and $x - 1$ are factors of the function. Since the graph is "S-shaped" at $x = 1$, $x - 1$ is a cubed factor. Thus, $f(x)$ has the form $k(x + 2)(x - 1)^3$. Since the graph points down when $x > 1$, $k$ represents a negative number. One possible equation is $f(x) = -(x + 2)(x - 1)^3$.

**Example 2** A rectangular dog pen is constructed using a barn wall as one side and 60 m of fencing for the other three sides. Find the dimensions of the pen that give the greatest area.

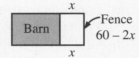

**Solution** **Method 1**

See Example 1 on text page 69.

**Method 2**

Use the symmetry of the graph of a quadratic equation. The $x$-coordinate of the vertex must be equidistant from the zeros of the function, as shown in the figure at the right.

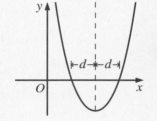

$A(x) = x(60 - 2x) = 2x(30 - x)$, so the zeros of $A(x)$ are 0 and 30. The $x$-coordinate of the vertex is the average of these zeros, 15. Thus, the dimensions ($x$ and $60 - 2x$) of the pen with greatest area are 15 m by 30 m.

## Exercises

**Give an equation for each polynomial graph shown. Since the $y$-axis has no indicated scale, more than one answer is possible.**

9.
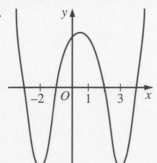

10.
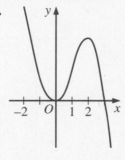

11. ***Application*** Calvin wants to make a gutter with a rectangular cross section by folding up strips of equal width on a long sheet of metal that is 26 cm wide. The larger the cross-sectional dimensions, the greater the capacity of the gutter. Find the dimensions that will maximize the gutter's capacity.

12. ***Writing*** Describe how you could find the approximate maximum volume in Example 2 on text page 69 using just a scientific calculator.

# Polynomial Equations

*Sections 2-5, 2-6, and 2-7*

**OVERVIEW**  Sections 2-6 and 2-7 present a variety of methods for solving polynomial equations including technology, factoring, and the rational root theorem. In Section 2-7 five important theorems about polynomial equations are stated and applied.

## KEY TERMS

**EXAMPLE/ILLUSTRATION**

**Polynomial equation that has a quadratic form** (p. 81)
an equation that can be written as $a(f(x))^2 + b(f(x)) + c = 0$, where $a \neq 0$

$$2x^4 - 5x^2 + 1 = 0$$
$$2(x^2)^2 - 5(x^2) + 1 = 0$$

$$3y^6 - y^3 - 5 = 0$$
$$3(y^3)^2 - y^3 - 5 = 0$$

## UNDERSTANDING THE MAIN IDEAS

### Using technology to solve a polynomial equation

- An excellent way to begin solving a polynomial equation is to graph the related polynomial function $y = P(x)$ if you have access to a graphing calculator or computer. A graph allows you to quickly determine the number of real roots and their approximate values. Use the trace, zoom, or rescale features to approximate the roots as shown in Example 1 on text page 75.

- The *location principle* allows you to identify the real roots of polynomial equations by using a table of values:

  If $P(a)$ and $P(b)$ have opposite signs, then $P(x) = 0$ has *at least* one *real* root between $a$ and $b$.

  Prepare a table of values for $P(x)$ using a computer or programmable calculator as in Example 2 on text page 76, or synthetic substitution. Look for changes in sign.

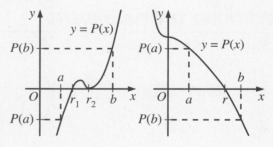

*Note:* Both graphing and tables of values can be used to approximate real roots, but *not* imaginary roots.

### Using algebra to solve a polynomial equation

Begin by writing the equation so that 0 is alone on one side.

- If the polynomial $P(x)$ is a cubic with four terms, try factoring by grouping. (See Example 1 on text page 80.)

- If the polynomial $P(x)$ has three terms, see if it has a quadratic form. If so, use substitution as in Example 2 on text page 81.

- If neither method above is helpful, use the *rational root theorem*:

  If $P(x) = 0$ has integral coefficients and the root $\frac{p}{q}$ (where $p$ and $q$ are nonzero integers with no common factor except 1), then $p$ must be a factor of the constant term and $q$ must be a factor of the coefficient of the term of highest degree.

---

*ADVANCED MATHEMATICS Student Resource Guide* **19**

Try the possible rational roots, beginning with the integral possibilities since these are easiest to evaluate. Once you know a root $r$, solve the *depressed equation* $Q(x) = 0$ where $Q(x)$ is the quotient when $P(x)$ is divided by $x - r$. If the depressed equation is *quadratic*, try factoring or the quadratic formula. If the depressed equation is *cubic*, try factoring, including factoring by grouping or use the rational roots theorem. (See Example 3 on text page 82.)

### General theorems about polynomial equations

- For a polynomial equation of positive degree $n$ with *real* coefficients:
    1. There are exactly $n$ roots (counting repeated roots individually).
    2. If there are any imaginary roots $a + bi$ or $a - bi$, then the complex conjugate is also a root.
    3. If $n$ is odd, then there is at least one real root.
- If the coefficients of a polynomial equation are *rational* and $a + \sqrt{b}$ is a root ($a$ and $b$ are rational and $\sqrt{b}$ is irrational), then $a - \sqrt{b}$ is also a root.
- If $ax^n + bx^{n-1} + cx^{n-2} + \ldots + k = 0$, with $a \neq 0$, then:
    1. the sum of the $n$ roots is $-\dfrac{b}{a}$.
    2. the product of the roots is $\dfrac{k}{a}$ if $n$ is even and $-\dfrac{k}{a}$ if $n$ is odd.

## CHECKING THE MAIN IDEAS

1. Use a graph to find the real roots of $x^4 - x^3 + 2x^2 - 5 = 0$ to the nearest tenth.

2. Use the location principle to find the real roots of $x^3 = 2x^2 + 1$ to the nearest tenth.

3. Solve $x^4 + 3x^3 + x^2 + 3x = 0$, giving all real and imaginary roots.

4. For which equation are $-3$, $\dfrac{1}{5}$, and $-\dfrac{2}{5}$ possible rational roots?

    **A.** $5x^3 - 7x + 6 = 0$      **B.** $5x^3 - 12x^2 - 1 = 0$
    **C.** $6x^3 - 7x + 5 = 0$      **D.** $10x^3 - x^2 - 3 = 0$

5. **a.** Find the maximum number and minimum number of real roots possible for a fifth-degree equation with real coefficients.
    **b.** *Critical Thinking*  Draw a sketch to illustrate your answers in part (a).

6. Find the sum and product of the roots of $2x^3 - 5x + 6 = 0$.

7. Find a quadratic equation with integral coefficients and roots $-1 \pm \sqrt{5}$. (See Example 1 on text page 87.)

8. *Critical Thinking*  Solve $x^2 + 2ix = 0$. Does your answer provide a counterexample for the Complex Conjugates theorem? Explain.

## USING THE MAIN IDEAS

**Example 1** Show that $P(x) = x^3 + 5x^2 + x - 3 = 0$ has no positive rational root, but that it does have a positive irrational root.

**Solution** The only possible positive rational roots are 1 and 3, and $P(1) = 4$ and $P(3) = 72$. Since neither value is zero, neither 1 nor 3 is a root of the equation. A graph or a table can be used to show that there *is* a real root between 0 and 1.

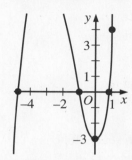

| $x$ | $P(x) = x^3 + 5x^2 + x - 3$ |
|---|---|
| 0 | $-3$ |
| 1 | 4 |
| 2 | 27 |

change of sign indicates a real root between $x = 0$ and $x = 1$.

The graph crosses the $x$-axis between $x = 0$ and $x = 1$, indicating a real root in that interval.

Since there is a positive real root but that root is not rational, the equation has a positive irrational root.

**Example 2** Find integers $c$ and $d$ such that the equation $x^3 - 3x^2 + cx + d = 0$ has $1 - i$ as one of its roots.

**Solution** Since $1 - i$ is a root and the equation has integral coefficients, $1 + i$ is also a root. Therefore, if the third root is $r$, use the sum of the roots.

$$(1 + i) + (1 - i) + r = -\frac{-3}{1} \longrightarrow 2 + r = 3, \text{ or } r = 1$$

To find the values of $c$ and $d$, use either of the following methods.

1. $(1 + i)(1 - i)(1) = -\frac{d}{1} \longrightarrow 2 = -d, \text{ or } d = -2$ ◀━━━ **Caution:** Remember to check the degree of the equation to determine the sign of the product of the roots.

   Since 1 is a root, $1^3 - 3(1^2) + c(1) - 2 = 0$
   $\longrightarrow c - 4 = 0, \text{ or } c = 4.$

2. Find a cubic equation with the roots $1 + i$, $1 - i$, and 1; then compare the coefficients of your equation with the equation above to find the values of $c$ and $d$.

## Exercises

9. Solve the equation in Example 1 to find the exact roots.

10. *Writing* Compare technological methods and algebraic methods of solving polynomial equations. What are the strengths and limitations of each method?

11. Show that the equation $y = 2x^3 - 5x^2 - 2$ has no rational roots, but that it does have an irrational root between $x = 2$ and $x = 3$.

12. Find integers $b$ and $c$ such that the equation $x^3 + bx^2 + cx - 4 = 0$ has $\sqrt{2}$ as one of its roots.

13. *Application* An open-topped box is made by cutting and discarding four identical squares from the corners of a square sheet of metal 10 in. on a side and then folding up the sides. Find each length $x$, to the nearest tenth, for which the volume of the box is 60 in.$^3$.

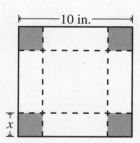

# CHAPTER REVIEW

## Chapter 2: Polynomial Functions

Complete these exercises before trying the Practice Test for Chapter 2. If you have difficulty with a particular problem, review the indicated section.

1. **a.** Give the zeros of $P(x) = 4x^3 - 4x^2 + x$.

   **b.** Use synthetic substitution to find $P\left(-\frac{1}{2}\right)$. *(Section 2-1)*

2. Find the quotient and the remainder when $x^4 - 2x^2 + 3x - 1$ is divided by $x + 1$. *(Section 2-2)*

3. Show that $x - 2$ is a factor of $x^{10} - 7x^7 - 9x^4 + 16$. *(Section 2-2)*

4. Sketch the graph of $y = -x(x - 2)^2$. *(Section 2-3)*

5. Let $f(x) = 2x^2 - 8x - 9$. State (**a**) whether the function has a maximum or a minimum value and (**b**) the value of $x$ at which the maximum or minimum occurs. *(Section 2-4)*

6. Describe how to find a maximum volume if you know the cubic function $V(x)$ that relates the volume to a certain dimension $x$. *(Section 2-4)*

7. Use two different methods to show that the equation $x^4 - 3x^3 + 1$ has a real root between 2 and 3. *(Section 2-5)*

8. Solve (a) $x^4 = 3x^2 + 4$ and (b) $x^3 + x^2 = 3x + 3$. *(Section 2-6)*

9. Use the rational root theorem to solve $3x^3 - x^2 - 8x - 4 = 0$. *(Section 2-6)*

10. Consider the equation $x^3 - x^2 - 7x + 15 = 0$.

    **a.** Find the sum and product of the roots.

    **b.** If $2 - i$ is a root, find the other roots. *(Section 2-7)*

---

If $P(x) = -4x^3 + 5x - 1$, use synthetic substitution to find each of the following.

1. $P(-2)$    2. $P\left(\frac{1}{3}\right)$    3. $P(i)$

4. One root of $3x^3 - 5x^2 - 3x + 2 = 0$ is 2. Find the remaining roots.

**For Exercises 5–7, let $P(x) = 8x^3 - 6x - 1$.**

5. Use synthetic division to find the quotient and remainder when $P(x)$ is divided by $x - \frac{1}{2}$.

6. List all the possible rational roots of the equation $P(x) = 0$.

7. Is $x - 1$ a factor of $P(x)$?

8. Sketch the graph of $P(x) = -x(x + 2)(x - 1)^2$.

9. Write the equation of the cubic polynomial whose graph is shown at the right. Since the $y$-axis has no indicated scale, more than one answer is possible.

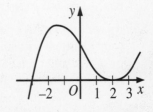

10. Find the remainder when $x^{10} - 1$ is divided by $x - 2$.

---

11. Show that $f(x) = x^3 + 3x + 1$ has no rational roots, but that it does have an irrational root between $x = -1$ and $x = 0$.

**For Exercises 12–14, let $P(x) = x^4 - 3x^2 - 14x - 12$.**

12. Without finding the zeros, what is (**a**) the sum of the zeros and (**b**) the product of the zeros of $P(x)$.

13. Find all the zeros of $P(x)$.

14. Use your answers to Exercise 13 to show that your answers to Exercise 12 are correct.

15. Two numbers have a sum of 40. Find the larger number if the product of twice the larger and four more than the smaller is a maximum.

---

**MIXED REVIEW**

*Chapters 1–2*

Let $P(x) = x^4 + 2x^3 + x^2 + 8x - 12$.

1. Find the quotient and remainder when $P(x)$ is divided by $x + 2$.

2. Show that $x - 1$ is a factor of $P(x)$.

3. Name the possible rational zeros of $P(x)$. Then find all the rational zeros.

4. Sketch the graph of $y = P(x)$.

5. Find $P(i)$.

6. Solve $P(x) = 0$, giving all real and imaginary roots.

**Line $l$ contains point $(2, -1)$ and has slope $\dfrac{3}{4}$.**

7. Find an equation for line $l$.

8. Find an equation of the line with $x$-intercept $-2$ that is perpendicular to $l$.

9. Find the point at which line $l$ intersects the line in Exercise 8.

10. Find any points of intersection of line $l$ and the parabola $y = x^2 + 4x + 3$. Sketch the graphs of the line and the parabola.

**Write each expression in the form $a + bi$.**

11. $\sqrt{-9} - \sqrt{-81}$    12. $(1 + i\sqrt{3}) - (2 - i\sqrt{3})$    13. $\dfrac{1 + i\sqrt{3}}{2 - i\sqrt{3}}$

14. Let $F(t)$ be a linear function such that $F(0) = 32$ and $F(10) = 50$. Find an equation for $F(t)$ and tell what the function represents.

**Solve. If necessary, round the real roots to the nearest tenth.**

15. $x^3 + 4x^2 = 1$    16. $x^4 + 8 = x^3 + 8x$    17. $x^2 - 8x = 1$

18. Show that $P(7, 3)$ is equidistant from $A(2, -2)$ and $B(0, 2)$.

19. If a ball is tossed upward, then $t$ seconds later its height in feet above the ground is $h(t)$. If $h(0) = 40$ (the ball is thrown upward from the top of a building), $h(2) = 120$, and $h(3) = 112$, find a quadratic model for the data. Then find the maximum height reached by the ball and the time at which it reaches the ground.

20. Find integers $b$ and $c$ such that the equation $x^3 + bx^2 + cx + 6 = 0$ has $-\sqrt{3}$ as one of its roots.

---

# Inequalities in One Variable

**OVERVIEW** These sections are concerned with solving and graphing various types of inequalities in one variable, including linear, absolute-value, and polynomial inequalities.

## KEY TERMS

| | EXAMPLE/ILLUSTRATION |
|---|---|
| **Linear or absolute-value inequality** (p. 95) an open sentence formed by replacing the equals sign in a linear or absolute-value equation by an inequality sign ($\neq$, $<$, $>$, $\leq$, or $\geq$) | $x > 5$ $\quad$ $\lvert x - 1 \rvert \leq 3$ |
| **Polynomial inequality** (p. 100) an inequality that can be written with a polynomial on one side and 0 on the other | $x^2 - 4x + 3 > 0$ |

## UNDERSTANDING THE MAIN IDEAS

### Solving a linear inequality

You can solve a linear inequality using much the same approach you would use to solve a linear equation: isolate the variable by performing the same operation to both sides. If the inequality involves fractions, begin by clearing the equation of fractions.

$$1 - 5x \geq 2(5 - x)$$
$$1 - 5x \geq 10 - 2x$$
$$1 - 3x \geq 10$$
$$-3x \geq 9$$
$$x \leq -3 \leftarrow \text{Reverse sign.}$$

*Note:* If you multiply or divide both sides of an inequality by a *negative* number, you must reverse the direction of the inequality sign. (See Example 1 on text page 95.)

Check: $1 - 5(-3) \overset{?}{=} 2(5 - (-3))$
$\quad\quad 1 + 15 = 2(8)$ ✔
$\quad\quad 1 - 5(-4) \overset{?}{\geq} 2(5 - (-4))$
$\quad\quad 1 + 20 \geq 2(9)$ ✔

### Solving a linear equation or inequality involving absolute value

| Sentence | Algebraic Method | Geometric Method |
|---|---|---|
| **1.** $\lvert x - k \rvert = c$ $\lvert x + 1 \rvert = 7$, or $\lvert x - (-1) \rvert = 7$ | $x - k = c$ or $x - k = -c$ $x + 1 = 7$ or $x + 1 = -7$ $x = 6$ or $x = -8$ | 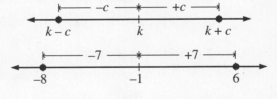 |
| **2.** $\lvert x - k \rvert < c$ $\lvert x + 1 \rvert < 7$, or $\lvert x - (-1) \rvert < 7$ | $-c < x - k$ and $x - k < c$ $-7 < x + 1$ and $x + 1 < 7$ $-8 < x < 6$ | 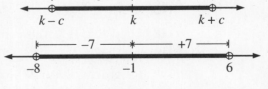 |
| **3.** $\lvert x - k \rvert > c$ $\lvert x + 1 \rvert > 7$, or $\lvert x - (-1) \rvert > 7$ | $x - k < -c$ or $x - k > c$ $x + 1 < -7$ or $x + 1 > 7$ $x < -8$ or $x > 6$ | 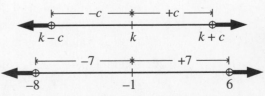 |

## Solving a polynomial inequality

- **Method 1**

  Find the zeros of the polynomial by factoring (use factoring by grouping or quadratic-form factoring if applicable) or by using synthetic substitution with the rational root theorem. Plot the zeros on a number line using open dots if the inequality sign is > or <, and using solid dots if the inequality sign is ≤ or ≥. Then use a test value for each interval of the number line to determine if the polynomial values in that interval are positive or negative. Select the intervals that are solutions. (See Examples 1 and 2 on text pages 100 and 101.)

- **Method 2**

  Use a graphing calculator or computer to graph the polynomial $P(x)$.
  In the sketch at the right:
  $P(x) = 0$ when $x = a$, $b$, or $c$,
  $P(x) > 0$ when $a < x < b$ or $x > c$,
  and $P(x) < 0$ when $x < a$ or $b < x < c$.
  *Note:* This method may give approximate rather than exact solutions.

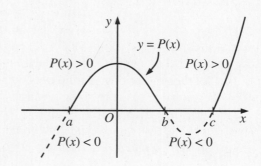

## CHECKING THE MAIN IDEAS

1. Solve $\dfrac{1-x}{4} \geq \dfrac{-1-2x}{2}$ and graph its solution.

2. *Complete:* If $|x - 7| > 3$, then ___?___ .

   **A.** the distance from $x$ to 3 is greater than 7

   **B.** $-3 < x - 7 < 3$

   **C.** $x - 7 < -3$ and $x - 7 > 3$

   **D.** the distance from $x$ to 7 is greater than 3

3. The graph of the inequality $|x + 2| \leq 3$ will look like which of the following?

   **A.**    **B.**

   **C.**   **D.**

**Solve each equation or inequality.**

4. $|x - 5| = 5$

5. $(x + 1)^2(x - 3) < 0$

6. $x^3 + 2x^2 - 8x \geq 0$

7. $|3 - 2x| \geq 5$

8. **Critical Thinking**  You should be able to solve the inequalities $|x - 7| \geq 0$ and $|x + 3| < -1$ by just looking at these inequalities and thinking about them. Solve each one and explain how you arrived at your answers.

## USING THE MAIN IDEAS

**Example 1** Solve $|2x + 5| \le 7$.

**Solution**  **Method 1: Algebraic method**
See the solution on text page 97.

**Method 2: Geometric method**
See the solution on text pages 97–98.

**Method 3**
Write and solve the related equation: $|2x + 5| = 7$.
Square both sides.    $|2x + 5|^2 = 7^2$
$$4x^2 + 20x + 25 = 49 \leftarrow |2x + 5|^2 = (2x + 5)^2$$
$$4x^2 + 20x - 24 = 0$$
$$x^2 + 5x - 6 = 0$$
$$(x + 6)(x - 1) = 0$$
$$x = -6 \text{ or } x = 1$$
Plot these solutions as solid dots on a number line.
Test a value between these solutions, say $x = 0$:
$$|2(0) + 5| \le 7 \checkmark$$
Since the test value results in a true inequality, shade
*between* $-6$ and $1$. The solution is $-6 \le x \le 1$.

**Example 2** Solve $3x^3 + 11x^2 \le 4$.

**Solution**  Rewrite the inequality with 0 alone on one side: $3x^3 + 11x^2 - 4 \le 0$

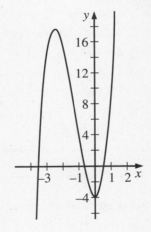

**Method 1**
Use a graphing calculator or a computer to graph $P(x) = 3x^3 + 11x^2 - 4$. Rescale or use the zoom feature to estimate the zeros to the nearest tenth: about $-3.6$, $-0.7$, and $0.6$. Since the inequality has the form $P(x) \le 0$, use the zeros and the graph to name the approximate intervals in which the graph is on or below the $x$-axis: $x \le -3.6$ or $-0.7 \le x \le 0.6$.

**Method 2**
The possible rational roots are $\pm 1$, $\pm 2$, $\pm 4$, $\pm\frac{1}{3}$, $\pm\frac{2}{3}$, and $\pm\frac{4}{3}$.

You can use synthetic substitution to evaluate these in turn. If you sketch the graph of $P(x) = 3x^3 + 11x^2 - 4$ (see figure), you notice that there are real roots between $-4$ and $-3$, between $-1$ and $0$, and between $0$ and $1$. The possible rational roots in these intervals are $\pm\frac{1}{3}$ and $\pm\frac{2}{3}$. Using synthetic substitution:

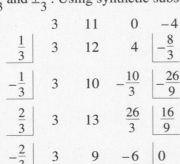

|  | 3 | 11 | 0 | −4 |
|---|---|---|---|---|
| $\frac{1}{3}$ | 3 | 12 | 4 | $-\frac{8}{3}$ |
| $-\frac{1}{3}$ | 3 | 10 | $-\frac{10}{3}$ | $\frac{26}{9}$ |
| $\frac{2}{3}$ | 3 | 13 | $\frac{26}{3}$ | $\frac{16}{9}$ |
| $-\frac{2}{3}$ | 3 | 9 | $-6$ | 0 |

The last line of the synthetic substitution shows that $P\left(-\frac{2}{3}\right) = 0$

and that $P(x) = 3x^3 + 11x^2 - 4 = \left(x + \frac{2}{3}\right)(3x^2 + 9x - 6) =$
$(3x + 2)(x^2 + 3x - 2)$.

Use the quadratic equation to find the remaining roots: $a = 1, b = 3$,
and $c = -2$.

$$x = \frac{-3 \pm \sqrt{3^2 - 4(1)(-2)}}{2 \cdot 1} = \frac{-3 \pm \sqrt{17}}{2}$$

Using the number line shown below, think about the signs of the
factors in each of the numbered intervals.

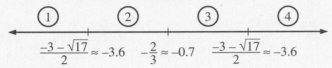

$$\frac{-3 - \sqrt{17}}{2} \approx -3.6 \qquad -\frac{2}{3} \approx -0.7 \qquad \frac{-3 - \sqrt{17}}{2} \approx -3.6$$

| Interval | sign of $3x + 2$ | sign of $x^2 + 3x - 2$ | sign of $P(x) = (3x + 2)(x^2 + 3x - 2)$ |
|---|---|---|---|
| 1 | − | + | − |
| 2 | − | − | + |
| 3 | + | − | − |
| 4 | + | + | + |

Thus, $P(x) \le 0$ in intervals 1 and 3, that is, when $x \le \dfrac{-3 - \sqrt{17}}{2}$ or

$-\dfrac{2}{3} \le x \le \dfrac{-3 + \sqrt{17}}{2}$.

## Exercises

**Solve the given equation or inequality and graph its solution.**

**9.** $\dfrac{x - 3}{4} - \dfrac{1 - x}{5} \ge \dfrac{2x + 3}{10}$

**10.** $|5 - 2x| = 1$

**11.** $|3x - 1| < 7$

**12.** $x^4 \le 3x^2 + 4$

**13.** *Writing* Write a paragraph in which you discuss the possibilities for
the solution of $(x - a)(x - b) < 0$, where $a$ and $b$ are constants.

**14.** *Application* The perimeter of a rectangular garden is 20 m. Find the
length of the longer side of the garden if its area is at least 22 m$^2$.

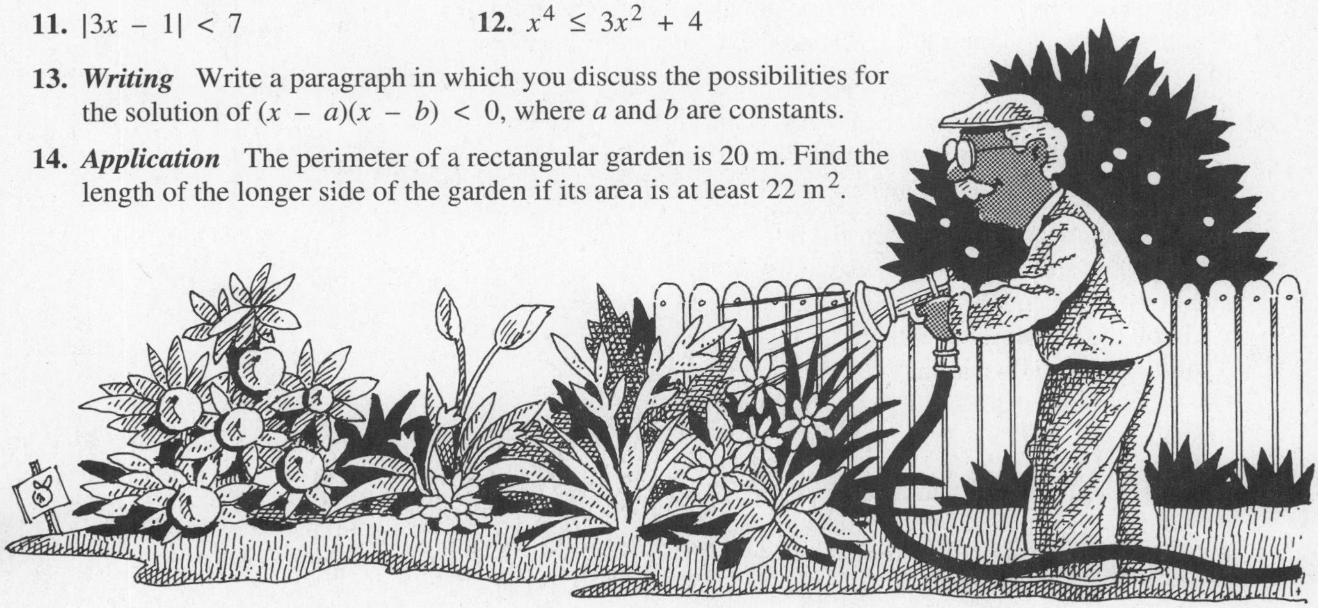

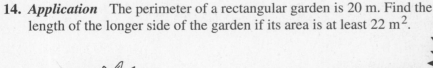

# Inequalities in Two Variables

## Sections 3-3 and 3-4

**OVERVIEW** | Section 3-3 presents methods for graphing various kinds of inequalities in two variables including linear, polynomial, and inequalities containing an absolute value. This basic skill is extended to graphing systems of inequalities. In Section 3-4, graphs of systems of inequalities are used to solve linear programming problems.

## KEY TERMS

**EXAMPLE/ILLUSTRATION**

**The graph of an inequality in two variables** (p. 104)
points, in the plane, whose coordinates make the inequality true

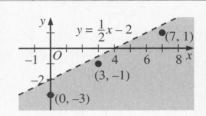

Each point in the shaded region makes the inequality $y < \frac{1}{2}x - 2$ true. The graph is the shaded region.

**Linear programming** (p. 108)
a method for solving certain decision-making problems in which a linear expression is to be maximized or minimized

See the Example on text page 111.

## UNDERSTANDING THE MAIN IDEAS

### Graphing an inequality

- To graph an inequality with the form $y > f(x)$ or $y < f(x)$, show the graph of $y = f(x)$ as a *dashed* line or curve. Shade the points *above* the curve to graph $y > f(x)$ and the points *below* the curve to graph $y < f(x)$.

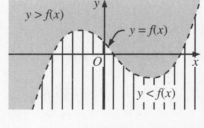

- To graph an inequality with the form $y \le f(x)$ or $y \ge f(x)$, use the same steps as above, but graph $y = f(x)$ as a *solid* line or curve to show that these points are also included in the graph.

- To graph an inequality in which $y$ is *not* alone on one side, replace the inequality symbol with an equals sign and solve for $y$. Graph the transformed equation using a solid or dashed line, as appropriate. Choose a test point to decide which region to shade.

$$5x - y \ge -5$$
$$\downarrow$$
$$5x - y = -5$$
$$5x + 5 = y$$

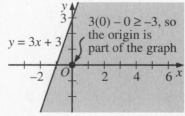

- To graph an inequality that involves $x$ or $y$ (but not both) in the coordinate plane, remember that $x = k$ is a vertical line through $(k, 0)$ and $y = k$ is a horizontal line through $(0, k)$.

$|x| > 1$ means $x < -1$ or $x > 1$: thus, the $x$-coordinate of each point in the graph is less than $-1$ or greater than 1.

### Graphing a system of inequalities

To graph a system consisting of two or more inequalities, find the points common to all the inequalities in the system. This is the region with overlapping shading (called the *feasible region* in a linear programming problem).

### Using linear programming to solve a word problem

- Define two variables, $x$ and $y$.
- Write a system of linear inequalitites in $x$ and $y$ to represent the facts given in the problem.
- Graph the system of inequaltites to determine the *feasible region*.
- Find the coordinates of each corner point of the feasible region.
- Write a function in the form $ax + by$ to be maximized or minimized.
- Evaluate the function $ax + by$ at each corner point.
- Select the required maximum or minimum value, which will occur at one of the corner points, and note the values of $x$ and $y$ at that point.

This process is illustrated in the Example on text page 111.

## CHECKING THE MAIN IDEAS

**1.** Choose the inequality whose graph is shown at the right.

    **A.** $2x - y > 6$         **B.** $2x - y \geq 6$

    **C.** $2x - y < 6$         **D.** $2x - y \leq 6$

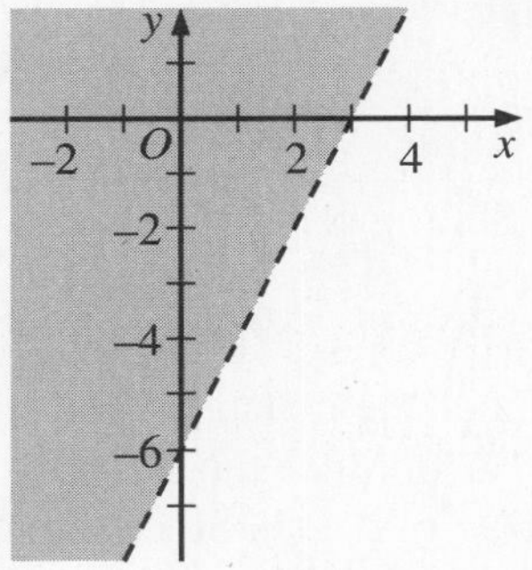

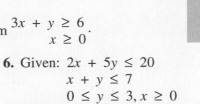

**Sketch the graph of each inequality.**

**2.** $y < 4x - x^2$ (See Example 1 on text page 105.)

**3.** $|y| \leq 2$

**4.** Graph the solution set of the system $\begin{array}{l} 3x + y \geq 6 \\ x \geq 0 \end{array}$.

**5.** Given: $3x + y \geq 6$
            $4x + 3y \geq 12$
            $x \geq 0, y \geq 0$
    Find the point $(x, y)$ that
    minimizes $C = 5x + 3y$.

**6.** Given: $2x + 5y \leq 20$
            $x + y \leq 7$
            $0 \leq y \leq 3, x \geq 0$
    Find the point $(x, y)$ that
    maximizes $P = 3x + 2y$.

**7.** *Critical Thinking* Consider a point $(a, b)$ on the line $y = mx + k$. If you choose a point $(a, c)$ that is above point $(a, b)$, why must $(a, c)$ satisfy the inequality $y > mx + k$?

## USING THE MAIN IDEAS

**Example 1** Sketch the graph of $1 < |x + 4| < 2$.

**Solution** Since $1 < |x + 4| < 2$ means $1 < |x + 4|$ and $|x + 4| < 2$, begin by solving $1 = |x + 4|$ and $|x + 4| = 2$.

    $x + 4 = 1$   or   $x + 4 = -1$   or   $x + 4 = 2$   or   $x + 4 = -2$

        $x = -3$           $x = -5$            $x = -2$           $x = -6$

---

    

Graph these values of $x$ on a number line:

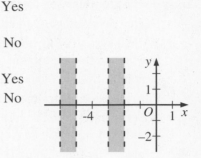

Test each interval.

| | Satisfies $1 < |x + 4| < 2$ ? |
|---|---|
| In interval A, $x > -2$ so $|x + 4| > 2$. | No |
| In interval B, $-3 < x < -2$. If, for example, $x = -2.5$, $|x + 4| = 1.5$, so $1 < |x + 4| < 2$. | Yes |
| In interval C, $-5 < x < -3$. If, for example, $x = -4$, $|x + 4| = 0$, so $|x + 4| < 1$. | No |
| In interval D, $-6 < x < -5$. If, for example, $x = -5.5$, $|x + 4| = 1.5$, so $1 < |x + 4| < 2$. | Yes |
| In interval E, $x < -6$, so $|x + 4| > 2$. | No |

Graph the dashed vertical lines $x = -6$, $x = -5$, $x = -3$, and $x = -2$ on a coordinate plane and shade the appropriate regions, as shown.

**Example 2** A pet store manager is preparing an order of two types of dog food, Pedigree Pellets and Mongrel Mash. These are packaged in bags of the same size. The store has display and storage space for a total of 800 bags. Past sales indicate that at least twice as much mash as pellets will be sold. The manager has decided to spend no more than $1500 on the order. The store's buying and selling prices per bag are as follows:

| | Pedigree Pellets ($x$) | Mongrel Mash ($y$) |
|---|---|---|
| Buying price | $4.50 | $1.50 |
| Selling price | $7.00 | $3.00 |

Write and explain five inequalities that must be satisfied.

**Solution** Since the manager cannot order a negative number of bags, $x \geq 0$ and $y \geq 0$. Since the store has display space for at most 800 bags, $x + y \leq 800$. Since the manager expects that at least twice as much mash as pellets will be sold, $y \geq 2x$. Since the total cost of the order must not exceed $1500, $4.5x + 1.5y \leq 1500$, or $3x + y \leq 1000$.

## Exercises

8. Sketch the graphs of (**a**) $|y - 2| > 3$ and (**b**) $3 < |y - 2| < 7$.

9. Graph the solution set of the system $y \geq x^2 + 3x - 4$ and $x + y \leq 8$. (See Example 2 on text page 105.)

10. Refer to Example 2 above.

   **a.** Write an expression in terms of $x$ and $y$ for the profit, $P$, earned on the order if all the bags are sold.

   **b.** How many bags of each type of dog food should be ordered to maximize the profit?

   **c.** *Writing* Write a short paragraph in which you describe a hidden restriction on the solution of the problem described in part (b).

   **d.** *Critical Thinking* Suppose the manager does not limit the cost of the order to $1500. Would the maximum profit on the order be the same as in part (b)? If not, what would it be?

# CHAPTER REVIEW

## Chapter 3: Inequalities

**Complete these exercises before trying the Practice Test for Chapter 3. If you have difficulty with a particular problem, review the indicated section.**

**1.** Solve (**a**) $3 - 2x > 5$ and (**b**) $|x| \leq 1.5$. *(Section 3-1)*

**2.** Solve (**a**) $x(x + 4) \geq 0$ and (**b**) $x^3 - 4x^2 + 4x > 0$. *(Section 3-2)*

**Sketch the graph of each inequality.** *(Section 3-3)*

**3.** $2x + 5y > 10$        **4.** $y \leq x^2 - 5x + 4$

**5.** Find the minimum value of $8x + 10y$ for the region shown at the right. *(Section 3-4)*

**6.** Write a system of inequalities to describe the following situation:
A magazine store has room for 200 football magazines. There are two major football magazines available: Magazine *A* and Magazine *B*. Customers generally buy at least 1.5 times as many issues of Magazine *A* as Magazine *B*. *(Section 3-4)*

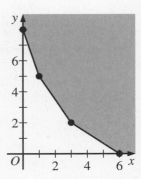

---

**Solve and graph each inequality on a number line.**

**1.** $3(x - 5) \geq 2x - 4(5 - 2x)$     **2.** $|2x - 3| \leq 7$

**3.** $|x + 2| < 3$                **4.** $2|x - 5| - 3 \geq 5$

**5.** $(3x - 2)(x + 1) < 0$      **6.** $(x - 2)(x + 1)^2(x + 3) \geq 0$

**Sketch the graph of each inequality in the coordinate plane.**

**7.** $4x - y \geq 4$            **8.** $|x| \geq 1$

**9.** $y \leq x^2 - 4$         **10.** $y \leq (x - 1)(x + 2)(x - 2)$

**Graph the solution set of each system.**

**11.** $x + y \geq 5$           **12.** $y \leq -x^2 + 6x$
      $x - y \geq -5$              $y \geq x$
      $x \leq 4$

**13.** Write a set of inequaltites that defines the region shown at the right.

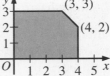

Let *R* be the region defined by the system $2x + 3y \leq 12, x + y \leq 5$, $x \geq 0$, and $y \geq 0$.

**14.** Graph *R*.

**15.** Determine the coordinates of all vertices of *R*.

**16.** What is the maximum value of $100x + 200y$ over the region *R*?

**17.** What is the maximum value of $300x + 200y$ over the region *R*?

**18.** What is the maximum value of $300x + 400y$ over the region *R*?

---

    

# Chapter 3

1. Sketch the graph of $y \leq -x^4 + 3x^3 + 4x^2 - 12x$.

2. Graph the system of equations $y = 2x + 1$ and $y = \frac{1}{2}(x + 2)^2 - 3$.

3. Use a calculator or computer to approximate, to the nearest tenth, the real roots of $x^4 + 2x^2 - x - 3 = 0$.

4. Write an equation of the line with $y$-intercept $-3$ and parallel to the line through $(2, 7)$ and $(6, 5)$.

**Solve.**

5. $|4x - 7| \geq 3$       6. $2x^2 - 8x + 3 = 0$     7. $6x^3 + 4x^2 = 7x - 5$

**Nails are measured in units called *pennies*, indicated by the letter *d*.**

| Nail size ($d$) | 2 | 4 | 6 | 8 |
|---|---|---|---|---|
| Nail length (in.) | 1 | 1.5 | 2 | 2.5 |

8. **a.** Find an equation that models how nail length is related to nail size.
   **b.** What kind of function is the model?

9. Find the nail size of a 3-inch long nail.

10. **a.** Find the midpoint $M$ of the line segment with endpoints $A(-5, -1)$ and $B(1, 4)$.
    **b.** Show that $M$ is equidistant from $A$ and $B$.

**For Exercises 11 and 12, let $P(x) = x^6 - 7x^4 - 5x^3 - 2x^2 - x - 6$.**

11. Show that $x - 3$ is a factor of $P(x)$.

12. Find the sum and product of the roots of $P(x) = 0$.

13. Write an equation for the line with undefined slope that contains the origin.

14. Simplify **(a)** $(2 + 3i)^2$ and **(b)** $\dfrac{2}{1+i}$.

15. Solve $x^4 - 10x^2 + 9 > 0$.

16. Find an equation of the line through $(-3, 4)$ and perpendicular to the line $2x - 5y = 10$.

17. Maximize $P = 2x + 3y$ subject to the constraints $x \geq 0$, $y \geq 0$, $x + 4y \leq 24$, and $5x + 4y \leq 40$.

**In a certain electric circuit, the available power $P$ (in watts), depends on the current, $I$ (in amperes), that is flowing, as shown in this chart.**

| $I$ | 1 | 2 | 3 |
|---|---|---|---|
| $P$ | 99 | 176 | 231 |

18. Find a quadratic function, $P(I)$, that models the relationship between the power and the current.

19. Find the zeros of $P$.

20. Find the maximum possible power.

---

# Sections 4-1 and 4-2

**OVERVIEW**  Section 4-1 introduces the general terminology and concepts related to functions. In Section 4-2, functions are combined in a variety of ways, including composition.

## KEY TERMS

**Function** (p. 119)
a relationship that pairs each member of one set (the *domain*) with *exactly one* member in another set (the *range*)

$-2 \rightarrow 3$
$0 \rightarrow 3$
$2 \rightarrow 1$
$4 \rightarrow -2$
Domain $= \{-2, 0, 2, 4\}$
Range $= \{-2, 1, 3\}$

**Graph of a function** (p. 120)
the set of all points in the plane determined by the function

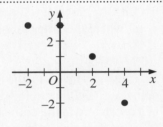

**Dependent/independent variables** (p. 121)
If $y$ is a function of $x$ (written $y = f(x)$), then $y$ is the dependent variable and $x$ is the independent variable, since the value of $y$ depends on the value of $x$.

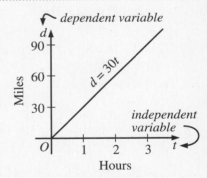

The distance traveled at 30 mi/h depends on the traveling time.

**Composite of functions** (p. 126)
The composite of functions $f$ and $g$ (written $f \circ g$) is the function whose value at $x$ is $f(g(x))$, provided $x$ is in the domain of $g$ and $g(x)$ is in the domain of $f$. (The operation that combines $f$ and $g$ this way is called *composition*.)

If $f(x) = x^2$ and $g(x) = 2x - 1$, then $g(3) = 2(3) - 1 = 5$ and $f(5) = 5^2 = 25$, so $f(g(3)) = 25$.

## UNDERSTANDING THE MAIN IDEAS

### Using the equation of a function

- To decide if an equation defines a function, solve the equation for $y$. See if there is only one value of $y$ for each value of $x$. If so, the equation defines a function.

- To find the domain of a function, examine its equation to determine any values of $x$ for which the function value is not a real number. Exclude these values from the domain. (See Example 1 on text page 120.)

- To find the zeros of a function, set the function equal to zero and solve the resulting equation for $x$.

### Using the graph of a function

- To decide if a graph specifies a function, imagine drawing vertical lines through the graph. If there is any vertical line that contains two or more points of the graph, then the graph does *not* specify a function. Otherwise, it does.

- To find the domain of a function, list the $x$-values of the graph individually or specify them using an inequality.

- To find the range of a function, list the $y$-values of the graph individually or specify them using an inequality.

- To find the zeros of a function, list the $x$-coordinate of each point where the graph crosses the $x$-axis.

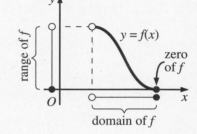

### Combining functions

- To find the sum $(f + g)(x)$ of two functions $f$ and $g$, add their equations (see Example 1(a) on text page 125) or use the graphs of $f$ and $g$ to add the corresponding $y$-values for each value of $x$. (See the diagram on text page 125).

- To find the difference $(f - g)(x)$ of two functions $f$ and $g$, subtract the equation for $g(x)$ from the equation for $f(x)$ or use the graphs of $f$ and $g$ to subtract the corresponding $y$-values for each value of $x$. (See the diagram on text page 125.)

- To find the product $(f \cdot g)(x)$ of two functions $f$ and $g$, multiply their equations.

- To find the quotient $\left(\dfrac{f}{g}\right)(x)$ of two functions $f$ and $g$, where $g(x) \neq 0$, divide the equation for $f$ by the equation for $g$. (See Example 1(b) on text page 125.)

- To find the composite $(f \circ g)(x)$ of two functions $f$ and $g$, substitute $g(x)$ for $x$ in the equation for $f$. (See Examples 2 and 3 on text page 127.)

## CHECKING THE MAIN IDEAS

1. Tell whether the graph at the right is that of a function. If it is, give the domain, the range, and the zeros.

2. Explain why the equation $x = -y^2$ does not define $y$ as a function of $x$.

3. Give the domain, range, and zeros of the function $f(x) = \sqrt{x^2 - 4}$.

4. *Critical Thinking* Describe the graph of $g(x) = \dfrac{x^2 - 4}{x + 2}$. Specify the range and zeros of the function.

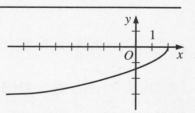

**5.** Refer to the graphs of $y = f(x)$ and $y = g(x)$ at the right.

   **a.** Find $(f + g)(1)$ and $(f - g)(1)$.

   **b.** For what values of $x$ is $(f - g)(x) \geq 0$?

   **c.** What is the maximum value of $(f + g)(x)$?

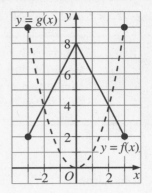

**Let $f(x) = x + 2$ and $g(x) = x^2 + x - 2$.**

  **6.** Find $(f + g)(x)$ and $(f - g)(x)$.

  **7.** Find $(f \cdot g)(x)$ and $\left(\dfrac{f}{g}\right)(x)$.

  **8.** Find $f(g(3))$ and $(f \circ g)(x)$.

  **9.** Find $g(f(3))$ and $(g \circ f)(x)$.

## USING THE MAIN IDEAS

**Example 1** Sketch the graph of $f(x) = \begin{cases} x^3 \text{ if } -2 \leq x < 1 \\ -x + 2 \text{ if } 1 \leq x \leq 4 \end{cases}$.

    Use the graph to find the range and zeros of $f$.

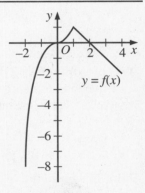

**Solution** The graph of $f$ is shown at the right. The range, or set of $y$-values on the graph, is $\{y \mid -8 \leq y \leq 1\}$. The zeros are 0 and 2.

**Example 2** Use a computer or graphing calculator to find the real solutions of $\sqrt{x + 4} = x$. Give answers to the nearest hundredth.

**Solution** **Method 1**

    Graph $y = \sqrt{x + 4}$ and $y = x$. Use the trace, zoom, or rescale feature to find that $x \approx 2.56$.

    **Method 2**

    Graph $y = \sqrt{x + 4} - x$. Each $x$-intercept is a solution. Use the trace, zoom, or rescale feature to find that $x \approx 2.56$.

## Exercises

**10.** Name two types of function rules for which the domain is *not* the set of real numbers.

**11. a.** Sketch the graph of $f(x) = \begin{cases} x \text{ if } x < 0 \\ -x \text{ if } x \geq 0 \end{cases}$.

   **b.** Use the graph to find the range and zeros of the function.

   **c.** Write a single function rule for $f$.

**12. *Application*** If \$1 is invested at an annual interest rate of $r$ (expressed as a decimal) compounded monthly, then the value of the investment in one year will be $A = \left(1 + \dfrac{r}{12}\right)^{12}$ dollars.

   **a.** Is $A$ a function of $r$? Explain.

   **b.** Give the domain and range of $A$.

**13. *Writing*** Write a paragraph in which you identify a property that holds for the addition of two real numbers but not for the composition of two functions. Include examples to support your answer.

**14.** Let $f(x) = \dfrac{1}{x}$. Show that $(f \circ f)(x) = x$ for all nonzero values of $x$.

*ADVANCED MATHEMATICS* Student Resource Guide **35**

# Graphs and Inverses of Functions

| OVERVIEW | Sections 4-3 and 4-4 investigate the effect on the graph of an equation when the equation is altered. Specific changes in an equation cause the graph of the equation to be reflected, stretched, shrunk, or slid. Two special functions, periodic functions and inverse functions, are explored. |
|---|---|

## KEY TERMS

**EXAMPLE/ILLUSTRATION**

**Reflection in a line** (p. 131)

flipping a figure over a *line of reflection*, which is located halfway between a point and its reflection

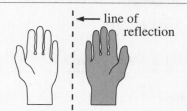

**Axis of symmetry** (p. 133)

a line that divides a figure into halves; if the figure were folded over this line, the halves would match exactly

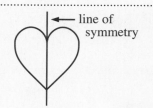

**Point of symmetry** (p. 133)

a point that is the center of a 180° rotation that produces a figure shaped exactly like the original figure

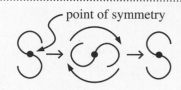

**Periodic function** (p. 138)

a function whose graph repeats or cycles over and over again

Graph of a periodic function

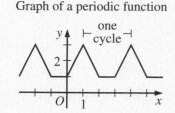

**Fundamental period of a periodic function** (p. 138)

the length of one complete cycle

period = 3

**Amplitude of a periodic function** (p. 139)

half the vertical distance between the maximum and minimum values of the function

amplitude $= \frac{1}{2}(3-1) = 1$

**Translating a graph** (p. 141)

sliding a graph vertically and/or horizontally to a new location

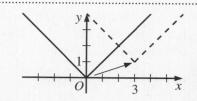

**Inverse functions** (p. 146)
two functions whose composition, in either order, results in the original input value

$$f(x) = 2x + 3, g(x) = \frac{x-3}{2}$$

$$f(g(7)) = f(2) = 7$$
$$g(f(7)) = g(17) = 7$$

$$f(g(x)) = f\left(\frac{x-3}{2}\right) = x$$
$$g(f(x)) = g(2x + 3) = x$$

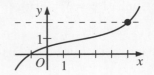

**One-to-one function** (p. 148)
a function whose graph has the property that no horizontal line intersects the graph in more than one point

## UNDERSTANDING THE MAIN IDEAS

### Sketching the graph of an altered equation

- To sketch the graph of $y = -f(x)$, reflect the graph of $y = f(x)$ in the $x$-axis (replace each point $(x, y)$ with $(x, -y)$).

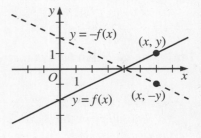

- To sketch the graph of $y = |f(x)|$, reflect in the $x$-axis any portion of the graph of $y = f(x)$ that is below the $x$-axis.

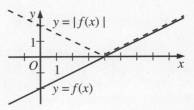

- To sketch the graph of $y = f(-x)$, reflect the graph of $y = f(x)$ in the $y$-axis (replace each point $(x, y)$ with $(-x, y)$).

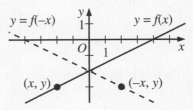

- To sketch the graph of $x = f(y)$, reflect the graph of $y = f(x)$ in the line $y = x$ (replace each point $(x, y)$ with $(y, x)$).

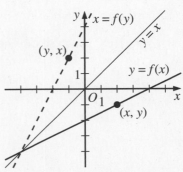

- To sketch the graph of $y = c \cdot f(x)$, replace each point $(x, y)$ on the graph of $y = f(x)$ with $(x, cy)$. If $c > 1$, the graph is stretched vertically. If $0 < c < 1$, the graph is shrunk vertically.

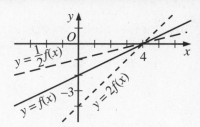

- To sketch the graph of $y = f(cx)$, replace each point $(x, y)$ on the graph of $y = f(x)$ with $\left(\dfrac{x}{c}, y\right)$. If $c > 1$, the graph is shrunk horizontally. If $0 < c < 1$, the graph is stretched horizontally.

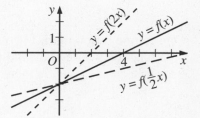

- To sketch the graph of $y - k = f(x - h)$, replace each point $(x, y)$ on the graph of $y = f(x)$ with $(x + h, y + k)$, shifting the graph $h$ units horizontally and $k$ units vertically.

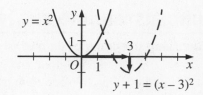

### Testing an equation for symmetry

Substitute:

|  |  |
|---|---|
| • Symmetry in the $x$-axis | $-y$ for $y$ |
| • Symmetry in the $y$-axis | $-x$ for $x$ |
| • Symmetry in the line $y = x$ | $x$ for $y$ and $y$ for $x$ |
| • Symmetry in the origin | $-x$ for $x$ and $-y$ for $y$ |

If the new equation is equivalent to the original equation, then the graph or the equation has the specified symmetry. (See the examples on text page 134.)

*Note:* You can use a graphing calculator to graph the original and new equations and see if their graphs are identical to test for symmetry.

### Periodic functions

If a periodic function $f$ has period $p$ and amplitude $A$, then:

- $y = c \cdot f(x)$ has period $p$ and amplitude $cA$.

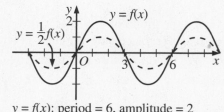

$y = f(x)$: period = 6, amplitude = 2
$y = \frac{1}{2}f(x)$: period = 6, amplitude = 1

- $y = f(cx)$ has period $\dfrac{p}{c}$ and amplitude $A$.

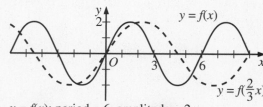

$y = f(x)$: period = 6, amplitude = 2
$y = f(\frac{2}{3}x)$: period = 9, amplitude = 2

**Inverses**

- A function $f(x)$ has an inverse $f^{-1}(x)$ if and only if $f$ is one-to-one. The graph of $f^{-1}$ is the reflection in the line $y = x$ of the graph of $f$, so that if $(a, b)$ is on the graph of $f$, then $(b, a)$ is on the graph of $f^{-1}$.
- To find a rule for an inverse function, $f^{-1}(x)$, let $x = f(y)$ and solve for $y$.

## CHECKING THE MAIN IDEAS

1. The graph of $y = f(x)$ is shown at the right. Match each graph with its equation.

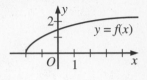

a. 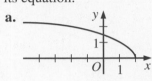  b. 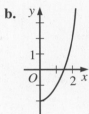  c.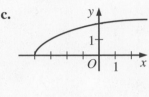

**A.** $y = |f(x)|$        **B.** $y = -f(x)$        **C.** $y = f(-x)$

**D.** $x = f(y)$          **E.** $y = f(x) + 2$      **F.** $y = f(x + 2)$

2. Refer to the graph of $y = f(x)$ shown in Exercise 1. Which equation listed in Exercise 1 has the same graph as $y = f(x)$?

3. Match each equation with its symmetry.

   **a.** $y = x^4 - x^2$    **b.** $y = x^3 + 2x$    **c.** $y^2 = x$    **d.** $y = 6 - x$

   **A.** symmetry in the $x$-axis         **B.** symmetry in the $y$-axis
   **C.** symmetry in the line $y = x$     **D.** symmetry in the origin

4. The graph of a periodic function $f$ is shown at the right. Give its fundamental period and amplitude. Then find $f(100)$ and $f(-100)$. (See Example 1 on text page 139.)

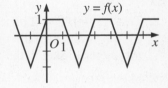

5. Use the graph of $f$ in Exercise 4 to sketch the graphs of $y = -2f(x)$ and $y = -2f(x) - 1$.

6. **a.** Use the graph of $g(x) = x^2 - 4x$ to show that $g$ has no inverse.
   **b.** How can the domain of $x$ be restricted so that $g$ will have an inverse?
   **c.** Using the restricted domain from part (b), sketch the graph of $g$ and $g^{-1}$ on a single set of axes.

7. Let $f(x) = \sqrt[3]{x + 1}$. Find $f^{-1}(x)$. (See Example 2 on text page 147.)

## USING THE MAIN IDEAS

**Example 1**  Sketch the parabola $x = -(y + 2)^2 + 4$. Label the vertex and axis of symmetry.

**Solution**  The required graph is the reflection of the graph of $y = -(x + 2)^2 + 4$ in the line $y = x$. That is, the required graph is obtained by reversing the coordinates of points on the parabola $y = -(x + 2)^2 + 4$ and reversing the roles of $x$ and $y$.

*ADVANCED MATHEMATICS Student Resource Guide* **39**

Since the vertex of the latter parabola is $(-2, 4)$, the parabola
$x = -(y + 2)^2 + 4$ has vertex $(4, -2)$. The axis of the parabola
$y = -(x + 2)^2 + 4$ is the line $x = -2$, so the axis of the parabola
$x = -(y + 2)^2 + 4$ is the line $y = -2$. The parabola
$y = -(x + 2)^2 + 4$ opens downward; since the $x$-values of
$x = -(y + 2)^2 + 4$ decrease from 4, the parabola opens to the left.

Prepare a table of values and sketch the graph.

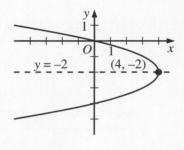

| $x$ | $y = -(x + 2)^2 + 4$ | $\Rightarrow$ | $x = -(y + 2)^2 + 4$ |
|---|---|---|---|
| $-5$ | $-5$ | | $(-5, -5)$ |
| $-4$ | $0$ | | $(0, -4)$ |
| $-3$ | $3$ | | $(3, -3)$ |
| $-2$ | $4$ | | $(4, -2)$ |
| $-1$ | $3$ | | $(3, -1)$ |
| $0$ | $0$ | | $(0, 0)$ |
| $1$ | $-5$ | | $(-5, 1)$ |

**Example 2** If $g(x) = -(x + 2)^2 + 4$, $x \geq -2$, find a rule for $g^{-1}(x)$.

**Solution** Exchange $x$ and $y$
in $y = -(x + 2)^2 + 4$: $x = -(y + 2)^2 + 4$, $y \geq -2$

Solve for $y$:  $(y + 2)^2 = 4 - x$, $y \geq -2$
$y + 2 = \pm\sqrt{4 - x}$, $y \geq -2$
$y = -2 + \sqrt{4 - x}$ since $y \geq -2$

Thus, $g^{-1}(x) = -2 + \sqrt{4 - x}$ where $x \leq 4$. The graph of $g^{-1}(x)$
is the "upper half" of the parabola shown in Example 1 above.

## Exercises

8. Test $|x| - |y| = 2$ to see if its graph has symmetry in (i) the $x$-axis,
   (ii) the $y$-axis, (iii) the line $y = x$, and (iv) the origin. Then use
   symmetry to sketch the graph of the equation.

9. Sketch the parabola $x = 2(y - 3)^2 - 1$. Label the axis of symmetry
   and the vertex.

10. *Writing* Write a paragraph in which you describe how to make various
    changes beginning with the graph of $y = |x|$ to obtain the graph of
    $y + 3 = -\frac{1}{2}|x - 2|$.

11. **a.** Sketch the graphs of $g(x) = 2(x - 3)^2 - 1$, $x \leq 3$, and $g^{-1}(x)$.
    **b.** Find a rule for $g^{-1}(x)$.

12. *Application* The volume of a gas at constant pressure, $V_t$, is affected
    by its temperature, $t$, such that $V_t = V_0\left(1 + \frac{1}{273}t\right)$, where $V_0$ is the
    volume of the gas at $0°C$ and $t$ is the temperature in degrees Celsius.
    **a.** Find the inverse conversion function that expresses $V_0$ in terms of $V_t$.
    **b.** What happens to the volume of a gas as the temperature decreases?

# Applications of Functions

> **OVERVIEW** Section 4-6 is concerned with graphs of functions of two variables in a two-dimensional coordinate system. Section 4-7 presents ways to minimize or maximize functions of two or more variables by expressing the functions in terms of only one variable.

## UNDERSTANDING THE MAIN IDEAS

### Functions of two variables

- Suppose $f$ is a function of variables $x$ and $y$ (written $f(x, y)$). To graph $f$ in a coordinate plane, hold $f$, $x$, or $y$ constant.

  Wages = hourly rate × hours worked, or $W = rt$

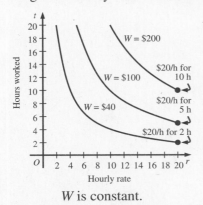

*W* is constant.

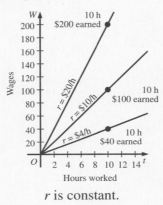

*r* is constant.

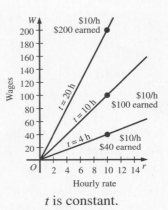

*t* is constant.

- Maximizing or Minimizing a Function
  To maximize or minimize a function of two variables, use a relationship among the variables to express the function in terms of only one variable. The relationship may follow from:
  (a) given information (See Example 1 on text page 158.),
  (b) a geometric property, such as the Pythagorean theorem or similarity of triangles (See Examples 2 and 3 on text pages 159–160.), or
  (c) a formula, such as $d = rt$ (See Example 2 on text page 159.).

## CHECKING THE MAIN IDEAS

**In the diagram, the segment of length $x$ is tangent to the circle.**

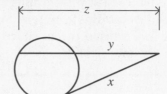

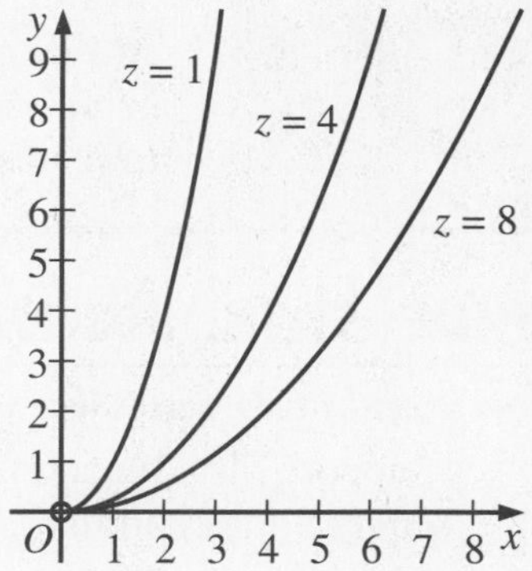

1. *Complete:* The graph shows that the length $y$ is a function of __?__ .

2. **a.** When $x = 2$ and $z = 1$, $y = $ __?__ .
   **b.** When $x = 2$ and $y = 1$, $z = $ __?__ .
   **c.** When $y = 2$ and $z = 1$, $x = $ __?__ .

3. Choose the correct rule for the function.

   **A.** $y = xz^2$     **B.** $y = x^2z$     **C.** $y = \dfrac{z^2}{x}$     **D.** $y = \dfrac{x^2}{z}$

4. *Critical Thinking* Explain the difference between $A(x, y)$ in the two diagrams below.

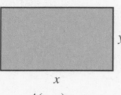

$A(x, y) = xy$

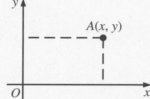

5. A cone is inscribed in a sphere with radius 4, as shown at the right.

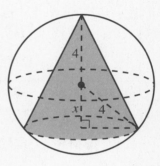

   **a.** Express the height of the cone in terms of $x$.
   **b.** Express the radius of the cone in terms of $x$.
   **c.** Express the volume $V$ of the cone as a function of $x$.

6. Refer to the diagram for Exercise 5. What is the domain of $x$?

## USING THE MAIN IDEAS

**Example 1** Let $p(n, r)$ be the number of ways that $r$ people out of a group of $n$ people can line up for a photograph. Then
$p(n, r) = n \cdot (n - 1) \cdot (n - 2) \cdot \ldots \cdot (n - r + 1)$.
   **a.** Show that $5 \cdot p(4, 2) = p(5, 3)$.
   **b.** Give several pairs $(n, r)$ for which $p(n, r) = 120$.

**Solution**   **a.** For $p(5, 3)$, $n = 5$ and $r = 3$, so $(n - r + 1) = 3$ and
$p(5, 3) = 5 \cdot 4 \cdot 3 = 60$.
For $p(4, 2)$, $n = 4$ and $r = 2$, so $(n - r + 1) = 3$ and
$5 \cdot p(4, 2) = 5(4 \cdot 3) = 60$.
Thus, $5 \cdot p(4, 2) = p(5, 3)$.
   **b.** Find consecutive whole numbers whose product is 120.
$5 \cdot 4 \cdot 3 \cdot 2 \cdot 1 = 120$, so $p(5, 5) = 120$,
$5 \cdot 4 \cdot 3 \cdot 2 = 120$, so $p(5, 4) = 120$, and
$6 \cdot 5 \cdot 4 = 120$, so $p(6, 3) = 120$.

**Example 2** Two sides of a rectangle lie along the axes as shown at the right. One vertex is a point on the line $3x + 4y = 24$.

  **a.** Express the area $A$ of the rectangle as a function of the $x$-coordinate of $P$.
  **b.** What is the domain of the area function?
  **c.** What is the maximum area $A$?

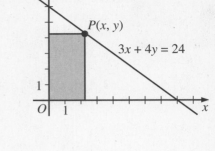

**Solution**
  **a.** Since $P$ is on the line $3x + 4y = 24$, $y = -0.75x + 6$. The rectangle has width $x$ and length $y$, so $A(x, y) = xy$ or $A(x) = x(-0.75x + 6)$.
  **b.** Since $P$ is in Quadrant I and on the line $3x + 4y = 24$, $0 < x < 8$.
  **c.** The area function is the quadratic function $A(x) = -0.75x^2 + 6x$,

  so the maximum value of $A$ occurs when $x = -\dfrac{6}{2(-0.75)} = 4$.

  $A(4) = -0.75(4^2) + 6(4) = 12$
  Thus, the maximum area is 12 square units.

## Exercises

**For Exercises 7 and 8, refer to the function $p(n, r)$ defined in Example 1.**

  **7. a.** Find $p(7, 4)$ and give a verbal description of $p(7, 4)$.
   **b.** Show that $3 \cdot p(7, 4) = p(7, 5)$.
   **c.** Find several pairs $(n, r)$ for which $p(n, r) = 720$.

  **8.** *Writing*  Write a few sentences describing the appearance of the graph of a constant curve such as $p(n, r) = 720$.

  **9.** $P(x, y)$ is an arbitrary point on the line $3x + 4y = 24$ as in Example 2.
   **a.** Express the distance from the origin to $P$ as a function of the $x$-coordinate of $P$.
   **b.** Find the minimum distance from the origin to $P$. (*Hint:* Find the minimum value of the square of the distance function.)

  **10.** *Application*  Use a computer or graphing calculator to find the maximum volume of the cone described in Exercise 5.

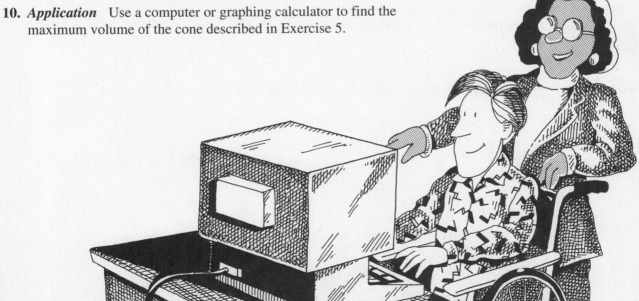

*ADVANCED MATHEMATICS Student Resource Guide* **43**

# CHAPTER REVIEW

## Chapter 4: Functions

Complete these exercises before trying the Practice Test for Chapter 4. If you have difficulty with a particular problem, review the indicated section.

1. Give the domain, range, and zeros of $f(x) = 4 - 3x - x^2$. *(Section 4-1)*

2. Let $f(x) = 2x - 1$ and $g(x) = \dfrac{1}{x}$. Find each of the following. *(Section 4-2)*

   **a.** $(f - g)(x)$      **b.** $(f \cdot g)(x)$      **c.** $\left(\dfrac{f}{g}\right)(x)$

   **d.** $(f \circ g)(2)$      **e.** $(f \circ g)(x)$      **f.** $(g \circ f)(x)$

3. Tell whether the graph of $x^3 + y^3 = 8$ has symmetry in (i) the $x$-axis, (ii) the $y$-axis, (iii) the line $y = x$, and/or (iv) the origin. *(Section 4-3)*

4. Refer to the graph of $y = g(x)$ at the right. Find the fundamental period of $g$, the amplitude of $g$, and $g(51)$. *(Section 4-4)*

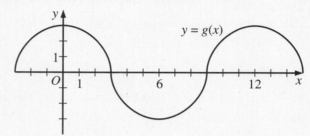

5. Let $f(x) = 1 - 2x$. Find $f^{-1}(3)$ and a rule for $f^{-1}(x)$. *(Section 4-5)*

6. Refer to the graph of $y = g(x)$ in Exercise 4. Tell whether $g$ has an inverse. *(Section 4-5)*

7. The surface area of a square prism is a function of the length of a side, $s$, of the square base and the height, $h$, of the prism.

   **a.** Give a rule for this surface area function.

   **b.** Find two pairs $(s, h)$ for which $A(s, h) = 100$. *(Section 4-6)*

8. Express the area, $A$, of an isosceles right triangle as a function of the length of the hypotenuse, $h$. *(Section 4-7)*

---

Give the domain of each function.

1. $h(x) = \dfrac{x - 1}{x^2 - 4x + 3}$      2. $f(x) = \sqrt{x^2 - 4}$

Sketch the graph of each function. Use the graph to find the range and zeros of the function.

3. $f(x) = x^2 + 6x + 5$      4. $s(x) = \sqrt{f(x)} = \sqrt{x^2 + 6x + 5}$

Let $f(x) = 2x - 1$, $g(x) = \sqrt[3]{x + 1}$, and $h(x) = x^2 + 2x$. Find each of the following.

5. $(f + h)(x)$      6. $\left(\dfrac{f}{g}\right)(x)$      7. $(f \circ g)(x)$      8. $g^{-1}(x)$

The surface area, $A$, of a cube is a function of the length of a side, $s$, of a face.

9. Express $A$ as a function of $s$.      10. Find $A(9)$.

---

**Use the graph of the periodic function**
**$f(x)$ shown at the right.**

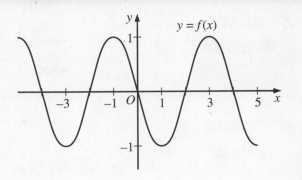

11. What is the period?

12. What is the amplitude?

13. Why does $f$ not have an inverse?

14. Sketch the graph of $y = |f(x)|$.

15. Sketch the graph of $y = f(x - 1)$.

16. Sketch the graph of $y = f(-x) + 2$.

---

**MIXED REVIEW**

*Chapters 1–4*

**Consider the points $A(-2, 1)$, $B(1, 5)$, and $C(2.5, 7)$.**

1. Show that $AB + BC = AC$.

2. Find an equation of the perpendicular bisector of $\overline{AB}$.

3. Find a cubic equation with integral coefficients and roots $\frac{1}{2}$ and $\sqrt{3} - 1$.

4. Find an equation of the form $f(x) = ax^2 + bx + c$ if $f(1) = 4$, $f(2) = 11$, and $f(4) = 31$.

**For Exercises 5–7, let $P(x) = x^4 - 3x^2 - 4$.**

5. Graph $y = P(x)$.       6. Find $P(1 - i)$.

7. Find the quotient and remainder when $P(x)$ is divided by $x + 1$.

8. The velocity, $v$, of a particle with respect to time, $t$, is given by the function $v(t) = t^2 - 4t + 5$. Find the minimum velocity of the particle.

9. Solve $3y^3 - 8y^2 + 6y + 3 > 0$.

10. Find the real root, to the nearest tenth, of $x^3 + x = 6$.

**Refer to the graph of $y = f(x)$ shown at the right.**

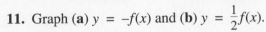

11. Graph (**a**) $y = -f(x)$ and (**b**) $y = \frac{1}{2}f(x)$.

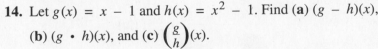

12. What kind of symmetry does the graph of $f$ have?

13. Give the fundamental period and amplitude of $f$. Then find $f(21.5)$.

14. Let $g(x) = x - 1$ and $h(x) = x^2 - 1$. Find (**a**) $(g - h)(x)$, (**b**) $(g \cdot h)(x)$, and (**c**) $\left(\dfrac{g}{h}\right)(x)$.

**The point $P(x, y)$ is a point in quadrant I on the line $x + 2y = 4$, as shown in the figure at the right.**

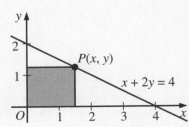

15. Express the area, $A$, of the shaded rectangle in terms of $y$.

16. Find the maximum possible area.

---

    

# Integral and Rational Exponents

## Sections 5-1 and 5-2

| OVERVIEW | These sections present the rules for evaluating and simplifying expressions involving integral exponents and rational exponents. |
|---|---|

## KEY TERMS

EXAMPLE/ILLUSTRATION

**Exponential growth** (p. 170)
a situation in which a fixed percent of growth occurs in a fixed amount of time

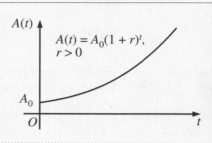

$A(t) = A_0(1 + r)^t$,
$r > 0$

**Exponential decay** (p. 170)
a situation in which a fixed percent of decay occurs in a fixed amount of time

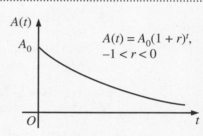

$A(t) = A_0(1 + r)^t$,
$-1 < r < 0$

## UNDERSTANDING THE MAIN IDEAS

### Laws of exponents

- Same bases

  1. $b^x \cdot b^y = b^{x+y}$ $\longrightarrow$ $2^3 \cdot 2 = (2 \cdot 2 \cdot 2)2 = 2^4 = 2^{3+1}$

  2. $\dfrac{b^x}{b^y} = b^{x-y}$ (provided $b \neq 0$) $\longrightarrow$ $\dfrac{2^4}{2^3} = \dfrac{2 \cdot 2 \cdot 2 \cdot 2}{2 \cdot 2 \cdot 2} = 2 = 2^1 = 2^{4-3}$

  3. $b^x = b^y$ if and only if $x = y$ $\longrightarrow$ If $2^x = 2^7$, then $x = 7$.
  (provided $b \neq 0, 1,$ or $-1$)      (Set the exponents equal.)

- Same exponents

  1. $(ab)^x = a^x b^x$ $\longrightarrow$ $(2 \cdot 5)^3 = 2^3 \cdot 5^3$ (or $10^3 = 8 \cdot 125$)

  2. $\left(\dfrac{a}{b}\right)^x = \dfrac{a^x}{b^x}$ (provided $b \neq 0$) $\longrightarrow$ $\left(\dfrac{5}{2}\right)^3 = \dfrac{5^3}{2^3} \left(\text{or } 2.5^3 = \dfrac{125}{8}\right)$

  3. $a^x = b^x$ if and only if $a = b$ $\longrightarrow$ If $(2t)^5 = 7^5$, then $2t = 7$.
  (provided $x \neq 0, a > 0,$ and $b > 0$)

- Power of a power: $(b^x)^y = b^{xy}$ $\longrightarrow$ $(3^5)^2 = 3^5 \cdot 3^5 = 3^{10} = 3^{5 \cdot 2}$

- Zero exponent: $b^0 = 1$ (provided $b \neq 0$) $\longrightarrow$ $(3x)^0 = 1$

- Negative exponent:

  $b^{-x} = \dfrac{1}{b^x}$ (provided $x > 0$ and $b \neq 0$) $\longrightarrow$ $2^{-5} = \dfrac{1}{2^5} = \dfrac{1}{32}$

- Rational exponent ($b > 0$; $b \neq 1$; $p, q$ integers; $q \neq 0$)

  $b^{1/q} = \sqrt[q]{b}$ $\longrightarrow$ $32^{1/5} = \sqrt[5]{32} = \sqrt[5]{2^5} = 2$

  $b^{p/q} = (\sqrt[q]{b})^p = \sqrt[q]{b^p}$ $\longrightarrow$ $32^{4/5} = (\sqrt[5]{32})^4 = 2^4 = 16$

To simplify an expression, rewrite it so it does not contain any negative exponents, powers of powers, fractional exponents in the denominator of a fraction, or a fraction divided by a fraction. Also the power should be as small as possible.

### Exponential growth and decay

To compute exponential growth and decay, substitute the initial value, $A_0$, and the growth or decay rate, $r$, written as a decimal in the model $A(t) = A_0(1 + r)^t$. (When $r > 0$, the model represents exponential *growth*; when $-1 < r < 0$, the model represents exponential *decay*.) Then substitute the required value(s) of $t$. A calculator may be helpful in evaluating the model. (See Examples 1 and 2 on text pages 169–170 and Examples 2 and 5 on text pages 176–177.)

## CHECKING THE MAIN IDEAS

**Match each expression with its value.**

**1.** $(2^{-4} + 4^{-2})^{-1}$ (See Example 4 on text page 171.)

    **A.** 32      **B.** $\dfrac{1}{32}$      **C.** $\dfrac{1}{8}$      **D.** 8

**2.** $(4^{7/2} - 49^{1/2})^{-1/2}$

    **A.** $\dfrac{1}{11}$      **B.** 11      **C.** 63      **D.** 71

**3.** $\left(\dfrac{1000}{27}\right)^{-2/3}$

    **A.** $\dfrac{9}{100}$      **B.** $\dfrac{100}{9}$      **C.** $\dfrac{10{,}000\sqrt{30}}{243}$      **D.** $-\dfrac{100}{9}$

**Simplify.**

**4.** $(-2x^{-1})^{-3}$      **5.** $(3^{-2}x)^2 \cdot x^{-5}$      **6.** $\dfrac{(3x^{-3})^5}{(3x^{-3})^8}$

**7.** $2t^{-4}(s^2t^3 - 3t^6)$      **8.** $\left(\dfrac{27^2}{81^3}\right)^{-1}$      **9.** $\dfrac{x^{-1}(x^{-1/3} - x^{2/3})}{x^{-4/3}}$

**10.** *Critical Thinking*  Explain why $(-25)^{-1/2}$ is not defined, but $-25^{1/2}$ and $-25^{-1/2}$ are defined.

**11.** The price of a new Panther sports car has been increasing at 5% per year. If the sports car costs $14,500 now, find the approximate cost **(a)** 18 months from now and **(b)** 8 months ago. (See Example 2 on text page 176.)

**12.** Suppose the value of a U.S. dollar has been decreasing at 7% per year.

    **a.** Write the value function, $V(t)$, to represent the situation.

    **b.** Find the value of a U.S. dollar three years ago.

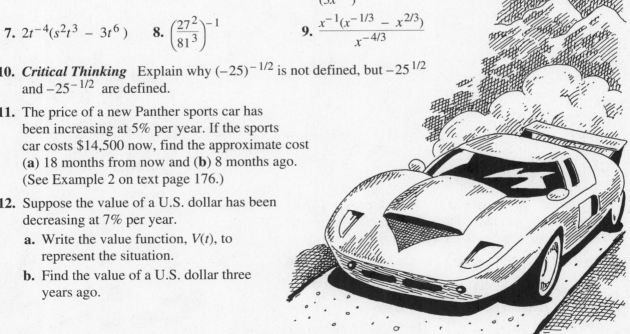

## USING THE MAIN IDEAS

**Example 1** Simplify $\sqrt{\dfrac{32^n}{4^{n+1} \cdot 8^{-3n}}}$ by using powers of the same base.

**Solution** Notice that the bases, 32, 4, and 8, are all powers of 2.

$$\sqrt{\frac{32^n}{4^{n+1} \cdot 8^{-3n}}} = \sqrt{\frac{(2^5)^n}{(2^2)^{n+1} \cdot (2^3)^{-3n}}}$$

$$= \sqrt{\frac{2^{5n}}{2^{2n+2} \cdot 2^{-9n}}} \qquad \leftarrow (b^x)^y = b^{xy}$$

$$= \sqrt{\frac{2^{5n}}{2^2 \cdot 2^{-7n}}} \qquad \leftarrow b^x \cdot b^y = b^{x+y}$$

$$= \sqrt{\frac{2^{12n}}{2^2}} \qquad \leftarrow \frac{b^x}{b^y} = b^{x-y}$$

$$= \frac{\sqrt{(2^{6n})^2}}{\sqrt{4}} \qquad \leftarrow \sqrt{\frac{a}{b}} = \frac{\sqrt{a}}{\sqrt{b}}, \; b^{xy} = (b^x)^y$$

$$= \frac{2^{6n}}{2} \qquad \leftarrow \sqrt{x^2} = x, x \geq 0$$

$$= 2^{6n-1} \qquad \leftarrow \frac{b^x}{b^y} = b^{x-y}$$

**Example 2** Solve $(1 - x)^{-3} = 125$.

**Solution** We want to get $x$ alone on one side.

Raise each side to the power $-\dfrac{1}{3}$ since $-3\left(-\dfrac{1}{3}\right) = 1$.

$$\left[(1 - x)^{-3}\right]^{-1/3} = 125^{-1/3} \qquad \leftarrow \text{If } a = b, \text{ then } a^x = b^x$$

$$(1 - x)^1 = \left(125^{1/3}\right)^{-1} \qquad \leftarrow (b^x)^y = b^{xy}$$

$$1 - x = \frac{1}{5} \qquad \leftarrow 5^{-1} = \frac{1}{5}$$

$$x = \frac{4}{5}$$

## Exercises

13. Simplify $\dfrac{\sqrt[3]{27^4}}{9^5 \cdot 81^{-2}}$ by using powers of the same base.

14. Write $\dfrac{(b^4)^n}{b \cdot b^{n-1}}$ as a single power of $b$.

15. *Application*  An artist sold a painting for \$400. Three years later the painting was sold for \$625. To the nearest percent, what was the annual growth rate in the value of the painting?

16. Solve $25^{x-2} = 125^3$. (See Example 3 on text page 176.)

17. Solve (**a**) $(4x)^{-2} = 100$ and (**b**) $(4 + x)^{-2} = 100$.

18. *Writing*  Explain how to simplify a product involving different roots such as $\sqrt[4]{x} \cdot \sqrt[3]{x} \cdot \sqrt[6]{x}$. Then simplify this expression.

# Exponential Functions

## Sections 5-3 and 5-4

**OVERVIEW** Section 5-3 presents irrational exponents and general exponential functions. Section 5-4 presents the number *e* and the natural exponential function. Both sections focus on applications such as doubling time, half-life, and compound interest.

## KEY TERMS

EXAMPLE/ILLUSTRATION

**Exponential function with base *b*** (p. 181)
a function of the form $f(x) = ab^x$ where *a* and *b* are positive numbers and $b \neq 1$

$y = 2 \cdot 3^x$

**e** (p. 186)
the value that $\left(1 + \dfrac{1}{n}\right)^n$ approaches as *n* becomes very large; a number that is approximately equal to 2.7183

| $n$ | $\left(1 + \dfrac{1}{n}\right)^n$ |
|---|---|
| 1000 | 2.7169 |
| 10,000 | 2.7181 |
| 100,000 | 2.7183 |
| 1,000,000 | 2.7183 |

**Natural exponential function** (p. 187)
the function $y = e^x$

$y = e^x$

## UNDERSTANDING THE MAIN IDEAS

### Exponential functions

- The graph of an exponential function is a smooth, unbroken curve. This means that an exponent can be a rational or an irrational number. The easiest way to evaluate an expression with an irrational exponent, such as $\pi^{\sqrt{7}}$, is to use your calculator: $\boxed{\pi}\ \boxed{y^x}\ \boxed{7}\ \boxed{\sqrt{x}}\ \boxed{=}$ [20.669736].

- You can use given information to specify an exponential function. (See Example 1 on text page 181.)

- If your calculator does not have an "*e*" button, use the $\boxed{1}$, $\boxed{\text{INV}}$, and $\boxed{\ln x}$ keys to find and then store a value for *e*, or simply store 2.7182818 in memory to use as an approximation for *e*.

### Applications

- When an exponential function is used to model exponential growth or decay, its equation is generally written in one of the following forms rather than in the form $f(t) = ab^t$:

1. $A(t) = A_0(1 + r)^t$ ← $A_0$ = amount at time $t = 0$
$r$ = growth/decay rate written as a decimal

2. $A(t) = A_0 b^{t/k}$ ← $A_0$ = amount at time $t = 0$
$k$ = time needed to multiply $A_0$ by $b$

Example: $A(t) = A_0\left(\dfrac{1}{2}\right)^{t/h}$ ← $h$ = half-life of a substance

- Use the "Rule of 72" to estimate how long it takes for a quantity to double:
  If a quantity is growing at $r\%$ per year (or other time unit), then the doubling time is about $(72 \div r)$ years (or other time unit).

- If $P_0$ dollars are invested at an annual rate $r$ (expressed as a decimal) and compounded $n$ times a year, then the amount in the account after $t$ years is given by $P(t) = P_0\left(1 + \dfrac{r}{n}\right)^{nt}$. If \$1 grows to $(1 + x)$ dollars, then the *effective annual yield* is $x$ (expressed as a percent.)

- If $P_0$ is an initial amount (dollars, population, and so on) that is compounded continuously at an annual rate $r$, then the amount at a future time $t$ is given by $P(t) = P_0 e^{rt}$.

## CHECKING THE MAIN IDEAS

**For Exercises 1–3, match each situation with the appropriate model.**

1. A cell population of $P_0$ doubles every 8 hours. What will the population be in $t$ hours?

2. $P_0$ dollars are invested at 8% annual interest compounded continuously. What would the investment be worth in $t$ years?

3. The value of a machine worth $P_0$ dollars now decreases in value by 8% per year. What will the value of the car be in $t$ years?

**A.** $P_0 \cdot e^{0.08t}$

**B.** $P_0 \cdot 8^{t/2}$

**C.** $P_0 \cdot 2^{t/8}$

**D.** $P_0(0.92)^t$

**E.** $P_0(1.08)^t$

4. Evaluate $\sqrt{3}^{\sqrt{2}}$ to the nearest tenth.

5. Find an exponential function $f$ such that $f(0) = 40$ and $f(2) = 90$. (See Example 1 on text page 181.)

6. If \$1000 is invested at 8% per year, in about how many years will the investment be worth \$2000?

7. *Critical Thinking* Explain why the value of an exponential function with positive base $b$ must be a positive number.

8. Evaluate $e^3$ and $3^e$ to the nearest tenth. Which value is greater?

9. Suppose that \$1000 is invested at 8% annual interest compounded monthly.
   a. Find the value of the investment after one year.
   b. Find the effective annual yield.

10. A cell population $t$ days from now is given by $A(t) = 8e^{0.5t}$.
   a. What is the cell population now?
   b. Find the cell population 4 days from now.

# USING THE MAIN IDEAS

**Example 1** The table below shows the amount $A(t)$, in grams, of carbon 11 (which decays exponentially) present after $t$ minutes.

| $t$ | 0 | 10 | 20 | 30 | 40 |
|---|---|---|---|---|---|
| $A(t)$ | 400 | 284 | 201 | 141 | 99 |

   **a.** Find an equation for $A(t)$.

   **b.** About how many grams will be left after $1\frac{1}{4}$ hours?

**Solution**   **a.** Notice that there is about half as much left, about 200 g, when $t = 20$ as when $t = 0$. Therefore, the half-life is about 20 min. The initial amount, $A_0$, is 400 g.

$$A(t) = A_0 \cdot \left(\frac{1}{2}\right)^{t/h} \longrightarrow A(t) = 400 \cdot \left(\frac{1}{2}\right)^{t/20}$$

    **b.** $1\frac{1}{4}$ h is 75 min; substitute 75 for $t$.

$$A(75) = 400 \cdot \left(\frac{1}{2}\right)^{75/20} = 400(0.5)^{3.75} \approx 30$$

    About 30 grams will be left after $1\frac{1}{4}$ hours.

**Example 2** How are the graphs of the functions $f(x) = e^x$ and $g(x) = -e^x$ related to each other? Tell whether or not $g$ has an inverse.

**Solution**  Notice that $g(x) = -f(x)$. Therefore, the graph of $g(x)$ is the reflection of the graph of $f(x)$ in the $x$-axis. The vertical-line test and horizontal-line test can be used to show that $g$ is a one-to-one function, so it has an inverse.

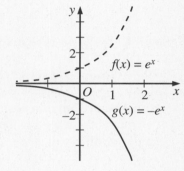

# Exercises

**11.** *Application*  The table shows the value of a house $V(t)$, in thousands of dollars, where $t$ is the number of years the Hernandez family has owned the house. Suppose $V(t)$ increases exponentially.

| $t$ | 0 | 4 | 8 | 12 | 16 |
|---|---|---|---|---|---|
| $V(t)$ | 102 | 122 | 144 | 170 | 203 |

   **a.** Find an equation for $V(t)$.

   **b.** Estimate the value of the house after 20 years of ownership.

**12.** The half-life of Na–24 is 15 hours. What percent of the amount present now will remain 24 hours from now?

**13.**  **a.** How are the graphs of the functions $f(x) = e^{-x}$ and $g(x) = \left(\frac{1}{e}\right)^x$ related to each other? Explain.

    **b.** How is the graph of $g$ related to the graph of $h(x) = e^x$?

**14.** Tell whether the function has an inverse.

   **a.** $f(x) = 1.6^{-x}$          **b.** $g(x) = e^{1/x}$

   **c.** $h(x) = x^e$           **d.** $k(x) = e^{x^2}$

**15.** *Writing*  If you invest \$1 at an annual rate of 6%, explain how the value of the investment after one year is affected by compounding more frequently. Also, explain if you could double the value of your one-year investment by compounding frequently.

        *ADVANCED MATHEMATICS Student Resource Guide*

# Logarithms

**OVERVIEW** | Sections 5-5, 5-6, and 5-7 focus on logarithms: evaluating and applying logarithms (including natural logarithms), and using laws of logarithms and the change-of-base formula. Section 5-7 also presents a technique for solving certain types of real-world problems by using logarithms to solve exponential equations.

## KEY TERMS

**EXAMPLE/ILLUSTRATION**

**Common logarithm of a *positive* number *x*** (p. 191)
the exponent you get when *x* is written as a power of 10

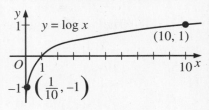

$$\log \frac{1}{10} = \log 10^{-1} = -1$$
$$\log 1 = \log 10^0 = 0$$
$$\log 10 = \log 10^1 = 1$$

**Logarithm of a *positive* number *x* to the base *b* where $b > 0$ and $b \neq 1$** (p. 193)
the exponent you get when *x* is written as a power of *b*

$\log_2 8 = 3$, since $2^3 = 8$

**Natural logarithm of a *positive* number *x*** (p. 193)
(written ln *x* or $\log_e x$) the exponent you get when *x* is written as a power of *e*

$\ln 4.0552 \approx 1.4$, since $e^{1.4} \approx 4.0552$

**Base *b* logarithmic function with $b > 0$ and $b \neq 1$** (p. 193)
the inverse of the base *b* exponential function

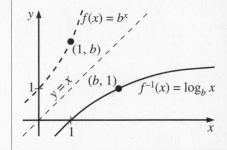

**Natural logarithmic function** (p. 193)
the inverse of the natural exponential function, $f(x) = e^x$

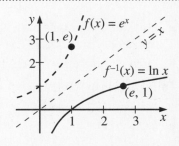

**Exponential equation** (p. 203)
an equation that contains a variable in the exponent

$4^{x-1} = 32$; $e^{2x} = 9$

---

# UNDERSTANDING THE MAIN IDEAS

## Logarithms

- A logarithm is an exponent.
  $x = b^a$ if and only if $\log_b x = a$
  $x = e^k$ if and only if $\ln x = k$

- Laws of Logarithms
  $M$, $N$, and $b$ are positive real numbers and $b \neq 1$.

  1. $\log_b MN = \log_b M + \log_b N$ ⟶ $\log_4 12 = \log_4 2 + \log_4 6$

  2. $\log_b \frac{M}{N} = \log_b M - \log_b N$ ⟶ $\log_4 \frac{5}{2} = \log_4 5 - \log_4 2$

  3. $\log_b M = \log_b N$ if and only if $M = N$ ⟶ If $\log x = \log 8$, then $x = 8$

  4. $\log_b M^k = k \log_b M$, for any real number $k$ ⟶ $\ln x^5 = 5 \ln x$

  5. change-of-base formula: $\log_b c = \dfrac{\log_a c}{\log_a b}$ ⟶ $\log_3 8 = \dfrac{\log 8}{\log 3} = \dfrac{\ln 8}{\ln 3}$

## Exponential equations

- If possible, write each side of the equation as a power of the same number. Then set the exponents equal.

  $\begin{aligned} 4^{x-1} &= 32 \\ (2^2)^{x-1} &= 2^5 \\ 2^{2x-2} &= 2^5 \\ 2x - 2 &= 5 \\ x &= 3.5 \end{aligned}$

- If you cannot write each side of the equation as a power of the same number, take the logarithm of each side and apply law 4 listed above.

  $\begin{aligned} 4^{x-1} &= 30 \\ \log 4^{x-1} &= \log 30 \\ (x-1)\log 4 &= \log 30 \\ x - 1 &= \frac{\log 30}{\log 4} \\ x &= 1 + \frac{\log 30}{\log 4} \approx 3.45 \end{aligned}$

# CHECKING THE MAIN IDEAS

**For Exercises 1–4, match each logarithm with a logarithm equal to it. Do not use a calculator.**

1. $\log_3 4$
2. $\log 64$
3. $\log 12$
4. $\log 0.004$

A. $\log 4 - 3$
B. $3 \log 4$
C. $\dfrac{\log 4}{\log 3}$
D. $\log 4 + \log 3$

5. Write $\ln 20 \approx 3$ in exponential form.

6. Find each logarithm. (Do not use a calculator; see Example 2 on text page 193.)

   a. $\log_3 81$    b. $\log_2 \frac{1}{16}$    c. $\log_2 4\sqrt{2}$    d. $\log_8 1$

7. Express $\left(\dfrac{\sqrt[3]{M}}{\log N}\right)^2$ in terms of $\log M$ and $\log N$. (See Examples 1 and 2 on text page 198.)

8. Simplify $\dfrac{1}{2}(\log_5 50 + \log_5 12.5)$. (See Example 3 on text page 198.)

9. *Critical Thinking*   Use your calculator to evaluate each of the expressions $\log 10^{8.2}$, $10^{\log 8.2}$, $\ln e^{8.2}$, and $e^{\ln 8.2}$. Then write two general laws suggested by these expressions and explain why each must be true.

10. Solve $(1.7)^x = 18$.

11. Solve $25^x = \dfrac{\sqrt{5}}{125^x}$.

## USING THE MAIN IDEAS

**Example 1**   Given $\log 4 \approx 0.6021$, find:
  **a.** $\log 40$          **b.** $\log 0.25$          **c.** $\log 2$

**Solution**   Write each logarithm in term of $\log 4$.
  **a.** $\log 40 = \log(4 \cdot 10) = \log 4 + \log 10 \approx 0.6021 + 1 = 1.6021$

  **b.** $\log 0.25 = \log \dfrac{1}{4} = \log 4^{-1} = -1(\log 4) \approx -0.6021$

  **c.** $\log 2 = \log \sqrt{4} = \log 4^{1/2} = \dfrac{1}{2}\log 4 \approx \dfrac{1}{2}(0.6021) \approx 0.3011$

**Example 2**   Simplify $10^{2 + 4 \log x}$.

**Solution**   $10^{2 + 4 \log x} = 10^2 \cdot 10^{4 \log x} \qquad \leftarrow b^{x+y} = b^x \cdot b^y$
$= 100 \cdot 10^{\log x^4} \qquad \leftarrow \log M^k = k \cdot \log M$
$= 100 \cdot x^4 \qquad\quad \leftarrow 10^x \text{ and } \log x \text{ are inverse functions.}$
$= 100x^4$

## Exercises

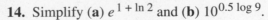

12. Given $\log 9 \approx 0.9542$, find:
  **a.** $\log \dfrac{1}{81}$          **b.** $\log 900$          **c.** $\log 27$

13. *Writing*   Suppose you use your calculator to approximate $\log 20$ and your calculator displays 2.30103. Write a convincing argument to explain why this value must be incorrect.

14. Simplify **(a)** $e^{1 + \ln 2}$ and **(b)** $10^{0.5 \log 9}$.

15. Express $y$ in terms of $x$ if $\log y = 2 \log x - 1$. (See Example 4 on text page 198.)

16. *Application*   If you invest \$100 at 6% annual interest compounded monthly, when will the investment be worth \$150? (See Example 2 on text page 204).

# CHAPTER REVIEW

## Chapter 5: Exponents and Logarithms

Complete these exercises before trying the Practice Test for Chapter 5. If you have difficulty with a particular problem, review the indicated section.

**1.** Simplify each expression. *(Section 5-1)*

    **a.** $-2^{-4}$         **b.** $(-3a^2)(2a^3)^2$     **c.** $\left(\dfrac{a}{b^3}\right)^{-3}$

**2.** Suppose the value of a U.S. dollar decreases 5% per year. Find the value of a U.S. dollar, to the nearest cent, in 4 years. *(Section 5-1)*

**3.** Simplify **(a)** $\left(\dfrac{1}{1000}\right)^{-1/3}$ and **(b)** $(81^{-3/2})^{-1/2}$. *(Section 5-2)*

**4.** Solve $16 = 2^{x+1}$. *(Section 5-2)*

**5.** Find an exponential function $f$ such that $f(0) = 8$ and $f(2) = 18$. *(Section 5-3)*

**6.** Write the formula for continuous compounding. Then use it to find the value of \$1 invested at 5% annual interest compounded continuously for one year. *(Section 5-4)*

**7.** Write the equation $\log_2 8 = 3$ in exponential form. *(Section 5-5)*

**8.** Find the value of $\log \dfrac{1}{10}$. *(Section 5-5)*

**9.** Simplify **(a)** $\dfrac{1}{2}\log_2 36 - \log_2 3$ and **(b)** $\ln \sqrt[3]{e}$. *(Section 5-6)*

**10.** Use a calculator to solve $3^x = 40$ for $x$ to the nearest hundredth. *(Section 5-7)*

Simplify each expression.

**1.** $\sqrt[4]{\dfrac{16^{x-1}}{16^{x-6}}}$     **2.** $\dfrac{\left(3x^{-5/3}\right)^2\left(2x^{-2/3}\right)}{3\left(x^{-2}\right)^3}$     **3.** $\log_b b^{2x}$

**4.** $(2^{-2} - 4^{-2})^{-1}$     **5.** $(2^{-2} \cdot 4^{-2})^{-1}$     **6.** $3^{(\log_3 5 + \log_3 10)}$

**7.** $\ln e^3$     **8.** $e^{\ln 5}$     **9.** $\log_6 8 + \log_6 9 - \log_6 2$

Express each of the following in terms of $\log A$ and $\log B$.

**10.** $\log \dfrac{10A}{B^2}$         **11.** $\log_{AB} 10$

The temperature $T$ (in °F) of a glass of lemonade left in a room for $t$ minutes is $T(t) = 95 - 60e^{-0.15t}$.

**12.** What is the temperature after 5 minutes?

**13.** How soon will the temperature be 80°F?

**Ten thousand dollars is invested at 10% compounded quarterly.**

**14.** What is its value of the investment after one year?

**15.** How long does it take for the investment to double in value?

    

**Use a calculator to find the value of *x* to the nearest hundredth.**

**16.** $e^x = 12$        **17.** $(1.02)^x = 2$        **18.** $(e^{-x})^3 = 0.055$

**19.** Express *y* in terms of *x* if $\log y = 4 \log x - 3 \log 3$.

**Solve. Express *x* as a logarithm if necessary.**

**20.** $e^{2x} - 5e^x + 4 = 0$        **21.** $4^{2x} - 8 \cdot 4^x + 12 = 0$

---

**MIXED REVIEW**

*Chapters 1–5*

**1.** Solve $|2x - 5| \le 4$ and graph the solution on a number line.

Let $f(x) = 2x^2 - x^4$.

**2.** Sketch the graph of $y \le P(x)$.

**3.** Find the domain, range, and zeros of *f*.

**4.** Describe the graph of $x = f(y)$. Tell whether the graph of $x = f(y)$ would represent a function.

**Solve each equation or system of equations.**

**5.** $25^{x-3} = 125^{4/3}$        **6.** $\begin{aligned} y &= 2x^2 - 5x - 3 \\ 3x + y &= 1 \end{aligned}$        **7.** $3x^2 + 2x + 1 = 0$

**Simplify.**

**8.** $3 - (1 - 2i)^2$        **9.** $\log 0.001 + 10^{2 \log 3}$        **10.** $(a^{-1} - b^{-1})(a + b)^{-1}$

$\triangle ABC$ **has vertices *A*(0, 0), *B*(10, 0) and *C*(2, 4).**

**11.** Find an equation of the line through *C* that contains the median to $\overline{AB}$.

**12.** Find the perimeter of $\triangle ABC$.

**13.** The sum of the roots of a cubic equation is −2 and the product is 12. If one root is −1, find the other two roots.

**14.** Tell whether the graph of the equation $|x| - |y| = 2$ has symmetry in (i) the *x*-axis, (ii) the *y*-axis, (iii) the line $y = x$, and/or (iv) the origin.

**15.** Suppose $1000 is invested at 6% per year. How much more will the investment be worth after 2 years if the interest is compounded continuously rather than monthly?

**16.** Solve $e^x = 100$ to the nearest hundredth.

**17.** Let $f(x) = \sqrt{x}$ and $g(x) = 2x - 1$. Find **(a)** $(f + g)(x)$, **(b)** $(f \circ g)(x)$, and **(c)** $(g \circ f)(x)$.

**18.** Express the area of an equilateral triangle in terms of its height, *h*.

**19.** Sketch the graph of a periodic function $f(x)$ with amplitude 3 and period 4. Then sketch the graph of $f(\frac{1}{2}x)$ on the same set of axes.

**20.** The sector of a circle shown at the right has radius *r* and perimeter 50 cm. Let $A(r)$ be the area function of the sector.

**a.** If $A(1) = 24$, $A(5) = 100$, and $A(10) = 150$, find a quadratic function for $A(r)$.

**b.** Find the maximum possible area of the sector.

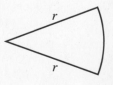

# Coordinate Proofs and Conic Sections

## Sections 6-1, 6-2, 6-3, and 6-4

**OVERVIEW**  Section 6-1 presents a method for proving theorems from geometry: coordinate, or analytic, geometry. Sections 6-2, 6-3, and 6-4 explore many of the conic sections: circles, ellipses, and hyperbolas. (The last type of conic section, parabolas, will be presented in Section 6-5.)

## KEY TERMS

**EXAMPLE/ILLUSTRATION**

**Conic section** (p. 213)
cross sections (circle, ellipse, hyperbola, parabola) resulting from slicing a double cone by a plane (When the result is a point, a line, or a pair of lines, the cross section is called a *degenerate* conic section. Each cross section has an equation of the form $Ax^2 + Bxy + Cy^2 + Dx + Ey + F = 0$.)

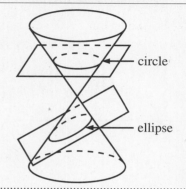

**Circle** (p. 219)
the set of all points $P(x, y)$ in the plane that are $r$ units from point $C(h, k)$

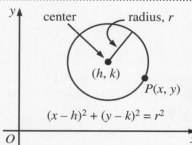

$$(x - h)^2 + (y - k)^2 = r^2$$

**Ellipse** (p. 226)
the set of all points $P(x, y)$ in the plane such that $PF_1 + PF_2 = 2a$, where $F_1$ and $F_2$ are fixed points called *foci* and $a$ is a positive real number

(See the figures under "Ellipse" in the Understanding the Main Ideas section.)

**Hyperbola** (p. 231)
the set of all points $P(x, y)$ in the plane such that $|PF_1 - PF_2| = 2a$, where $F_1$ and $F_2$ are fixed points called *foci* and $a$ is a positive real number (The *asymptotes* of a hyperbola are lines that the hyperbola approaches more and more closely.)

(See the figures under "Hyperbola" in the Understanding the Main Ideas section.)

## UNDERSTANDING THE MAIN IDEAS

### Coordinate proofs

To prove a theorem from geometry by using coordinates:
1. introduce a coordinate system in which the figure has as many zero coordinates as possible; and
2. choose an appropriate method from the summary on text page 217.

## Circles

- Remember that the circle with center $(h, k)$ and radius $r$ has equation $(x - h)^2 + (y - k)^2 = r^2$ (center-radius form). If the equation of a circle is not in this form, rewrite the equation. (See Example 1(b) on text page 220.)

- To find the coordinates of the points where a line and circle intersect, either graph the equations by hand, with a graphing calculator, or with a computer, or find them algebraically by following the steps listed in the summary on text page 221.

## Ellipses

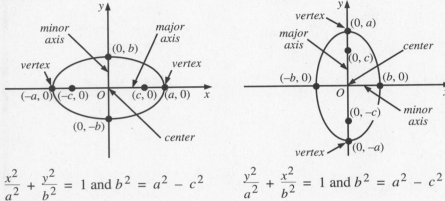

$$\frac{x^2}{a^2} + \frac{y^2}{b^2} = 1 \text{ and } b^2 = a^2 - c^2 \qquad \frac{y^2}{a^2} + \frac{x^2}{b^2} = 1 \text{ and } b^2 = a^2 - c^2$$

When you graph an ellipse of one of these forms, find the intercepts and a few first-quadrant points. Then use the symmetry of the ellipse to locate points in the other quadrants.

- To find the intersection points of a line and an ellipse, use the same methods as for a line and a circle: graph the equations or solve algebraically by following the steps in the summary on text page 221.

## Hyperbolas

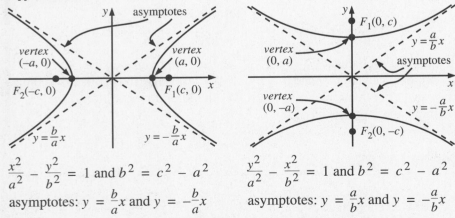

$$\frac{x^2}{a^2} - \frac{y^2}{b^2} = 1 \text{ and } b^2 = c^2 - a^2 \qquad \frac{y^2}{a^2} - \frac{x^2}{b^2} = 1 \text{ and } b^2 = c^2 - a^2$$

asymptotes: $y = \frac{b}{a}x$ and $y = -\frac{b}{a}x$ asymptotes: $y = \frac{a}{b}x$ and $y = -\frac{a}{b}x$

To graph a hyperbola of one of these forms, graph the intercepts and the asymptotes. Also sketch a rectangle with dimensions $2a$ and $2b$, as shown on text pages 232 and 233. Then find some first-quadrant points. Use the symmetry of the hyperbola to locate points in the other quadrants.

- To find the intersection points of a line and a hyperbola, use the same methods as for a line and a circle: graph the equations or solve algebraically by following the steps in the summary on text page 221.

## CHECKING THE MAIN IDEAS

**Match the required proof with the method you would use.**

1. Prove: $\overline{AX}$ is a median of $\triangle ABC$.
2. Prove: $\triangle XYZ$ is isosceles.
3. Prove: $\overline{AP}$ is an altitude of $\triangle ABC$.

**A.** distance formula

**B.** slope formula

**C.** midpoint formula

**Refer to isosceles trapezoid *PQRS* shown at the right.**

4. Give the coordinates of points $R$ and $S$ without introducing new variables.

5. Prove that the line segments joining the midpoints of successive sides of any isosceles trapezoid form a rhombus.

6. Write an equation of the circle with center $(-2, 5)$ that is tangent to the $y$-axis (that is, intersects the $y$-axis in exactly one point).

7. Write the equation $x^2 + y^2 - 8x - 2y - 8 = 0$ in center-radius form. Give the center and radius. (See Example 1(b) on text page 220.)

8. Sketch the ellipse $4x^2 + y^2 = 36$. Find the coordinates of its vertices and foci.

9. An ellipse has vertex $(-10, 0)$, focus $(-8, 0)$, and center at the origin. Find an equation of the ellipse. (See the Example on text page 227.)

10. Sketch the hyperbola $9x^2 - 16y^2 = 144$ and its asymptotes. Give equations for the asymptotes.

11. Sketch the hyperbola $(x - 3)y = 6$.

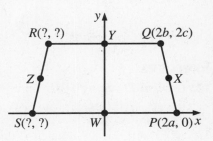

## USING THE MAIN IDEAS

**Example 1** Sketch the graphs of $4x^2 + 16y^2 = 16$ and $3x - 2y = -2$ on a single set of axes. Then determine algebraically where they meet.

**Solution** The graph suggests that there are two intersection points and that $(0, 1)$ is one of them.

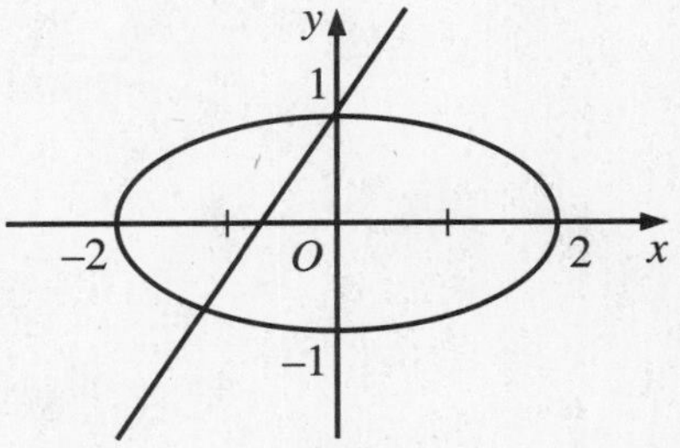

1. Solve $3x - 2y = -2$ for $y$:

$$y = \frac{3x + 2}{2} = 1.5x + 1$$

2. Substitute for $y$ in $4x^2 + 16y^2 = 16$:

$$4x^2 + 16(1.5x + 1)^2 = 16$$

3. Solve for $x$: $4x^2 + 16(2.25x^2 + 3x + 1) = 16$

$$40x^2 + 48x + 16 = 16$$
$$40x^2 + 48x = 0$$
$$5x^2 + 6x = 0$$
$$x(5x + 6) = 0$$
$$x = 0 \text{ or } x = -1.2$$

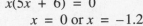

4. Find the *y*-value for each *x*-value.

$x = 0: y = 1.5(0) + 1 = 1 \quad x = -1.2: y = 1.5(-1.2) + 1 = -0.8$

The intersection points are $(0, 1)$ and $(-1.2, -0.8)$.

**Example 2** Give the domain and the range of the function $y = -\sqrt{x^2 - 1}$. Then graph the function.

**Solution** Since $-\sqrt{x^2 - 1}$ is defined only if $|x| \geq 1$, the domain is $\{x \mid x \leq -1 \text{ or } x \geq 1\}$. Notice that $y = 0$ when $x = \pm 1$ and $y < 0$ when $x < -1$ or $x > 1$. Therefore, the range is $\{y \mid y \leq 0\}$. Squaring both sides of $y = -\sqrt{x^2 - 1}$ gives $y^2 = x^2 - 1$, or $x^2 - y^2 = 1$. This shows that the graph of $y = -\sqrt{x^2 - 1}$ is the bottom half of the hyperbola $x^2 - y^2 = 1$.

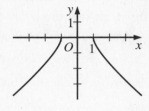

## Exercises

12. ***Critical Thinking*** Explain how the graph of $x^2 + y^2 = r^2$ is related to the graph of $(x - h)^2 + (y - k)^2 = r^2$. Then explain how the graph of $\frac{x^2}{25} + \frac{y^2}{36} = 1$ is related to the graph of $\frac{(x-2)^2}{25} + \frac{(y+4)^2}{36} = 1$.

13. ***Writing*** Discuss some of the advantages of using coordinate methods to prove a geometric theorem.

14. Give the domain and the range of the function $y = \sqrt{4 - (x - 2)^2}$. Then graph the function.

**Sketch the graphs of the given equations on a single set of axes. Then determine algebraically where the graphs intersect.**

15. $9x^2 + 16y^2 = 144$
$x + y = 0$

16. $2x + y = 6$
$xy = -8$

17. Find an equation of the hyperbola with center at the origin, a vertex at $(0, 4)$, and an asymptote with equation $y = 2x$.

18. ***Application*** Suppose you drive 100 miles for *t* hours at a speed of *r* miles per hour. Describe the set of points $P(r, t)$.

# More About Conics

## Sections 6-5, 6-6, and 6-7

**OVERVIEW**

Section 6-5 presents the geometric definition of a parabola. Section 6-6 illustrates some methods of solving a system of second-degree equations in two variables. The last section of Chapter 6, Section 6-7, explores the properties common to all the conic sections.

## KEY TERMS

**EXAMPLE/ILLUSTRATION**

**Parabola** (p. 238)

the set of all points $P(x, y)$ that are the same distance from a fixed point $F$ (called the *focus*) and a fixed line $d$ that does not contain $F$ (called the *directrix*)

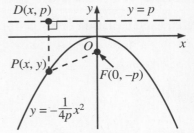

(See the other figures on text page 238 also.)

**Eccentricity of a conic section** (p. 247)

the positive constant $e$ such that $PF : PD = e$, where $F$ is a fixed point (*focus*) not on a fixed line $d$ (*directrix*), and $PD$ is the perpendicular distance from $P$ to $d$

In the diagram above, $PF = PD$, so $PF : PD = 1$ and so $e = 1$. For *every* parabola, $e = 1$.

## UNDERSTANDING THE MAIN IDEAS

### Parabolas

Recall that the vertex of a parabola is the point at which the axis of symmetry intersects the parabola (see text page 37). A parabola whose vertex is at the origin and whose directrix is either horizontal or vertical has an equation of the form $y = \pm\frac{1}{4p}x^2$ or $x = \pm\frac{1}{4p}y^2$. (Refer to the diagrams and examples on text pages 238–239.)

### Systems of second-degree equations

A system of two second-degree equations in two variables can be solved by algebraic methods (substitution or elimination), graphical methods such as a graphing calculator or computer graphing software, or a combination of algebraic and graphical methods. These approaches are illustrated in Examples 1–3 on text pages 242–244.

### Conic sections

- A specific value or range of values for the eccentricity $e$ is associated with each type of conic section.
  If $e = 0$, the conic is a circle.
  If $0 < e < 1$, the conic is an ellipse.
  If $e = 1$, the conic is a parabola.
  If $e > 1$, the conic is a hyperbola.

When you know the focus, directrix, and value of $e$ for a conic, you can derive an equation for that conic. (See Example 1, text page 248.)

- You can identify a nondegenerate conic section with the equation $Ax^2 + Bxy + Cy^2 + Dx + Ey + F = 0$ ($A, B, C$ not all zero) by calculating the value of $B^2 - 4AC$. (See Example 2 on text page 248.)
  If $B^2 - 4AC < 0$, the graph is a circle ($B = 0, A = C$), or an ellipse.
  If $B^2 - 4AC = 0$, the graph is a parabola.
  If $B^2 - 4AC > 0$, the graph is a hyperbola.

## CHECKING THE MAIN IDEAS

**Match each equation or value for $e$ with the appropriate conic section.**

1. $16x^2 + 8xy + y^2 + 16x - 4y - 2 = 0$      **A.** circle

2. $2x^2 + 2y^2 + 8x + 16y + 15 = 0$      **B.** ellipse

3. $e = \dfrac{4}{3}$      **C.** hyperbola
        **D.** parabola

4. $e = \dfrac{3}{4}$

5. Find the vertex, focus, and directrix of each parabola whose equation is given.

   **a.** $y = \dfrac{1}{2}x^2$      **b.** $y - 2 = \dfrac{1}{2}x^2$      **c.** $x = -y^2 - 3$

**Solve each system.**

6. $x^2 + y^2 = 10$         7. $x^2 + 16y^2 = 64$
   $(x - 2)^2 + (y + 1)^2 = 25$     $x^2 - y = -4$

8. *Critical Thinking* Discuss some of the limitations of using a graphical approach to solve a system of second-degree equations.

9. Given that the distance from point $P(x, y)$ to point $F(4, 0)$ is $\dfrac{4}{5}$ the distance from $P$ to line $d$ with equation $x = 6.25$, write an equation to show that $P$ lies on an ellipse. (See Example 1 on text page 248.)

10. Identify the graph of $y = 2x + \dfrac{1}{x}$ if the graph is not degenerate. (See Example 2 on text page 248.)

## USING THE MAIN IDEAS

**Example 1** Find an equation of the parabola with focus $(0, -2.5)$ and directrix $y = 2.5$.

**Solution** The vertex is equidistant from the focus and directrix, so the vertex is at the origin. The directrix is horizontal, the focus is on the $y$-axis, and the vertex is $(0, 0)$, so the equation has the form $y = \pm\dfrac{1}{4p}x^2$. Since the directrix is *above* the focus, the parabola opens downward. The equation has the form $y = -\dfrac{1}{4p}x^2$. $F(0, -p) = (0, -2.5)$, so $p = 2.5$ and $4p = 10$. The required equation is $y = -\dfrac{1}{10}x^2$.

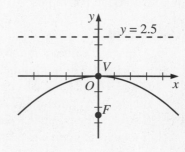

**Example 2** Describe the graph of $x^2 + xy = 0$.

**Solution**
$x^2 + xy = 0$
$x(x + y) = 0$
$\quad x = 0$ or $y = -x$

The graph of $x = 0$ is a line, the $y$-axis.

The graph of $y = -x$ is also a line, passing through the origin.

Therefore, the graph of $x^2 + xy = 0$ is two intersecting lines.

## Exercises

**Find an equation of each parabola and describe its graph.**

**11.** focus, $(2, 0)$; directrix: $x = -2$

**12.** vertex, $(0, 0)$; focus, $(0, 1)$

**13.** Solve $9x^2 + 25y^2 = 225$ and $x^2 + y^2 = 16$ simultaneously.
(See Example 3 on text page 244.)

**14.** *Writing* Compare the graphs of $4x^2 + y^2 = 0$ and $4xy = 0$.

**15.** *Application* When an equation of an ellipse is given in the
form $\dfrac{x^2}{a^2} + \dfrac{y^2}{b^2} = 1$, where $b^2 = a^2 - c^2$, the eccentricity

$e$ is equal to $\dfrac{c}{a}$. Mercury's orbit is an ellipse with

the sun almost at one focus. The closest and
farthest distances of Mercury from the sun are
about $2.86 \times 10^7$ miles and $4.30 \times 10^7$ miles,
respectively. Find the eccentricity of this
orbit to the nearest tenth.

ORBIT OF MERCURY

*ADVANCED MATHEMATICS Student Resource Guide* **63**

# CHAPTER REVIEW

## Chapter 6: Analytic Geometry

Complete these exercises before trying the Practice Test for Chapter 6. If you have difficulty with a particular problem, review the indicated section.

1. Study the coordinates of equilateral $\triangle OPQ$. Then give the coordinates of point $Q$ in terms of $a$. *(Section 6-1)*

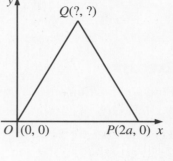

2. How would you show that point $Z\left(a, \dfrac{a\sqrt{3}}{3}\right)$ is equidistant from $O$, $P$, and $Q$? *(Section 6-1)*

3. Find an equation of the circle with center $(-3, 2)$ and radius 4. *(Section 6-2)*

4. Sketch the ellipse $\dfrac{x^2}{25} + \dfrac{y^2}{36} = 1$. Find the coordinates of the vertices and foci. *(Section 6-3)*

5. Sketch the hyperbola $\dfrac{y^2}{16} - \dfrac{x^2}{9} = 1$. Find the coordinates of the vertices and foci, and the equations of the asymptotes. *(Section 6-4)*

6. Tell whether the parabola $x = \dfrac{1}{16}y^2$ opens up, down, left, or right. Find the vertex and focus, and the equation of the directrix. *(Section 6-5)*

7. Solve the system $x^2 + y^2 = 25$ and $x^2 + (y + 5)^2 = 64$. *(Section 6-6)*

8. The figure at the right shows $F(4, 0)$, directrix $x = 1$, and $P(x, y)$. The distance from $P(x, y)$ to $F$ is twice the distance from $P$ to the directrix. Show that $P$ lies on a hyperbola. (*Hint:* Begin with the equation $PF = 2 \cdot PD$.) *(Section 6-7)*

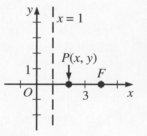

1. Prove that the line segments joining the midpoints of the sides of a square form a square.

**Write an equation for each figure described.**

2. circle with center $(2, -3)$ and radius 5

3. ellipse with center $(0, 0)$, vertex $(3, 0)$, and focus $(2, 0)$

4. hyperbola with asymptotes $y = \pm 2x$ and vertex $(0, 2)$

5. parabola with vertex $(2, -3)$ and focus $(2, 0)$

**Sketch the graph of each conic section. Then give the coordinates of the center, foci, and vertices where applicable. If the conic is a circle, also give the radius.**

6. $x^2 + 4y^2 - 16 = 0$        7. $x^2 - 4y^2 - 16 = 0$

8. $x^2 + y^2 - 4x + 8y + 11 = 0$        9. $2y^2 - x - 4y + 5 = 0$

**Graph each system and label the points of intersection.**

10. $x^2 + y^2 - 25 = 0$        11. $x^2 - y^2 - 1 = 0$
    $x^2 + y^2 - 2x + 14y + 25 = 0$        $x^2 - y - 3 = 0$

---

**Solve each system algebraically.**

**12.** $x^2 + y^2 - 25 = 0$
$x^2 + y^2 - 2x - 14y + 25 = 0$

**13.** $x^2 - y^2 - 1 = 0$
$x^2 - y - 3 = 0$

**Identify the graphs of the following nondegenerate conic sections.**

**14.** $x^2 - 2xy - y^2 + y - 5 = 0$

**15.** $x^2 - 3xy + 5y^2 - 2x - 7 = 0$

**A parabolic arch spans a road that is 50 feet wide.**

**16.** How high is the arch if the middle 30 feet of road has a minimum clearance of 12 feet?

**17.** Where is the focus of the arch in relation to the vertex?

---

**MIXED REVIEW**

*Chapters 1–6*

**Solve.**

**1.** $81^{5/4} = 27^{x/6}$

**2.** $x^e = 50$

**3.** $x^4 - 9x^2 + 8 = 0$

**4.** When a ball is thrown upward, its approximate height in meters $t$ seconds later is given by $h(t) = 36 + 24t - 5t^2$. Find the domain, range, and zeros of $h$.

**5.** Show that $x^3 + 2x^2 + 4 = 0$ has no rational root, but does have an irrational root. Then approximate the irrational root to the nearest tenth.

**6.** Let $d(s)$ be the total distance in feet traveled by a car moving at $s$ mi/h from the time the driver applies the brakes to the time the car stops. If $d(20) = 41$, $d(30) = 76$, and $d(40) = 123$, find a quadratic model for $d$. Then find the distance traveled by a car moving 50 mi/h.

**7.** Simplify $\log_2 48 - \frac{1}{2}\log_2 9$.

**8.** The half-life of strontium 90 is about 28 years. A laboratory has 1 kg of strontium 90. How many grams will remain a century from now?

**9.** Solve $2x^3 + x^2 < 8x + 4$.

**10.** Let $f(x) = \frac{1}{2}x - 5$. Find $f^{-1}(x)$ and show that $(f \circ f^{-1})(x) = (f^{-1} \circ f)(x)$.

**11.** Describe the graph of $y^2 - 4x^2 = 4$. Then tell if the graph has symmetry in (a) the $x$-axis, (b) the $y$-axis, (c) the line $y = x$, and/or (d) the origin.

**12.** Find an equation of the line through $(3, 5)$ that is perpendicular to the line through $(-3, -1)$ and $(5, 3)$.

**13.** Use the graph of $y = f(x)$ to sketch the graphs of
(a) $y = f(x + 2)$ and (b) $y = f(2x)$.

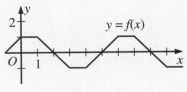

**14.** Simplify $(1 + x^{-1})^{-1}$.

**15.** Describe the graph of $x^2 + y^2 + 2x - 6y = 39$.

**16.** Solve $|4x - 1| < 2$.

**17.** Simplify (a) $(1 + i)^2$ and (b) $\dfrac{3 + i}{1 + i}$.

---

# Angles, Arcs, and Sectors

## Sections 7-1 and 7-2

| OVERVIEW | Section 7-1 presents a basic topic: the measurement of angles in degrees and in radians. In Section 7-2, this topic is applied to arc length, the area of a sector of a circle, and apparent size. |
|---|---|

## KEY TERMS

EXAMPLE/ILLUSTRATION

**Angle** (p. 257)

the rotation about a point of an initial ray to its terminal ray (The arrow shows the direction of the rotation.)

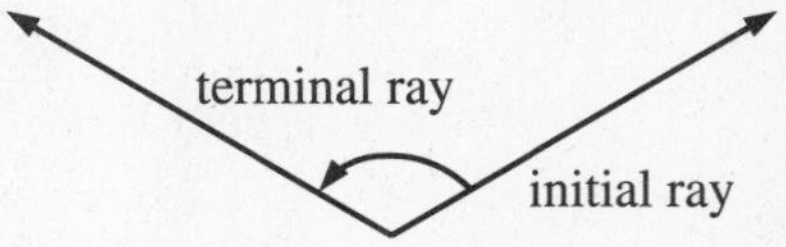

**Revolution** (p. 257)

a complete circular motion, or 360 *degrees* (360°), or $2\pi$ (about 6.28) *radians*

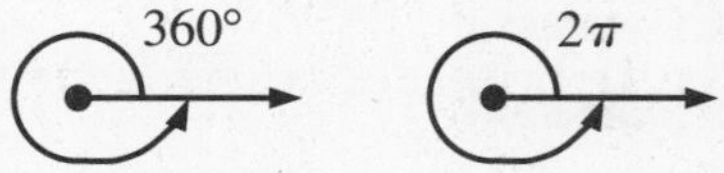

**Radian measure** (p. 258)

the radian measure $\theta$ of central angle $AOB$ is the number of radius units in the length $s$ of arc $AB$

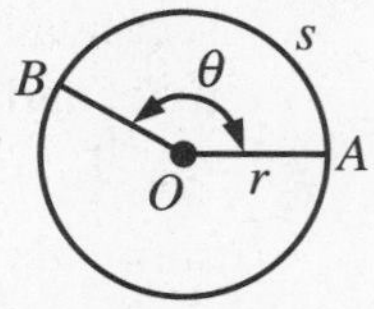

$$\theta = \frac{s}{r}$$

**Standard position of an angle** (p. 259)

an angle with vertex at the origin and initial ray along the positive $x$-axis

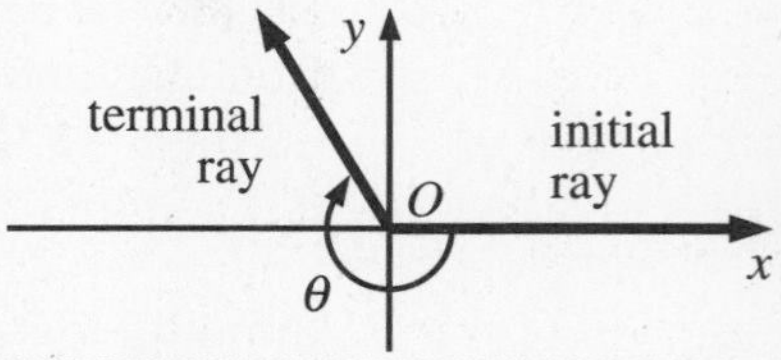

**Quadrantal angle** (p. 259)

an angle in standard position whose terminal ray lies along an axis; an angle whose measure is a multiple of 90° or $\frac{\pi}{2}$

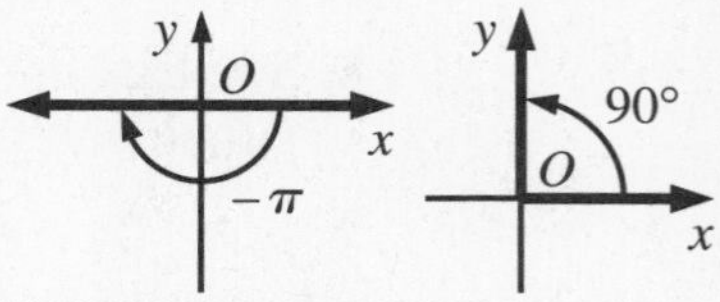

**Coterminal angles** (p. 260)

angles in standard position with the same terminal ray

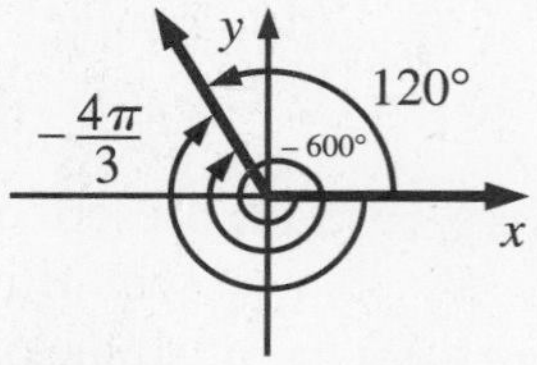

$$\theta = (120 + 360n)°$$
$$\theta = \frac{2\pi}{3} + 2n\pi$$
$\left.\right\}$ $n$ is an integer.

**Sector of a circle** (p. 263)

a region bounded by a central angle and the intercepted arc

**Apparent size of an object** (p. 264)

the measure of an angle subtended at a person's eye by an object being observed

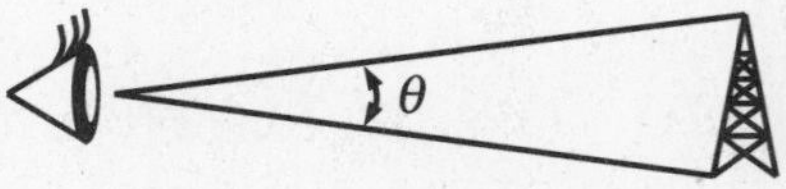

## UNDERSTANDING THE MAIN IDEAS

### Angle measurement

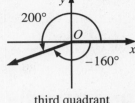

third quadrant
coterminal angles

- Angles can be measured in radians, degrees, or revolutions. A counterclockwise rotation represents a positive angle measure and a clockwise rotation represents a negative angle measure.
- An angle in standard position is classified according to the quadrant in which its terminal ray lies. That is, if an angle is not a quadrantal angle, then it is a first-quadrant, second-quadrant, third-quadrant, or fourth-quadrant angle.
- To find a radian measure $\theta$, use the formulas:

  1. $\theta = \dfrac{s}{r}$ where $s$ is the arc length and $r$ is the radius, or

  2. $1 \text{ degree} = \dfrac{\pi}{180}$ radians (when the degree measure of $\theta$ is known)

- To find a degree measure when you know the radian measure, use the formula $1 \text{ radian} = \dfrac{180}{\pi}$ degrees. Note that an angle measure without a degree symbol, such as 1.6 or $\dfrac{\pi}{4}$, indicates radian measure. Learn the radian and degree measures of the special angles shown in the diagrams near the top of text page 259.
- To find an angle coterminal with a given angle, add or subtract $360°$ (or $2\pi$) from the given angle measure.

### Sectors of circles

- If $\theta$ is the *degree* measure of a central angle of a circle:

$$s = \frac{\theta}{360} \cdot 2\pi r$$

$$K = \frac{\theta}{360} \cdot \pi r^2$$

Use the degree mode setting on your calculator.

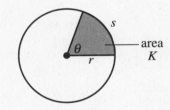

- If $\theta$ is the *radian* measure of a central angle of a circle:

$$s = r\theta$$

$$K = \frac{1}{2}r^2\theta \text{ and } K = \frac{1}{2}rs$$

Use the radian mode setting on your calculator.

Arc length $s$ is measured in *linear* units, area $K$ in *square* units.

### Apparent size

To estimate the diameter, $d$, of a planet, star, or other distant object, use the formula $d \approx r\theta$ where $r$ is the distance of the planet or star from the observer's eye and $\theta$ is the apparent size.

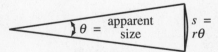

$r$ = object's distance from the eye

## CHECKING THE MAIN IDEAS

**Choose the measure that is equivalent to the given one.**

1. $36° \ 36' \ 36''$    **A.** $36.6°$    **B.** $36.66°$    **C.** $36.61°$

2. $\dfrac{11\pi}{4}$    **A.** $315°$    **B.** $-315°$    **C.** $495°$

3. Convert $200°$ to radians. Leave your answer in terms of $\pi$.

4. Convert **(a)** $3\pi$ and **(b)** $-\dfrac{5\pi}{3}$ to degrees.

   *ADVANCED MATHEMATICS Student Resource Guide* **67**

5. Give the radian measure of a central angle $\theta$ with radius 4 and arc length 6.

6. Find two angles, one positive and one negative, that are coterminal with the angle $-\dfrac{\pi}{2}$. (See Example 2 on text page 260.)

7. A sector of a circle has radius 8 cm and central angle 1.75 radians. Find its arc length and area.

## USING THE MAIN IDEAS

**Example 1** Convert 1.8 radians to decimal degrees (to the nearest tenth) and to degrees and minutes (to the nearest ten minutes).

**Solution** **Method 1**

$$1.8 \text{ radians} = 1.8 \times \frac{180}{\pi} \qquad \leftarrow \text{Use your calculator to evaluate.}$$
$$\approx 103.1324°$$
$$= (103 + 0.1324)°$$
$$= 103° + 0.1324(60)'$$
$$\approx 103°10' \text{ (to the nearest ten minutes)}$$

**Method 2**

Use a proportion to determine the correct conversion form:

$$\frac{180°}{\pi \text{ radians}} = \frac{x°}{1.8 \text{ radians}}$$
$$x = 1.8 \times \frac{180}{\pi}$$

Proceed as in Method 1 above.

**Example 2** A sector of a circle has perimeter 14 cm and area 6 cm$^2$. Find all possible radii.

**Solution** Since $P = 2r + s, 2r + s = 14$ or $s = 14 - 2r$

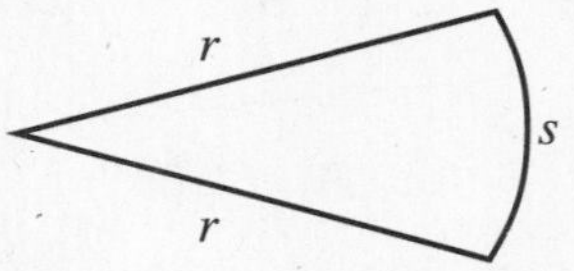

$$K = \frac{1}{2}rs$$
$$6 = \frac{1}{2}r(14 - 2r)$$
$$12 = -2r^2 + 14r$$
$$r^2 - 7r + 6 = 0$$
$$(r - 6)(r - 1) = 0$$
$$r = 6 \text{ cm} \quad \text{or} \quad r = 1 \text{ cm}$$

## Exercises

8. Convert 305.9° to radians (to the nearest hundredth).

9. Convert 4.06 radians to decimal degrees (to the nearest tenth) and to degrees and minutes (to the nearest ten minutes).

10. A sector of a circle has perimeter 16 cm and area 12 cm$^2$. Find all possible arc lengths.

11. *Application* The apparent size of a mountain 60 mi away is 0.04 radians. Find the height of the mountain to the nearest hundred feet.

# The Sine and Cosine Functions

## Sections 7-3 and 7-4

**OVERVIEW**  Sections 7-3 and 7-4 focus on the two most important trigonometric functions: the sine and cosine. Skills developed include evaluating the sine and cosine functions, sketching their graphs, and solving simple trigonometric equations.

## KEY TERMS

**Sine of an angle $\theta$ (written sin $\theta$)** (p. 268)

the ratio $\frac{y}{r}$, where $P(x, y)$ is a point on circle $O$ with equation $x^2 + y^2 = r^2$ and $\theta$ is an angle in standard position with terminal ray $OP$

**Cosine of an angle $\theta$ (written cos $\theta$)** (p. 268)

the ratio $\frac{x}{r}$, where $P(x, y)$ is a point on circle $O$ with equation $x^2 + y^2 = r^2$ and $\theta$ is an angle in standard position with terminal ray $OP$

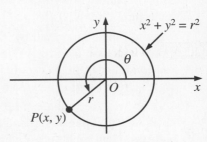

$$\sin \theta = \frac{y}{r} \,;\, \cos \theta = \frac{x}{r}$$

**Unit circle** (p. 269)

the circle $x^2 + y^2 = 1$

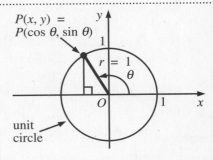

$$\left. \begin{array}{l} \sin \theta = y \\ \cos \theta = x \end{array} \right\} \text{Note:} \ r = 1$$

**Reference angle for an angle $\theta$** (p. 275)

the acute positive angle $\alpha$ formed by the terminal ray of $\theta$ and the $x$-axis

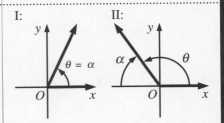

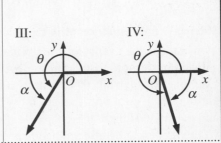

## UNDERSTANDING THE MAIN IDEAS

### Evaluating sin θ and cos θ

- *Notice:* For any angle θ, $-1 \le \cos\theta \le 1$ and $-1 \le \sin\theta \le 1$

- If θ is a quadrantal angle, use its coordinates on the unit circle, as shown, to evaluate sin θ and cos θ. If θ is not a quadrantal angle, you can determine the sign of sin θ and cos θ by determining the signs of *x* and *y* in the quadrant containing the terminal ray of θ.

- A scientific calculator allows you to approximate sin θ and cos θ for any angle θ. *Caution:* Be careful to use the appropriate angle setting (degree or radian).

- Another way to evaluate sin θ and cos θ is to use trigonometric tables (see text pages 800 and 805–821). If θ is a first-quadrant angle, you can look up sin θ and cos θ easily. If θ is not a first-quadrant angle, find the value of sin α or cos α, where α is the reference angle for θ, since $\sin\alpha = |\sin\theta|$ and $\cos\alpha = |\cos\theta|$. Then use the diagram above to determine the appropriate sign for sin θ or cos θ.

- If θ is a special first-quadrant angle, you can remember the exact values of sin θ and cos θ by recalling the pattern in the values shown in the table at the right. If θ is a multiple of a special first-quadrant angle, use the table to find the value of the sine or cosine of the reference angle and determine the appropriate sign.

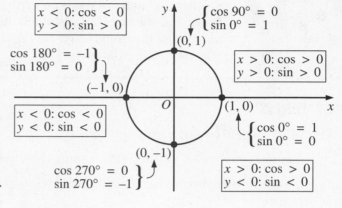

$$\begin{cases} \cos 90° = 0 \\ \sin 0° = 1 \end{cases}$$

$$\cos 180° = -1 \\ \sin 180° = 0 \end{Bmatrix}$$

$x < 0: \cos < 0$
$y > 0: \sin > 0$

$x > 0: \cos > 0$
$y > 0: \sin > 0$

$x < 0: \cos < 0$
$y < 0: \sin < 0$

$$\begin{cases} \cos 0° = 1 \\ \sin 0° = 0 \end{cases}$$

$x > 0: \cos > 0$
$y < 0: \sin < 0$

$$\cos 270° = 0 \\ \sin 270° = -1 \end{Bmatrix}$$

| θ (°) | θ (rad.) | sin θ | cos θ |
|---|---|---|---|
| 0° | 0° | $\frac{\sqrt{0}}{2}$ | $\frac{\sqrt{4}}{2}$ |
| 30° | $\frac{\pi}{6}$ | $\frac{\sqrt{1}}{2}$ | $\frac{\sqrt{3}}{2}$ |
| 45° | $\frac{\pi}{4}$ | $\frac{\sqrt{2}}{2}$ | $\frac{\sqrt{2}}{2}$ |
| 60° | $\frac{\pi}{3}$ | $\frac{\sqrt{3}}{2}$ | $\frac{\sqrt{1}}{2}$ |
| 90° | $\frac{\pi}{2}$ | $\frac{\sqrt{4}}{2}$ | $\frac{\sqrt{0}}{2}$ |

### Graphing sin θ and cos θ

Become familiar with the general graphs of sin θ and cos θ shown on text page 278. To graph these functions yourself, you can use a calculator to find ordered pairs (θ, sin θ) and (θ, cos θ), or you can plot values corresponding to special angles, remembering that $\frac{\sqrt{2}}{2} \approx 0.7$ and $\frac{\sqrt{3}}{2} \approx 0.9$. Keep in mind that the sine and cosine functions repeat their values (and graphs) every 360° or $2\pi$ radians:

$\sin(\theta \pm 360°) = \sin\theta$ and $\sin(\theta \pm 2\pi) = \sin\theta$
$\cos(\theta \pm 360°) = \cos\theta$ and $\cos(\theta \pm 2\pi) = \cos\theta$

### Solving simple trigonometric equations

You can use the graphs of sin θ and cos θ to help you solve some simple trigonometric equations. For example, to solve sin θ = 1, draw a horizontal line at sin θ = 1. This shows that sin θ = 1 when θ = 90° and every 360° thereafter; θ = 90° + *n* · 360°, where *n* is an integer. Similarly, sin θ = 0.5 when θ = 30° + *n* · 360° or θ = 150° + *n* · 360°, where *n* is an integer.

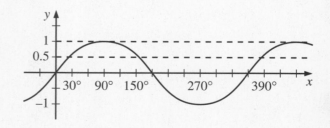

## CHECKING THE MAIN IDEAS

1. Which expression is *not* equal to 0?

   **A.** $\sin(-270°)$  **B.** $\cos 90°$  **C.** $\sin(-2\pi)$  **D.** $\cos\left(-\dfrac{\pi}{2}\right)$

2. In which quadrant(s) do $\sin\theta$ and $\cos\theta$ have the same sign?

   **A.** I only  **B.** I and II  **C.** I and III  **D.** I and IV

3. Which expression is *not* equal to $\sin 210°$?

   **A.** $\sin(-120°)$  **B.** $\cos 120°$  **C.** $\sin(-30°)$  **D.** $\cos 240°$

4. ***Critical Thinking*** Solve the equation $\cos\theta = 1.5$ for all $\theta$ in radians. Justify your answer.

5. Without using a calculator or table, state whether each expression is positive, negative, or zero.

   **a.** $\cos 141°$  **b.** $\sin 308°$  **c.** $\sin\left(-\dfrac{6\pi}{5}\right)$  **d.** $\cos\dfrac{13\pi}{4}$

6. Express each of the following in terms of a reference angle. (See Example 1 on text page 275.)

   **a.** $\sin 284°$  **b.** $\cos 284°$  **c.** $\sin 500°$  **d.** $\sin(-318°)$

7. Use a calculator or table to find the value of each expression to four decimal places.

   **a.** $\sin(-17.2°)$  **b.** $\cos 48°$  **c.** $\cos(-1.3)$  **d.** $\sin 4$

8. Give the exact value of each expression in simplest radical form.

   **a.** $\cos(-225°)$  **b.** $\sin 210°$  **c.** $\cos\dfrac{11\pi}{6}$  **d.** $\sin\left(-\dfrac{2\pi}{3}\right)$

## USING THE MAIN IDEAS

**Example 1** If $\theta$ is a third-quadrant angle and $\cos\theta = -\dfrac{1}{3}$, find $\sin\theta$.

**Solution** **Method 1**

Since $\cos\theta = \dfrac{x}{r} = \dfrac{-1}{3}$, let $x = -1$ and $r = 3$.

Sketch a circle with radius 3. Locate the third-quadrant point $P$ on the circle with $x$-coordinate $-1$. Since $P(x, y)$ is a point on the circle $x^2 + y^2 = 9$,

$$(-1)^2 + y^2 = 9$$
$$y^2 = 8$$
$$y = \pm 2\sqrt{2}$$

Since $P$ is in Quadrant III, $y < 0$.

Thus, $y = -2\sqrt{2}$ and $\sin\theta = \dfrac{y}{r} = \dfrac{-2\sqrt{2}}{3} = -\dfrac{2\sqrt{2}}{3}$.

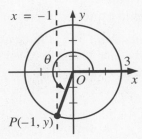

**Method 2**

For every angle $\theta$, $(\sin\theta)^2 + (\cos\theta)^2 = 1$.
(See Class Exercise 7 on text page 271.)

Therefore, $(\sin\theta)^2 + \left(-\dfrac{1}{3}\right)^2 = 1$.

---

*ADVANCED MATHEMATICS Student Resource Guide*

$$(\sin \theta)^2 = 1 - \frac{1}{9} = \frac{8}{9}$$

$$\sin \theta = \pm\sqrt{\frac{8}{9}} = \pm\frac{\sqrt{8}}{\sqrt{9}} = \pm\frac{2\sqrt{2}}{3}$$

Since $\theta$ is in Quadrant III, $\sin \theta < 0$.

Therefore, $\sin \theta = -\dfrac{2\sqrt{2}}{3}$.

**Example 2** Sketch the graph of $y = \sin(-\theta)$. Then use the graph to express $\sin(-\theta)$ in terms of $\sin \theta$.

**Solution** The graph of $y = f(-x)$ is obtained by reflecting the graph of $y = f(x)$ in the $y$-axis. Reflect the graph of $y = \sin \theta$ in the $y$-axis, as shown. Comparing the graph of $y = \sin(-\theta)$ to the graph of $y = \sin \theta$, we see that the graph of $y = \sin(-\theta)$ is the reflection of $y = \sin \theta$ in the $x$-axis. Therefore, $\sin(-\theta) = -\sin \theta$.

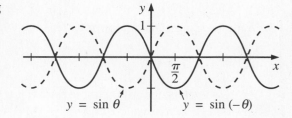

$y = \sin \theta \qquad y = \sin(-\theta)$

## Exercises

9. Use the figure at the right to find $\sin \theta$ and $\cos \theta$. (See Example 1 on text page 268.)

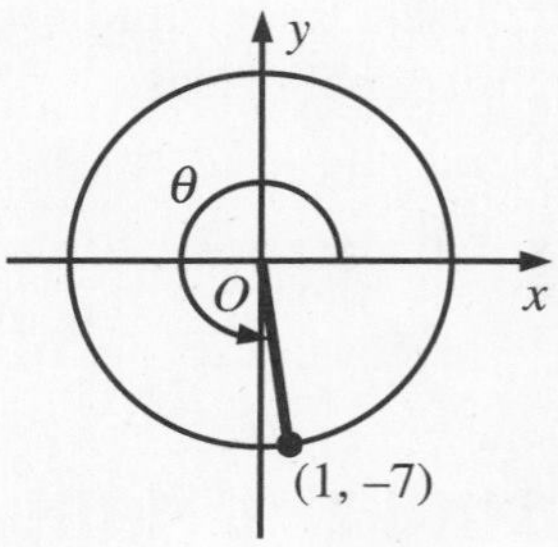

10. If $\theta$ is a second-quadrant angle and $\sin \theta = \dfrac{21}{29}$, find $\cos \theta$.

11. *Writing*

    **a.** Use the unit-circle definition of $\sin \theta$ and $\cos \theta$ to describe how to find the values of $\theta$ between $0°$ and $360°$ for which $\sin \theta = -\cos \theta$.

    **b.** Name the values of $\theta$ between $0°$ and $360°$ for which $\sin \theta = -\cos \theta$.

    **c.** Verify your answers in part (b) by evaluating $\sin \theta$ and $\cos \theta$ for each value of $\theta$.

    **d.** Describe how you could use the graphs of $\sin \theta$ and $\cos \theta$ to find these values of $\theta$.

$(1, -7)$

12. Sketch the graph of $y = \cos(-\theta)$. Then use the graph to express $\cos(-\theta)$ in terms of $\cos \theta$.

13. *Application* The latitude of Bombay, India is about $19°$N. Using 3963 mi for the radius of Earth, about how far from Bombay is the North Pole? (See Example 3 on text page 277.)

# More Trigonometric Functions and Inverse Trigonometric Functions

## Sections 7-5 and 7-6

**OVERVIEW**  Section 7-5 is concerned with evaluating and graphing the four other trigonometric functions: tangent, cotangent, secant, and cosecant. In Section 7-6, definitions of the inverse trigonometric functions are developed and applied.

### KEY TERMS

### EXAMPLE/ILLUSTRATION

If $P(x, y)$ is a point on circle $O$ with equation $x^2 + y^2 = r^2$ and $\theta$ is an angle in standard position with terminal ray $OP$:

**Tangent of $\theta$ (written tan $\theta$) (p. 282)**

the ratio $\dfrac{y}{x}$, $x \neq 0$

**Cotangent of $\theta$ (written cot $\theta$) (p. 282)**

the ratio $\dfrac{x}{y}$, $y \neq 0$

**Secant of $\theta$ (written sec $\theta$) (p. 282)**

the ratio $\dfrac{r}{x}$, $x \neq 0$

**Cosecant of $\theta$ (written csc $\theta$) (p. 282)**

the ratio $\dfrac{r}{y}$, $y \neq 0$

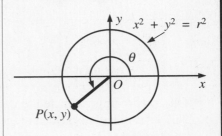

$\tan \theta = \dfrac{y}{x}, x \neq 0$

$\cot \theta = \dfrac{x}{y}, y \neq 0$

$\sec \theta = \dfrac{r}{x}, x \neq 0$

$\csc \theta = \dfrac{r}{y}, y \neq 0$

**Inverse tangent function (written Tan$^{-1}$ $x$) (p. 286)**

$y = \text{Tan}^{-1} x$ means that $\tan y = x$ and $-\dfrac{\pi}{2} < y < \dfrac{\pi}{2}$.

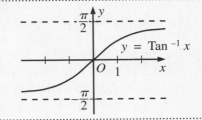

**Inverse sine function (written Sin$^{-1}$ $x$) (p. 287)**

$y = \text{Sin}^{-1} x$ means that $\sin y = x$ and $-\dfrac{\pi}{2} \leq y \leq \dfrac{\pi}{2}$.

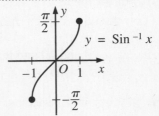

**Inverse cosine function (written Cos$^{-1}$ $x$) (p. 288)**

$y = \text{Cos}^{-1} x$ means that $\cos y = x$ and $0 \leq y \leq \pi$.

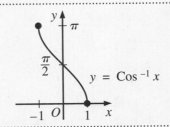

# UNDERSTANDING THE MAIN IDEAS

## The other trigonometric functions

- Let $P(x, y)$ be a point on the unit circle $x^2 + y^2 = 1$, and let $\theta$ be an angle in standard position with terminal ray $OP$.

| Function | $\sin \theta = y$ | $\cos \theta = x$ | $\tan \theta = \dfrac{y}{x}$ $= \dfrac{\sin \theta}{\cos \theta}$ | $\cot \theta = \dfrac{x}{y}$ $= \dfrac{\cos \theta}{\sin \theta}$ | $\sec \theta = \dfrac{1}{x}$ $= \dfrac{1}{\cos \theta}$ | $\csc \theta = \dfrac{1}{y}$ $= \dfrac{1}{\sin \theta}$ |
|---|---|---|---|---|---|---|
| Domain | all $\theta$ | all $\theta$ | $\theta \neq \dfrac{\pi}{2} + n\pi$ | $\theta \neq n\pi$ | $\theta \neq \dfrac{\pi}{2} + n\pi$ | $\theta \neq n\pi$ |
| Range | $-1 \leq \sin \theta \leq 1$ | $-1 \leq \cos \theta \leq 1$ | all reals | all reals | $\sec \theta \geq 1$ or $\sec \theta \leq -1$ | $\csc \theta \geq 1$ or $\csc \theta \leq -1$ |
| Period | $2\pi$ | $2\pi$ | $\pi$ | $\pi$ | $2\pi$ | $2\pi$ |

- One way to approximate $\tan \theta$, $\cot \theta$, $\sec \theta$, or $\csc \theta$ is to use a scientific calculator. Be sure the calculator is in the appropriate angle setting (degree or radian). If you get an error message when you try to evaluate a particular function value, it is likely that the function is not defined for that angle measure. If your calculator does not have cot, csc, and sec keys, use the tan, sin, and cos keys, respectively, with the $\boxed{1/x}$ key. (For example, evaluate $\cot 67.2°$ by finding $\tan 67.2°$ and then pressing the $\boxed{1/x}$ key.)

- Another way to evaluate trigonometric functions is to use trigonometric tables (see text pages 800 and 805–821). If $\theta$ is a first-quadrant angle, you can look up the values easily or give exact values if the angle is a special angle (see the diagrams near the top of text page 259). If $\theta$ is not a first-quadrant angle, you must find the associated reference angle, determine the value of the trigonometric function of the reference angle, and determine the appropriate sign based on the quadrant containing the terminal ray of the angle (see the diagram at the right).

- You should become familiar with the graphs of $\tan \theta$ and $\sec \theta$ shown on text page 284 and of $\cot \theta$ and $\csc \theta$ (see Written Exercises 9 and 10 and text page 285). Keep in mind the domain, range, and period of each function when you draw its graph.

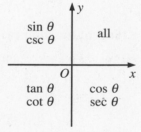

Quadrants in which values are positive

## Inverse trigonometric functions

- 

| Function | $y = \operatorname{Sin}^{-1} x$ | $y = \operatorname{Cos}^{-1} x$ | $y = \operatorname{Tan}^{-1} x$ |
|---|---|---|---|
| Domain | $-1 \leq x \leq 1$ | $-1 \leq x \leq 1$ | all reals |
| Range | $-\dfrac{\pi}{2} \leq y \leq \dfrac{\pi}{2}$ | $0 \leq y \leq \pi$ | $-\dfrac{\pi}{2} < y < \dfrac{\pi}{2}$ |

**Caution:** The range of $y = \operatorname{Cos}^{-1} x$ is unlike the ranges of the other two inverse trigonometric functions.

- To evaluate an inverse trigonometric function using a calculator, choose the angle setting (degrees or radians), enter the value, and then push the appropriate key(s), such as $\boxed{\text{INV}}$, $\boxed{\text{TAN}}$, $\boxed{\text{ARCTAN}}$, or $\boxed{\tan^{-1}}$.

- To evaluate an inverse trigonometric function without using a calculator, use the function definition.

## CHECKING THE MAIN IDEAS

1. Find the value of each expression to four significant digits.

   **a.** sec 195°  **b.** tan 325°  **c.** cot 1.11  **d.** csc 2.01

2. Express each of the following in terms of a reference angle.

   **a.** cot 107°  **b.** csc 250°  **c.** tan 4  **d.** sec 6

3. Give the values of $x$, in radians, for which sec $x$ is **(a)** undefined, **(b)** 0, **(c)** 1, and **(d)** −1.

4. *Critical Thinking*  Explain how you could translate and reflect the graph of $y = \tan x$ to obtain the graph of $y = \cot x$.

5. Find the value of each expression to the nearest tenth of a degree.

   **a.** $\text{Sin}^{-1}\, 0.8$  **b.** $\text{Cos}^{-1}\left(-\dfrac{2}{5}\right)$  **c.** $\text{Tan}^{-1}\, 4$

6. Find the value of each expression to the nearest hundredth of a radian.

   **a.** $\text{Cos}^{-1}\,(-0.66)$  **b.** $\text{Sin}^{-1}\dfrac{7}{8}$  **c.** $\text{Tan}^{-1}\,(-0.48)$

## USING THE MAIN IDEAS

**Example 1**  If $\theta$ is a fourth-quadrant angle and $\csc \theta = -\dfrac{17}{8}$, find the values of the other five trigonometric functions.

**Solution**  **Method 1**

Since $\csc \theta = -\dfrac{17}{8}$, then $\dfrac{r}{y} = \dfrac{17}{-8}$, $r = 17$, and $y = -8$.

Sketch a circle with radius 17. Locate the fourth-quadrant point $P$ on the circle with $y$-coordinate $-8$.

Since $P(x, -8)$ is a point on the circle $x^2 + y^2 = 17^2$, then $x^2 + (-8)^2 = 289$, $x^2 = 225$, and $x = \pm 15$.

$P$ is in Quadrant IV, so $x > 0$. Therefore, $x = 15$.

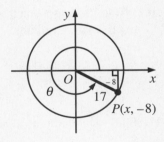

$$\sin \theta = \frac{y}{r} = -\frac{8}{17} \qquad \cos \theta = \frac{x}{r} = \frac{15}{17} \qquad \tan \theta = \frac{y}{x} = -\frac{8}{15}$$

$$\sec \theta = \frac{r}{x} = \frac{17}{15} \qquad \cot \theta = \frac{x}{y} = -\frac{15}{8}$$

**Method 2**

Since $\csc \theta = \dfrac{r}{y} = -\dfrac{17}{8}$ and $\sin \theta = \dfrac{y}{r}$, $\sin \theta = -\dfrac{8}{17}$.

From Exercise 7 on text page 271, $(\sin \theta)^2 + (\cos \theta)^2 = 1$.

$$\left(-\frac{8}{17}\right)^2 + (\cos \theta)^2 = 1$$

$$(\cos \theta)^2 = 1 - \frac{64}{289} = \frac{225}{289}$$

$$\cos \theta = \frac{15}{17} \leftarrow \cos \theta > 0 \text{ in Quadrant IV}$$

Thus, we have: $\sin \theta = -\dfrac{8}{17}$; $\cos \theta = \dfrac{15}{17}$; $\sec \theta = \dfrac{1}{\cos \theta} = \dfrac{17}{15}$;

$$\tan \theta = \frac{\sin \theta}{\cos \theta} = \frac{-\dfrac{8}{17}}{\dfrac{15}{17}} = -\frac{8}{15}; \cot \theta = \frac{1}{\tan \theta} = -\frac{15}{8}.$$

**Example 2** Find (**a**) the approximate value and (**b**) the exact value of

$$\cos\left(\text{Sin}^{-1}\frac{3}{4}\right).$$

**Solution**   **a.** To find the approximate value, set your calculator for radian measures and input 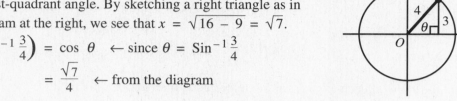 to find that $\cos\left(\text{Sin}^{-1}\frac{3}{4}\right) \approx 0.66$.

**b.** To find the exact value, let $\theta = \text{Sin}^{-1}\frac{3}{4}$. Then $\sin\theta = \frac{3}{4}$ and $\theta$ is a first-quadrant angle. By sketching a right triangle as in the diagram at the right, we see that $x = \sqrt{16 - 9} = \sqrt{7}$.

$$\cos\left(\text{Sin}^{-1}\frac{3}{4}\right) = \cos\theta \quad \leftarrow \text{since } \theta = \text{Sin}^{-1}\frac{3}{4}$$

$$= \frac{\sqrt{7}}{4} \quad \leftarrow \text{from the diagram}$$

# Exercises

**7.** If $\theta$ is a third-quadrant angle and $\sec\theta = -\frac{61}{11}$, find the values of the other five trigonometric functions.

**8.** If $\theta$ is a first-quadrant angle and $\cot\theta = \frac{1}{2}$, find the values of the other five trigonometric functions.

**9.** *Writing*   The graphs of which of the six trigonometric functions have asymptotes? Explain you answer.

**10.** Without using a calculator or table, find the exact value of each expression in radians.

**a.** $\text{Sin}^{-1}\dfrac{\sqrt{3}}{2}$
    **b.** $\text{Sin}^{-1}\left(-\dfrac{\sqrt{3}}{2}\right)$

**c.** $\text{Cos}^{-1}\left(-\dfrac{\sqrt{3}}{2}\right)$
    **d.** $\text{Tan}^{-1}\left(-\sqrt{3}\right)$

**11.** Find (**a**) the approximate value and (**b**) the exact value of

$$\sec\left(\text{Sin}^{-1}\left(-\tfrac{2}{5}\right)\right).$$

**12.** *Application*   The diagram at the right shows a cube.
  **a.** Find $AB$ and $BC$.
  **b.** Express the measure of $\angle ACB$ in terms of an inverse trigonometric function. Draw a diagram to support your answer.
  **c.** Evaluate the expression in part (b) to find the measure of $\angle ACB$ to the nearest tenth of a degree.

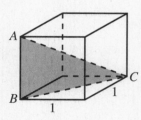

## Chapter 7: Trigonometric Functions

**QUICK CHECK**

*Chapter 7*

**Complete these exercises before trying the Practice Test for Chapter 7. If you have difficulty with a particular problem, review the indicated section.**

1. Convert 330° to radians, leaving your answer in terms of $\pi$. *(Section 7-1)*

2. Convert $3\pi$ to degree measure. *(Section 7-1)*

3. A sector of a circle has radius 10 cm and central angle 2 radians. Find its arc length and area. *(Section 7-2)*

4. Find the values of sin $(-\pi)$ and cos $(-\pi)$ without using a calculator or table. *(Section 7-3)*

5. If $\theta$ is a second-quadrant angle and sin $\theta = \dfrac{21}{29}$, find cos $\theta$. *(Section 7-3)*

6. Express sin 305° in terms of a reference angle and find its value to four decimal places. *(Section 7-4)*

7. Find the exact value of cos $\left(-\dfrac{5\pi}{6}\right)$ in simplest radical form. *(Section 7-4)*

8. Express tan 6 in terms of a reference angle and then find its value to four significant digits. *(Section 7-5)*

9. If tan $x = \dfrac{8}{15}$ and $0 < x < \dfrac{\pi}{2}$, find the values of the other five trigonometric functions. *(Section 7-5)*

10. Without using a calculator or table, find the value of $\text{Sin}^{-1}\left(-\dfrac{\sqrt{3}}{2}\right)$ in terms of $\pi$. *(Section 7-6)*

**PRACTICE TEST**

*Chapter 7*

**Convert each degree measure to radians, both in terms of $\pi$ and also to the nearest hundredth.**

1. 210°

2. $-100°$

**Convert each radian measure to degrees, both to the nearest ten minutes and also to the nearest tenth of a degree.**

3. $\dfrac{5\pi}{3}$

4. 7.28

**Name two angles, one positive and one negative, that are coterminal with each given angle.**

5. 110°

6. 2.3

**A sector of a circle has a radius of 4 feet and a central angle of 80°.**

7. What is the arc length of the sector?

8. What is the area of the sector?

**An 8-inch blade on an electric fan is rotating at 1800 rpm.**

9. Through how many radians does the blade turn in one minute?

10. About how many miles does the tip of the blade travel in one minute?

Given $\sin \theta = 0.6$ and $\cos \theta < 0$, find the value of each expression without using a calculator.

**11.** $\cos \theta$

**12.** $\tan \theta$

Find the values of the other five trigonometric functions.

**13.** $\sec x = 3, \dfrac{3\pi}{2} < x < 2\pi$

**14.** $\cot x = -\dfrac{3}{2}, \dfrac{\pi}{2} < x < \pi$

Give the exact value of each expression.

**15.** $\sin 270°$

**16.** $\cos 300°$

**17.** $\text{Sin}^{-1}(0.5)$

**18.** $\text{Tan}^{-1}(-1)$

**19.** Find $\tan(2 \text{ Sin}^{-1}(-0.5))$ without using a calculator.

**20.** Find $\tan(2 \text{ Sin}^{-1}(0.5))$ using a calculator.

---

## MIXED REVIEW

*Chapters 1–7*

**1.** Sketch the graph of $3x - 2y = -9$. Find the slope and intercepts of the line.

**2.** Find the value of the discriminant of the equation $4.5x^2 - 6x + 2 = 0$. What does this value tell you about the equation?

Graph the solution on a number line.

**3.** $7 + 2x < 9$

**4.** $|7 + 2x| < 9$

Evaluate.

**5.** $25^{-3/2} - 25^0$

**6.** $\log_3 \dfrac{1}{27}$

**7.** $\log \sqrt[3]{100}$

**8.** $\sec(-420°)$

**9.** $\text{Tan}^{-1}(-1)$

**10.** $e^{2 \ln 5}$

**11.** If $f(x) = 2x$ and $g(x) = x^3$, find $(f \circ g)(x)$ and $(g \circ f)(x)$.

**12.** If $\angle A$ is obtuse and $\cos A = -\dfrac{1}{3}$, find $\sin A$ and $\tan A$.

**13.** If $\log M = 0.2$ and $\log N = 0.5$, find $\log M^3\sqrt{N}$.

Solve each system.

**14.** $2x + y = 8$
$3x + 2y = -6$

**15.** $x = 2y^2 - 3$
$x^2 + y^2 = 9$

**16.** $4x^2 - y^2 = 16$
$x^2 + 4y^2 = 1$

**17.** Find an equation of the parabola with focus $(2, 0)$ and directrix $x = 0$.

**18.** $\triangle ABC$ has vertices $A(0, 0)$, $B(4, 8)$, and $C(4, -2)$. Use two different methods to show that $\triangle ABC$ is a right triangle.

**19.** Find the real zeros of $P(x) = x^4 - x^3 + x^2 - 1$ to the nearest tenth.

**20.** Solve $64^{x-1} = 16^{x+1}$.

**21.** A person's target heart rate, $h$, during exercise is a function of age, $a$. If $h(30) = 161$, $h(35) = 157$, and $h(40) = 153$, find a model for $h(a)$ and use it to approximate the target heart rate for a person who is 20.

**22.** Tell whether the graph of $y = \cos x$ has symmetry in (a) the $x$-axis, (b) the $y$-axis, (c) the line $y = x$, and/or (d) the origin.

# Equations and Applications of Sine Waves

## Sections 8-1, 8-2, and 8-3

| OVERVIEW | Sections 8-1, 8-2, and 8-3 cover solutions to basic trigonometric equations and graphs of general sine waves. In addition, trigonometric equations and functions are applied to topics in analytic geometry and to the modeling of periodic phenomena. |
|---|---|

## KEY TERMS

EXAMPLE/ILLUSTRATION

| | |
|---|---|
| **Trigonometric equation** (p. 295)<br>an equation in which the variable is "inside" a trigonometric function | $6 \cos 3x = 7$ |
| **Inclination of a line** (p. 296)<br>the angle $\alpha$, where $0° \leq \alpha < 180°$, that is measured from the positive $x$-axis to the line | 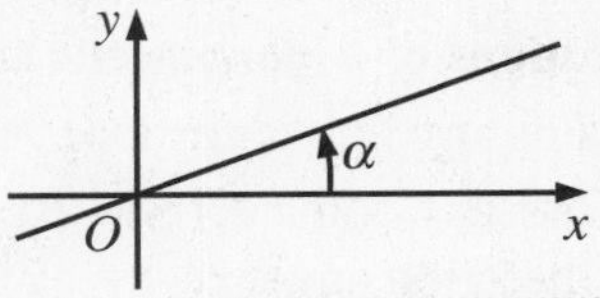 |
| **Direction of a conic** (p. 297)<br>the angle $\alpha$, where $0 < \alpha < \dfrac{\pi}{2}$, that is measured from the positive $x$-axis to an axis of the conic | |
| **Sine wave** (p. 309)<br>a graph obtained by stretching/shrinking, reflecting, or translating the graph of $y = \sin x$ or $y = \cos x$ | |

## UNDERSTANDING THE MAIN IDEAS

### Solving trigonometric equations

- Approach a trigonometric equation of the form
$$a \sin x + b = c$$
as you would the linear equation $ax + b = c$. $\longrightarrow$

- Isolate the trigonometric function. $\longrightarrow$

- Find the general solution by using the inverse trigonometric function, reference angles, and your knowledge of the function's period.

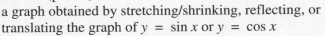

$3 \sin \theta + 7 = 9$
$3 \sin \theta = 2$

$\sin \theta = \dfrac{2}{3}$

$\theta = \sin^{-1}\left(\dfrac{2}{3}\right)$

$\theta \approx 41.8° + n \cdot 360°$ and
$\quad 138.2° + n \cdot 360°$

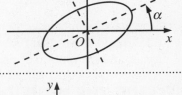
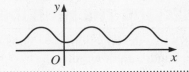
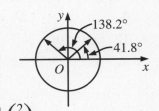

### Equations and graphs of general sine waves

*Note:* The graphs at the top of the next page are drawn for $A > 0$. The informationsummarized by the graphs is true for cosine functions as well as for the sine functions.

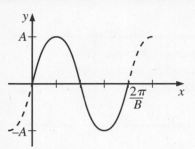

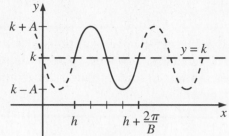

$y = A \sin Bx \quad (A \neq 0, B > 0)$

amplitude $= |A|$

period $= \dfrac{2\pi}{B}$

The graph of $y - k = A \sin B(x - h)$ is the graph of $y = A \sin Bx$ translated $h$ units horizontally and $k$ units vertically.

### Applications of trigonometric equations and functions

- For any line with slope $m$ and inclination $\alpha \neq 90°$, $m = \tan \alpha$.
- For any conic $Ax^2 + Bxy + Cy^2 + Dx + Ey + F = 0$ with direction angle $\alpha$,

$$\alpha = \frac{\pi}{4} \text{ if } A = C \text{ or } \tan 2\alpha = \frac{B}{A - C} \text{ if } A \neq C, 0 < 2\alpha < \pi.$$

- Sine waves can be used to model periodic phenomena such as tides, radio waves, sunrises and sunsets, weather patterns, musical tones, and the motion of pendulums.

## CHECKING THE MAIN IDEAS

1. Solve the equation $7 = 5 \cos x + 3$. Give answers to the nearest hundredth of a radian.

2. Identify the conic whose equation is $5x^2 + 2xy - y^2 = 6$ and find its direction angle $\alpha$ to the nearest degree. (See Example 4 on text page 297.)

3. The inclination of the line $5x - y = 2$ is:
   **A.** $179°$      **B.** $-79°$      **C.** $79°$      **D.** $101°$

4. The period of $y = 3 \sin 4x$ is :
   **A.** $\dfrac{2\pi}{3}$      **B.** $\dfrac{\pi}{2}$      **C.** $4\pi$      **D.** $6\pi$

5. Redraw the left diagram at the top of this page for $y = A \sin Bx$, $A < 0$.

## USING THE MAIN IDEAS

**Example 1** Without using tables or a calculator, solve $2 \cos 3x = 1$ for $0 \le x < 2\pi$.

**Solution** $y = 2 \cos 3x$ has amplitude 2 and period $\dfrac{2\pi}{3}$. A quick sketch shows that the graph of $y = 2 \cos 3x$ intersects the horizontal line $y = 1$ at six points. The equation $2 \cos 3x = 1$ will have six solutions.

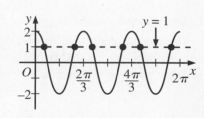

$$2 \cos 3x = 1$$

$$\cos 3x = \frac{1}{2}$$ ◀▰▰▰ **Caution:** Since $\cos 3x \neq 3 \cos x$, do not divide both sides of the equation by 3. Instead, use the inverse cosine function to solve for 3x.

$$3x = \cos^{-1}\left(\frac{1}{2}\right)$$

$$3x = \frac{\pi}{3}, \frac{5\pi}{3}$$ ◀▰▰▰ **Caution:** Don't stop after listing the first two values of $\cos^{-1}\left(\frac{1}{2}\right)$. If $0 \leq x < 2\pi$,
$$\frac{7\pi}{3}, \frac{11\pi}{3} \Big\} +2\pi$$ then $0 \leq 3x < 6\pi$ — therefore, list all
$$\frac{13\pi}{3}, \frac{17\pi}{3} \Big\} +2\pi$$ values of $\cos^{-1}\left(\frac{1}{2}\right)$ between $0$ and $6\pi$.

$$x = \frac{\pi}{9}, \frac{5\pi}{9}, \frac{7\pi}{9}, \frac{11\pi}{9}, \frac{13\pi}{9}, \frac{17\pi}{9}$$

**Example 2** Graph $y + 2 = -3 \sin \frac{\pi}{2}(x - 1)$. Show at least one full period.

**Solution** Identify the values of $A$, $B$, $h$ and $k$ from the general equation $y - k = A \sin B(x - h)$:

$$y - (-2) = -3 \sin \frac{\pi}{2}(x - 1) \longleftarrow \begin{cases} k = -2, \ A = -3, \\ B = \frac{\pi}{2}, h = 1 \end{cases}$$

Analyze how each number affects the graph of $y = \sin x$:

- $A = -3$ reflects the sine wave in the $x$-axis and makes the amplitude $|-3| = 3$.

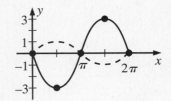

- $B = \frac{\pi}{2}$ gives the sine wave a period of $2\pi \div \frac{\pi}{2} = 2\pi \cdot \frac{2}{\pi}$, or 4. Notice that the $x$-axis is now marked in whole number units, rather than in units of $\pi$.

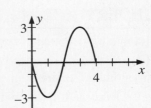

- The values $h = 1$ and $k = -2$ translate the sine wave 1 unit to the right and 2 units down. The cycle which took place for $x = 0$ to $x = 4$ now takes place for $x = 1$ to $x = 5$. The axis of the sine wave is now $y = -2$.

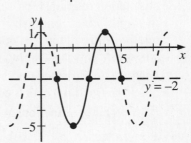

## Exercises

6. Solve $2 \sin 4x = \sqrt{3}$ for $0 \leq x \leq 2\pi$.

7. Use a graphing calculator to determine the number of solutions of the equation $5 \cos 2x = x + 1$.

8. *Writing* Write a paragraph in which you explain how each constant and sign in the equation $y = 2 - 4 \cos \left(x - \frac{\pi}{6}\right)$ affects the graph of $y = \cos x$.

9. Sketch the graph of $y = \frac{1}{2} \sin \pi(x + 1)$. Show at least one full period.

# Identities and Equations

## Sections 8-4 and 8-5

| OVERVIEW | Section 8-4 develops fundamental trigonometric relationships: reciprocal relationships, relationships with negatives, Pythagorean relationships, and cofunction relationships. These relationships are used to simplify trigonometric expressions and to prove trigonometric identities. In Section 8-5, trigonometric equations are solved using algebra, trigonometric identities, and technology. |
|---|---|

## KEY TERMS

**EXAMPLE/ILLUSTRATION**

| | |
|---|---|
| **Cofunctions** (p. 318)<br><br>the three pairs of functions: sine and cosine, tangent and cotangent, and secant and cosecant | $\sin 30° = \cos 60°$<br>$\cos 30° = \sin 60°$<br>$\tan 30° = \cot 60°$<br>$\cot 30° = \tan 60°$<br>$\sec 30° = \csc 60°$<br>$\csc 30° = \sec 60°$ |
| **Trigonometric identity** (p. 318)<br><br>an equation that is true for all values of the variable for which both sides of the equation are defined | $\tan \theta = \dfrac{\sin \theta}{\cos \theta}$<br><br>$(\sin \theta)^2 + (\cos \theta)^2 = 1$<br>$\qquad \downarrow \qquad\qquad \downarrow$<br>$\sin^2 \theta + \cos^2 \theta = 1$ |

## UNDERSTANDING THE MAIN IDEAS

### Simplifying trigonometric expressions

Here are some strategies you can consider when simplifying a trigonometric expression.

- Become familiar with the reciprocal, negative, Pythagorean, and cofunction relationships on text pages 317–318. Look for a way to simplify a trigonometric expression by applying one of these relationships. (This was done in Example 1 on text page 319, when $\cos^2 x$ was substituted for $1 - \sin^2 x$.)

- Perform an indicated operation; for example, multiply factors or combine fractions over a common denominator.

- Express all the functions in terms of sine and cosine; for example, express $\tan x$ as $\dfrac{\sin x}{\cos x}$, $\cot x$ as $\dfrac{\cos x}{\sin x}$, $\sec x$ as $\dfrac{1}{\cos x}$, and $\csc x$ as $\dfrac{1}{\sin x}$.

- See whether factoring will help you simplify the expression.

### Solving trigonometric equations

- If a trigonometric equation contains just one trigonometric function, solve for it directly. Factoring may be helpful. (See Example 1 on text page 323.)

- If a trigonometric equation contains more than one trigonometric function, try factoring (see Example 3 on text page 324) or try substituting to get an equation that involves just one function (see Examples 2 and 5 on text pages 324 and 325). Another way to get a single-function equation is to divide both sides of the equation by the same expression. Be very careful that this does not cause a root to be lost. (See Example 4 on text page 325.)
- Check all four quadrants to avoid overlooking solutions.

## CHECKING THE MAIN IDEAS

**1.** Which expression *cannot* be simplified to $\sin \theta$?

   **A.** $\dfrac{1}{\csc \theta}$     **B.** $\sin (-\theta)$     **C.** $\cos \left(\dfrac{\pi}{2} - \theta\right)$     **D.** $\cos \theta \tan \theta$

**2.** The relationship $1 + \tan^2 \theta = \sec^2 \theta$ is valid for:

   **A.** all angles $\theta$          **B.** $\theta \neq \dfrac{\pi}{2} + n\pi$

   **C.** $\theta \neq n\pi$            **D.** $\theta \neq \dfrac{n\pi}{4}$

**Simplify.**

**3.** $\sin^2 \theta + 2\cos^2 \theta - 1$          **4.** $\dfrac{\sin x}{\cot (90° - x)}$

**5.** $\dfrac{\sec^2 A - \tan^2 A}{\csc (-A)}$          **6.** $\dfrac{\sec \theta - 1}{1 - \cos \theta}$

**7.** *Critical Thinking* Explain when it is appropriate to divide both sides of an equation by $\cos \theta$ and when it is not. Give examples to illustrate your answer.

**Solve for $0° \leq \theta < 360°$. Give answers to the nearest tenth of a degree.**

**8.** $2\tan^2 \theta - 1 = 5$          **9.** $3\cos^2 \theta + 4\cos \theta = -1$

**10.** $\cos^2 \theta - \sin \theta = 1$          **11.** $\tan^2 x = 2\tan x$

**Solve for $0 \leq x < 2\pi$.**

**12.** $\sin x - \cos x = 0$          **13.** $4\sin x = \csc x$

## USING THE MAIN IDEAS

**Example 1** Prove $\tan \theta + \cot \theta = \sec \theta \csc \theta$.

**Solution** Show that both sides are equal to the same expression.

$$\tan \theta + \cot \theta = \frac{\sin \theta}{\cos \theta} + \frac{\cos \theta}{\sin \theta}$$

$$= \frac{\sin^2 \theta}{\sin \theta \cos \theta} + \frac{\cos^2 \theta}{\sin \theta \cos \theta}$$

$$= \frac{\sin^2 \theta + \cos^2 \theta}{\sin \theta \cos \theta}$$

$$= \frac{1}{\sin \theta \cos \theta}$$

        

$$\sec \theta \csc \theta = \frac{1}{\cos \theta} \cdot \frac{1}{\sin \theta}$$

$$= \frac{1}{\sin \theta \cos \theta}$$

Since each side is equal to $\dfrac{1}{\sin \theta \cos \theta}$, the identity has been proven.

**Example 2** Solve $4 \cot^2 \theta + 7 = 8 \csc \theta$ for $0 \le x < 2\pi$. Give answers to the nearest hundredth of a radian when necessary.

**Solution** **Method 1** (Algebraic)

$$4 \cot^2 \theta + 7 = 8 \csc \theta$$
$$4(\csc^2 \theta - 1) + 7 = 8 \csc \theta \qquad \leftarrow 1 + \cot^2 \theta = \csc^2 \theta$$
$$4 \csc^2 \theta - 8 \csc \theta + 3 = 0 \qquad \leftarrow \textit{Note:} \text{ If you can't factor,}$$
$$(2 \csc \theta - 1)(2 \csc \theta - 3) = 0 \qquad \text{use the quadratic formula.}$$

$$\csc \theta = \frac{1}{2} \qquad\qquad \text{or} \qquad\qquad \csc \theta = 1.5$$
$$\downarrow \qquad\qquad\qquad\qquad\qquad\qquad \downarrow$$

No solutions:
$|\csc \theta| \ge 1$ for all $\theta$

Use the inverse sine (and reciprocal
function) or inverse cosecant
function of your calculator
(set in radian mode). $\csc \theta > 0$ in
quadrants I and II; $\theta \approx 0.73$ or
$\theta \approx \pi - 0.73 \approx 2.41$

**Method 2** (Graphical)

Graph $y = 4 \div (\tan x)^2 + 7 - 8 \div \sin x$ for
$0 \le x < 6.3$ on a graphing calculator. A quick
check shows that the two solutions in the domain
$0 \le x < 2\pi$ occur in the intervals $0 < x < 1$
and $2 < x < 3$. Using the zoom or trace features,
you obtain the approximate solutions 0.73 and 2.41.

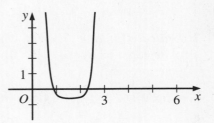

# Exercises

**14.** Prove $\dfrac{1 + \tan^2 \theta}{1 + \cot^2 \theta} = \tan^2 \theta$.

**15.** Prove $\sec \theta - \cos \theta = \sin \theta \tan \theta$.

**16.** *Writing* Do you agree or disagree with the statement "cofunctions of complementary angles are equal?" Explain.

**Solve for $0 \le x < 2\pi$. Give answers to the nearest hundredth of a radian when necessary.**

**17.** $2 \cos^2 x + \sin x + 1 = 0$      **18.** $\sin^2 x + 3 \sin x - 1 = 0$

**19.** *Application*

   **a.** What is the area of a square inscribed in a circle with radius 1?

   **b.** The formula for the area of a regular polygon with $n$ vertices inscribed in a circle with radius $r$ is $A = nr^2 \sin \theta \cos \theta$. Write the trigonometric formula obtained by substituting appropriate values for $A$, $n$, and $r$ from part (a).

   **c.** Describe two methods for solving the trigonometric equation in part (b).

   **d.** Use one of the methods in part (c) to find the value of $\theta$.

# CHAPTER REVIEW

## Chapter 8: Trigonometric Equations and Applications

Complete these exercises before trying the Practice Test for Chapter 8. If you have difficulty with a particular problem, review the indicated section.

1. Solve $\cos \theta = -0.4$ for $0° \le \theta < 360°$. Give answers to the nearest tenth of a degree. *(Section 8-1)*

2. Find the inclination of a line with slope $\frac{5}{2}$. *(Section 8-1)*

3. Give the amplitude and period of $y = -2 \sin \frac{1}{2}x$. Then sketch its graph. *(Section 8-2)*

4. Describe how to obtain the graph of $y + 1 = 3 \cos 4(x - 2)$ from the graph of $y = \cos x$. *(Section 8-3)*

5. Simplify (a) $\dfrac{1}{\sin \left( \dfrac{\pi}{2} - x \right)}$ and (b) $(1 + \csc \theta)(1 - \csc \theta)$. *(Section 8-4)*

6. Solve $\sin^2 \theta - \sin \theta = 0$ for $0° \le \theta < 360°$. *(Section 8-5)*

---

**Solve for $0° \le \theta < 360°$. Give answers to the nearest tenth of a degree.**

1. $4 \tan \theta - 1 = 0$  2. $\sin \theta \cos \theta = \sin \theta$  3. $(2 \sin \theta - 1)(3 \cos \theta - 1) = 0$

**Solve for $0 \le x < 2\pi$. Give answers to the nearest hundredth of a radian.**

4. $1 + \cos 4x = 0$  5. $\tan^3 x = 3 \tan x$  6. $(\sin x)(-1 + \sin x) = \cos^2 x$

7. Find the inclination for the line $8x + 3y = 10$.

**Write an equation in the specified form for the graph shown at the right.**

8. as a sine function

9. as a cosine function

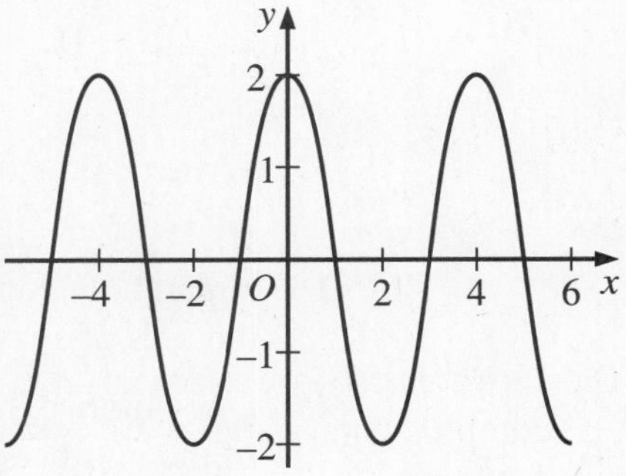

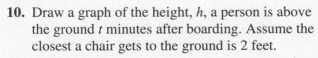

**A giant ferris wheel is 60 feet in diameter. The ferris wheel completes a revolution every 5 minutes.**

10. Draw a graph of the height, $h$, a person is above the ground $t$ minutes after boarding. Assume the closest a chair gets to the ground is 2 feet.

11. Write a function for the height, $h$, a person is above the ground $t$ minutes after boarding.

12. How high above the ground is a person 13 minutes after boarding the ferris wheel?

**Give the amplitude and period of each function. Then sketch its graph.**

13. $y = 1.5 \sin 2x$  14. $y = 2 - 3 \cos \pi x$

**Simplify.**

15. $\sec x \tan (-x) \cos x$  16. $\cos (90° - A)(\csc A - \sin A)$

---

*ADVANCED MATHEMATICS Student Resource Guide* **85**

**17.** $\dfrac{\cot x + \tan x}{\tan x}$

**18.** $\dfrac{\cos^2 x}{1 - \sin x}$

---

**MIXED REVIEW**

*Chapters 1–8*

**1.** If $\sec x = \dfrac{41}{9}$ where $\dfrac{3\pi}{2} < x \le 2\pi$, find the values of the other five trigonometric functions without using a table or a calculator.

**2.** If $f(x)$ is an exponential function such that $f(1) = 20$ and $f(2) = 4$, evaluate $f(-1)$.

**Solve.**

**3.** $x^4 + 3x^2 < 4$

**4.** $x^{-1.25} = 32$

**5.** $3x^3 + 12 = x^2 + 13x$

**6.** $4x + y = 10$
   $xy = 6$

**7.** $125^{x-1} = \dfrac{1}{25}$

**8.** $|x - 2| \ge 4$

**9.** *ABCD* is a parallelogram. Give the coordinates of $C$ in terms of $a$, $b$, and $c$. Then prove that if the lengths of the diagonals are equal, then the parallelogram is a rectangle.

**10.** If $f(x) = e^x$, find $f^{-1}(x)$ and its zeros, domain, and range.

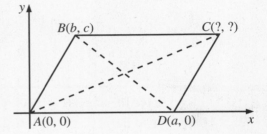

**Sketch the graph of each equation.**

**11.** $y = 4 \sin 2(x - \pi)$

**12.** $x^2 - y^2 = 4$

**Simplify.**

**13.** $\tan \left( \mathrm{Sin}^{-1} \left( -\dfrac{3}{5} \right) \right)$

**14.** $3 \log 20 - \dfrac{1}{2} \log 64$

**15.** $2 - i + \dfrac{1}{2 - i}$

**16.** $\log_a b \cdot \log_b a$

**17.** Write an equation of the line through $(-2, -1)$ that is perpendicular to the line $x + 3y = 6$.

**18.** A parabola contains the points $(-4, 30)$, $(-3, 22.5)$, and $(-2, 16)$. Find the axis of symmetry, vertex, and intercepts of the parabola.

**19.** One number is 8 more than another number. Find the minimum product of the numbers.

**20.** Let $f(x) = \sqrt{x - 1}$ and $g(x) = 3x^2 + 2$. Find $(f \circ g)(x)$ and $(g \circ f)(x)$. Then tell which function, $f$ or $g$, does *not* have an inverse.

**21.** Find an equation of the circle with diameter $\overline{PQ}$ where $P = (-8, -1)$ and $Q = (4, 5)$.

**22.** Solve $2 \sin^2 x = 3(1 - \cos x)$ for $0 \le x < 2\pi$.

**23.** Find the sum and product of the roots of $5x^3 - 32x^2 + 62x - 20 = 0$. Then find the roots and verify their sum and product.

**24.** What kind of symmetry does the graph of $y = \tan x$ have? Explain.

**25.** Identify the graph of $x^2 + 2xy + 3y^2 = 6$.

---

# Right Triangle Trigonometry and Areas of Triangles

## Sections 9-1 and 9-2

| OVERVIEW | In Section 9-1, trigonometry is used to find unknown sides and angles of right triangles. In Section 9-2, the lengths of two sides of a triangle and the measure of the included angle are used to compute the area of the triangle. |
|---|---|

## KEY TERMS

**Solving a triangle** (p. 331)
using the known sides and angles of a triangle to find the unknown sides and angles

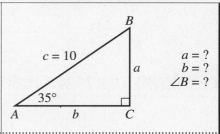

**Segment of a circle** (p. 340)
a region bounded by an arc of the circle and the chord connecting the endpoints of the arc

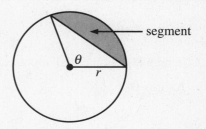

## UNDERSTANDING THE MAIN IDEAS

### Solving a right triangle

If you know the lengths of two sides of a right triangle, or the length of one side and the measure of one acute angle, then you can find the measures of the remaining sides and angles using the trigonometric functions defined below.

sine of $\angle A$: $\sin A = \dfrac{\text{opposite}}{\text{hypotenuse}} = \dfrac{a}{c}$

cosine of $\angle A$: $\cos A = \dfrac{\text{adjacent}}{\text{hypotenuse}} = \dfrac{b}{c}$

tangent of $\angle A$: $\tan A = \dfrac{\text{opposite}}{\text{adjacent}} = \dfrac{a}{b}$

cosecant of $\angle A$: $\csc A = \dfrac{\text{hypotenuse}}{\text{opposite}} = \dfrac{c}{a}$

secant of $\angle A$: $\sec A = \dfrac{\text{hypotenuse}}{\text{adjacent}} = \dfrac{c}{b}$

cotangent of $\angle A$: $\cot A = \dfrac{\text{adjacent}}{\text{opposite}} = \dfrac{b}{a}$

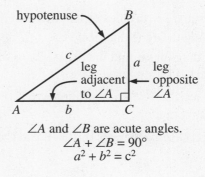

$\angle A$ and $\angle B$ are acute angles.
$\angle A + \angle B = 90°$
$a^2 + b^2 = c^2$

It is convenient to use a capital letter to label an angle of a triangle and the corresponding lower-case letter to label the length of the side opposite that angle.

### Area of a triangle

If you know the lengths of two sides of a triangle and the measure of the included angle, you can compute the area $K$ of the triangle using the formula $K = \frac{1}{2} \cdot$ (one side) $\cdot$ (another side) $\cdot$ (sine of included angle).

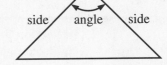

In terms of the triangle shown:

$K = \frac{1}{2}ab \sin C$

$K = \frac{1}{2}ac \sin B$

$K = \frac{1}{2}bc \sin A$

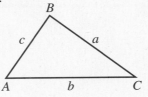

## CHECKING THE MAIN IDEAS

1. Which equation is *not* correct?

   **A.** $\csc 50° = \frac{y}{12}$

   **B.** $\tan 40° = \frac{x}{12}$

   **C.** $\cos 50° = \frac{x}{y}$

   **D.** $\sin 40° = \frac{12}{y}$

2. Which equation could you use to find the measure of $\angle D$?

   **A.** $\cos D = \frac{12}{13}$

   **B.** $\cos D = \frac{13}{12}$

   **C.** $\sin D = \frac{12}{13}$

   **D.** $\sin D = \frac{13}{12}$

   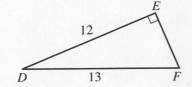

3. Refer to $\triangle DEF$ to the right of Exercise 2. In the following statement, replace the question mark with a trigonometric function to form a true statement: $\tan D = $ __?__ $F$.

4. In $\triangle XYZ$, $\angle X = 90°$, $\angle Z = 21°$, and $y = 32$. Find $x$ and $z$ to three significant digits.

5. Find the measures of the acute angles of a 5–12–13 right triangle. Give your answers to the nearest tenth of a degree.

6. **a.** Sketch a square with sides of length 1 and draw a diagonal.

   **b.** Find the length of the diagonal.

   **c.** Use your diagram to find the exact values of the six trigonometric functions of 45°.

7. In $\triangle ABC$, $a = 12$, $b = 15$, and $\angle C = 65°$. Find the area of $\triangle ABC$ to three significant digits.

8. The area of $\triangle DEF$ is 28 cm². If $d = 8$ and $f = 35$, find all possible measures of $\angle E$.

9. *Critical Thinking*

   **a.** If you know the lengths of two sides of a triangle and the measure of the included angle, how many different areas are possible? Explain.

   **b.** If you know the lengths of two sides of a triangle and the area of the triangle, how many different measures for the included angle are possible? Explain.

# USING THE MAIN IDEAS

**Example 1** From the top of a building 72 m tall, the angle of depression to the base of a building across the street is 82°. Find the width of the street to three significant digits.

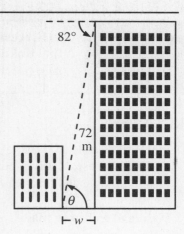

**Solution** Read the problem carefully. Recall that an angle of depression is formed by the horizontal and a line of sight. Draw a diagram that shows the given information. Since the angle of depression and $\theta$ are alternate interior angles formed by two horizontal lines and the line of sight, $\theta = 82°$. You can use either of these equations:

$\tan 82° = \dfrac{72}{w}$ or $\cot 82° = \dfrac{w}{72}$

Either gives $w \approx 10.1$; the street is about 10.1 m wide.

**Example 2** Find the approximate area of a regular pentagon inscribed in a circle of radius 8 cm.

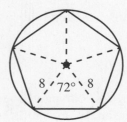

**Solution** The measure of each central angle is $\dfrac{360°}{5} = 72°$. The circle has radius 8 cm. The radii divide the pentagon into 5 congruent triangles, so $K = 5\left(\dfrac{1}{2} \cdot 8 \cdot 8 \cdot \sin 72°\right) \approx 152 \text{ cm}^2$.

# Exercises

10. Each leg of an isosceles triangle is 32 cm long. The measure of the included angle is 108°. Find the length of the third side to three significant digits.

11. *Application*   The pilot of a rescue helicopter flying at an altitude of 50 ft finds that the angle of depression to a stranded hiker is 42°. About how far must the helicopter fly at the same altitude to be directly over the hiker?

12. Find the exact area of a regular hexagon inscribed in a circle of radius 1.

13. Find the exact area of a regular hexagon circumscribed about a circle of radius 1.

14. *Writing*   Approximate the areas in Exercises 12 and 13 to three significant digits. Does $\pi$ lie between these values? Why?

# Law of Sines, Law of Cosines, and Applications

| OVERVIEW | In Sections 9-3 and 9-4, the law of sines and the law of cosines are used to find unknown sides and angles of triangles. In Section 9-5, problems from navigation and surveying are solved by using trigonometry. |
|---|---|

## KEY TERMS

**EXAMPLE/ILLUSTRATION**

**Ambiguous case** (p. 346)

the SSA case, that is, when you are given the lengths of two sides of a triangle and the measure of a *nonincluded* angle

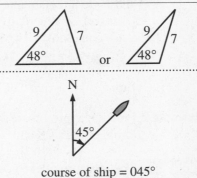

**Course of a ship or plane** (p. 359)

the angle measured clockwise from the north direction to the direction of the ship or plane

course of ship = 045°

**Compass bearing of one location from another** (p. 359)

the angle measured clockwise from the north direction (with respect to the first location) to the second location

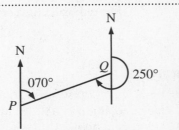

The bearing of $Q$ from $P$ is 070°.
The bearing of $P$ from $Q$ is 250°.

## UNDERSTANDING THE MAIN IDEAS

### Solving a triangle

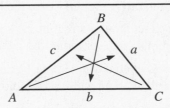

- The Law of Sines

  In any $\triangle ABC$: $\dfrac{\sin A}{a} = \dfrac{\sin B}{b} = \dfrac{\sin C}{c}$

  Notice that the law of sines involves the sine of an angle and the length of the side opposite that angle.

- The Law of Cosines

  In any $\triangle ABC$:

  $c^2 = a^2 + b^2 - 2ab \cos C$ or $\cos C = \dfrac{a^2 + b^2 - c^2}{2ab}$

  $b^2 = a^2 + c^2 - 2ac \cos B$ or $\cos B = \dfrac{a^2 + c^2 - b^2}{2ac}$

  $a^2 = b^2 + c^2 - 2bc \cos A$ or $\cos A = \dfrac{b^2 + c^2 - a^2}{2bc}$

*Note:* If an angle is obtuse, its cosine is negative.

- When you use given information to solve a triangle, remember the following:

  1. Begin by sketching and labeling a triangle. This will help you identify which case (ASA, AAS, SSA, SAS, SSS) applies.

  2. If you know the measures of two angles of the triangle, find the measure of the third angle by using the fact that the sum of the measures of the angles of a triangle is 180°.

  3. Use the chart on text page 352 if you are not sure how to proceed. Be careful when you work with the ambiguous case, SSA. This case may produce 0, 1, or 2 triangles. If the law of sines indicates that the sine of an angle is greater than 1, then no triangle exists. When you use the law of sines, remember that every angle $\alpha$ and its supplement, $180° - \alpha$, have the same sine value. Finally, remember that the angle measures and lengths are in the same order; that is, in $\triangle ABC$, if $a < b < c$, then $\angle A < \angle B < \angle C$ and if $\angle A < \angle B < \angle C$, then $a < b < c$.

### Applications to navigation and surveying

- Keep in mind the way that angles are measured in navigation problems and in surveying problems. In navigation problems, north is 0° and the bearing or course is measured clockwise from the north line. In surveying problems, an angle measure is expressed as an acute angle measured east or west of the north-south line (see text pages 359–360).

- Begin a problem by making a diagram. Then use the table on text page 352 if you are not sure whether to apply the law of sines or the law of cosines.

## CHECKING THE MAIN IDEAS

**Match each triangle with the case associated with it.**

**1.**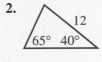

**2.**

**3.**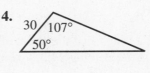

**4.** 30 / 107°  50°

   **A.** SSS     **B.** SAS     **C.** SSA     **D.** ASA     **E.** AAS

**Solve each $\triangle ABC$. Give angle measures to the nearest tenth of a degree and lengths to three significant digits.**

**5.** $\angle A = 150°, \angle B = 10°, b = 18$

**6.** $\angle B = 35°, b = 8, c = 10$

**7.** You are told that $\angle T = 70°$ and that $r = \frac{6}{5}t$ in $\triangle RST$. Show that no such $\triangle RST$ is possible.

**Solve each triangle. Give angle measures to the nearest tenth of a degree and lengths to three significant digits.**

   **8.** $x = 7, y = 13, z = 15$           **9.** $r = 15, t = 8, \angle S = 125°$

10. Town $T$ has a compass bearing of 207° from City $C$. What is the bearing of $C$ from $T$?

11. A ship sails for 1 h at a speed of 15 knots on a course of 300°. To the nearest nautical mile, how far west of its starting point is it then?

12. Two hikers travel 8 mi southeast and then 6 mi east. Find the hikers' distance and bearing from their starting point at this time.

## USING THE MAIN IDEAS

**Example 1** In $\triangle ABC$, $\cos A = \frac{1}{3}$, $\cos B = -\frac{\sqrt{2}}{2}$, and $b = 6$. Find the value of $a$ in simplest radical form.

**Solution** Recall the Pythagorean identity $\sin^2 \theta + \cos^2 \theta = 1$ for every angle $\theta$. Since $\angle A$ and $\angle B$ are angles of a triangle, the measure of each is between 0° and 180°. Therefore, $\sin A > 0$ and $\sin B > 0$.

$$\sin A = \sqrt{1 - \left(\frac{1}{3}\right)^2} = \frac{2\sqrt{2}}{3} \text{ and } \sin B = \sqrt{1 - \left(-\frac{\sqrt{2}}{2}\right)^2} = \frac{\sqrt{2}}{2}.$$

Using $\frac{\sin A}{a} = \frac{\sin B}{b}$: $\dfrac{\frac{2\sqrt{2}}{3}}{a} = \dfrac{\frac{\sqrt{2}}{2}}{6}$

$$\frac{\sqrt{2}}{2}a = 4\sqrt{2}$$
$$a = 8$$

**Example 2** An isosceles trapezoid has sides of lengths 8 cm, 6 cm, 8 cm, and 10 cm. Find the approximate length of each diagonal.

**Solution** The sketch of trapezoid $ABCD$ shows that $\cos A = \frac{2}{8} = \frac{1}{4}$.

In $\triangle ABD$, $(BD)^2 = 8^2 + 10^2 - 2(8)(10)\left(\frac{1}{4}\right) = 124$
So, $BD \approx 11.14$
The diagonals are about 11.14 cm long.

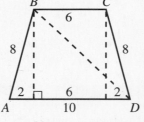

## Exercises

13. At a restaurant, Estella found that the angle of elevation to the top of Mt. Adams was 25°. After driving 2 km toward the mountain, the angle of elevation was 38°. Find the approximate height of Mt. Adams.

14. In $\triangle ABC$, $\tan A = \frac{1}{2}$, $\tan B = -3$, and $a = 20$. Find $b$ in simplest radical form.

15. A triangle has sides of length 6, 10, and 12. Find the length of the median to the longest side. Give your answer in simplest radical form.

16. *Writing* Explain the difference in the way an angle in standard position is measured and the way the course of a ship is measured.

17. From the north corner of Park Avenue, a surveyor walked N38°E for 128 m and then S48°E for 150 m. Find the surveyor's distance and compass reading from his starting point.

# CHAPTER REVIEW

## Chapter 9: Triangle Trigonometry

Complete these exercises before trying the Practice Test for Chapter 9. If you have difficulty with a particular problem, review the indicated section.

1. Find the measures to the nearest tenth of a degree, of the acute angles of a 5–12–13 right triangle. *(Section 9-1)*

2. Find the area of $\triangle ABC$ if $a = 20$, $b = 18$, and $\angle C = 65°$. Give your answer to the nearest whole number. *(Section 9-2)*

3. Why is the SSA situation called the *ambiguous case*? *(Section 9-3)*

4. Solve $\triangle RST$ if $\angle R = 50°$, $\angle S = 102°$, and $r = 36$. Give lengths to three significant digits. *(Section 9-3)*

5. Name the two cases (SSS, SAS, ASA, AAS, SSA) which require the use of the law of cosines. *(Section 9-4)*

6. Find the measures of the angles of a triangle with sides of lengths 4, 7, and 10. Give your answers to the nearest tenth of a degree. *(Section 9-4)*

7. Ship A sights ship B on a compass bearing of 210°. What is the bearing of ship A from ship B? *(Section 9-5)*

Where appropriate, give angle measures to the nearest tenth of a degree and lengths of sides in simplest radical form or to three significant digits.

$\triangle ABC$ is a right triangle with right angle $C$.

1. If $a = 2$ and $b = 7$, find $\angle A$.    2. If $\angle A = 40°$ and $c = 10$, find $b$.

An isosceles triangle has sides of length 10, 10, and 15.

3. Find the measures of the angles.    4. Find the area of the triangle.

Determine whether there are 0, 1, or 2 triangles possible for each of the following sets of measurements.

5. $b = 5$, $c = 10$, $\angle B = 20°$    6. $b = 10$, $c = 5$, $\angle B = 20°$

Find the measure of the largest angle in $\triangle ABC$.

7. $b = 10$, $c = 5$, $\angle B = 20°$    8. $a = 12$, $b = 10$, $c = 15$

Find the area of $\triangle ABC$.

9. $a = 10$, $b = 17$, $\angle C = 100°$    10. $a = 10$, $b = 17$, $c = 15$

11. Why should you avoid using the law of sines to find the measure of the largest angle?

12. Why is $\sin C = \sin (A + B)$ in any $\triangle ABC$?

Two ships leave port at 2 P.M. One travels due east at 20 knots and the other on a course of 200° at 30 knots.

13. How far apart are the ships at 4 P.M.?

14. If the slower ship continues on its course, on what course must the faster ship travel to meet the slower ship at midnight?

15. Where will they meet at midnight?

16. In quadrilateral $ABCD$, $AD = 12$, $DC = 10$, $BC = 20$, $\angle D = 90°$, and $\angle C = 130°$. Find the area to the nearest square unit.

17. What is the area of the segment of a circle defined by $x^2 + y^2 \leq 25$ and $y \geq 3$.

## MIXED REVIEW

Chapters 1–9

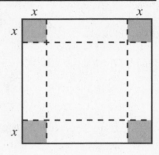

1. Let $P(x) = 2x^4 - x^3 + 13x^2 - 8x - 24$. Use synthetic division to show that $x + 1$ is a factor of $P(x)$. Then find all the real and imaginary roots of $P(x) = 0$.

2. An open box is formed from an 80 cm by 80 cm piece of metal by cutting four identical squares from the corners and folding up the sides. Express the volume of the box in terms of $x$. Then describe how you could find the maximum possible volume.

**Simplify.**

3. $(-1 + i)^3$

4. $e^{3 \ln 2 - 1}$

5. $\log_2 \sqrt[5]{512}$

6. If the value of a machine depreciates 16% each year, about how long will it take for a new machine to be worth about half its original value?

**Solve.**

7. $4x^3 + 12x^2 < x + 3$

8. $|3x - 1| = 5$

9. $(5 - x)^{-3} = 8$

10. $2x + 3y = 12$
    $3x + 5y = 15$

11. $x + y = 3$
    $x^2 + 9y^2 = 9$

12. $81\sqrt{3} = 9^x$

13. A parallelogram has sides of length 8 cm and 10 cm and an included angle of 120°. Find the exact lengths of its diagonals and its exact area.

14. Find an equation of the line with $x$-intercept $-3$ that is parallel to the line through $(0, 5)$ and $(8, -1)$.

15. Find the domain, range, and zeros of the function $f(x) = -2x^2 + x + 4$.

16. Solve $5 \sin x = -4 \cos x$ for $0 \leq x < 2\pi$. Give answers to the nearest hundredth of a radian.

17. Let $f(x) = 7 - 2x$. Find $f^{-1}(x)$ and show that $(f \circ f^{-1})(x) = (f^{-1} \circ f)(x)$.

18. Find an equation of an ellipse with center at the origin, vertex $(0, 15)$, and focus $(0, 12)$.

19. If $\theta$ is a third-quadrant angle such that $\csc \theta = -7$, find the other five trigonometric functions of $\theta$.

20. If $\triangle ABC$ is a right triangle with right angle $C$, explain why $\sin A = \cos B$.

21. Use the figure at the right to solve $\triangle RST$.

22. Sketch the graphs of $y = x^2 - 3$ and $x^2 + y^2 = 9$ on the same set of axes. Then solve the system algebraically.

23. Prove that the diagonals of a rectangle bisect each other.

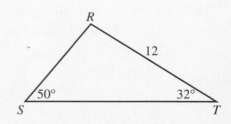

# Trigonometric Addition Formulas

## Sections 10-1, 10-2, 10-3, and 10-4

**OVERVIEW**  This chapter focuses on the sum, difference, double-angle, and half-angle formulas for sine, cosine, and tangent. These formulas are used to evaluate the sine, cosine, and tangent of certain angles, to simplify expressions, and to prove identities. In Section 10-4, trigonometric equations and inequalities are solved using algebra, trigonometric identities, and technology.

## UNDERSTANDING THE MAIN IDEAS

### Formulas for cos ($\alpha \pm \beta$) and sin ($\alpha \pm \beta$)

- When using these formulas, remember to reverse the signs for cosine and match the signs for sine:

$$\cos(\alpha + \beta) = \cos\alpha\cos\beta - \sin\alpha\sin\beta \qquad \cos(\alpha - \beta) = \cos\alpha\cos\beta + \sin\alpha\sin\beta$$

$$\sin(\alpha + \beta) = \sin\alpha\cos\beta + \cos\alpha\sin\beta \qquad \sin(\alpha - \beta) = \sin\alpha\cos\beta - \cos\alpha\sin\beta$$

  Note that there are many correct ways to use these formulas to find the exact value of an expression. (See Example 1 on text page 370.)

- The formulas for rewriting a sum or difference of sines or cosines as a product are listed on text page 372.

### Formulas for tan ($\alpha \pm \beta$)

- Notice that the formulas for tan ($\alpha + \beta$) and tan ($\alpha - \beta$) contain the sum or difference of tan $\alpha$ and tan $\beta$ in the numerator and their product in the denominator:

$$\tan(\alpha + \beta) = \frac{\tan\alpha + \tan\beta}{1 - \tan\alpha\tan\beta} \qquad \tan(\alpha - \beta) = \frac{\tan\alpha - \tan\beta}{1 + \tan\alpha\tan\beta}$$

  Each identity is defined only when tan $\alpha$, tan $\beta$, and the specified value, tan ($\alpha + \beta$), are *all* defined. Notice that in each formula the signs in the numerator and denominator are different, and that the sign in the numerator matches the sign in the quantity ($\alpha \pm \beta$).

- An angle $\theta$ formed by intersecting lines $l_1$ and $l_2$ with respective slopes $m_1$ and $m_2$ can be found by using the formula

$$\tan\theta = \frac{m_1 - m_2}{1 + m_1 m_2}.$$

### Double-angle and half-angle formulas

The table on text page 383 summarizes the double-angle and half-angle formulas. Notice that there are three formulas for cos $2\alpha$ and for $\tan\frac{\alpha}{2}$. When choosing a formula for cos $2\alpha$ or for $\tan\frac{\alpha}{2}$, think about which form will be most helpful. Also, notice the $\pm$ signs in the formulas for $\sin\frac{\alpha}{2}$, $\cos\frac{\alpha}{2}$, and $\tan\frac{\alpha}{2}$; when using these formulas, choose the appropriate sign, + or –, depending on the quadrant of $\frac{\alpha}{2}$ (*not* the quadrant of $\alpha$).

|          | I | II | III | IV |
|----------|---|----|----|----|
| sin, csc | + | +  | –  | –  |
| cos, sec | + | –  | –  | +  |
| tan, cot | + | –  | +  | –  |

### Solving trigonometric equations

A good first step in solving a trigonometric equation $f(x) = g(x)$ is to draw a quick sketch of $y = f(x)$ and $y = g(x)$ on the same set of axes. This will often help you identify the number of solutions in the interval $0 \le x < 2\pi$ and their approximate values. This is especially helpful if you are using a scientific calculator to identify solutions because it helps you determine the quadrants in which solutions occur. The table on text page 386 describes two methods for solving trigonometric equations: a graphical method and an algebraic method.

## CHECKING THE MAIN IDEAS

**Match each expression with an expression having the same value. Do *not* use a calculator.**

1. $-\sqrt{\dfrac{1 + \cos 225°}{2}}$

    **A.** $\dfrac{2 \tan 56.25°}{1 - \tan^2 56.25°}$

2. $2 \sin 100° \cos 100°$

    **B.** $-\sqrt{1 - \sin^2 20°}$

3. $\dfrac{1 - \cos 225°}{\sin 225°}$

    **C.** $-\cos 67.5°$

4. $1 - 2 \sin^2 100°$

    **D.** $\sin 50° \cos 150° + \cos 50° \sin 150°$

5. Prove $\cos (\pi - x) = -\cos x$. (See Example 4 on text page 372.)

6. Find the exact value of $\cos 255°$.

7. If $\tan \alpha = 2$ and $\tan \beta = -\dfrac{1}{4}$, find $\tan (\alpha + \beta)$ and $\tan (\alpha - \beta)$.

8. Find the exact value of each expression.

    **a.** $\dfrac{\tan 170° - \tan 50°}{1 + \tan 170° \tan 50°}$
        **b.** $\sin \dfrac{5\pi}{12} \cos \dfrac{\pi}{6} - \cos \dfrac{5\pi}{12} \sin \dfrac{\pi}{6}$

**Simplify the given expression.**

9. $\cos^2 2x - \sin^2 2x$
    10. $\sqrt{\dfrac{1 - \cos 70°}{2}}$
    11. $\cos 3x \cos x + \sin 3x \sin x$

12. $\dfrac{\sin 4\alpha}{1 + \cos 4\alpha}$
    13. $2 \cos^2 \dfrac{\alpha}{2} - 1$
    14. $\cos (\pi + x) + \cos (\pi - x)$

15. ***Critical Thinking*** Explain why the domain of a trigonometric function is generally restricted to $0° \le x < 360°$ or $0 \le x < 2\pi$?

16. Solve $\cos 2x + \cos x = 0$ for $0° \le x < 360°$.

## USING THE MAIN IDEAS

**Example 1** If $\cos A = \dfrac{7}{25}$ and $\angle A$ is acute, find $\sin 2A$ and $\cos \dfrac{A}{2}$.

**Solution**   Method 1

    Since $\cos A = \dfrac{7}{25}$, we can draw the diagram shown. By the Pythagorean theorem, $y = \sqrt{25^2 - 7^2} = \sqrt{576} = 24$.

    Therefore, $\sin A = \dfrac{24}{25}$.

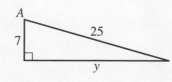

$$\sin 2A = 2 \sin A \cos A = 2 \cdot \frac{24}{25} \cdot \frac{7}{25} = \frac{336}{625}$$

Since $\angle A$ is acute, $\angle \frac{A}{2}$ is also acute and thus $\cos \frac{A}{2} > 0$.

$$\cos \frac{A}{2} = \sqrt{\frac{1 + \cos A}{2}} = \sqrt{\frac{1 + \frac{7}{25}}{2}} = \sqrt{\frac{32}{50}} = \frac{4\sqrt{2}}{5\sqrt{2}} = \frac{4}{5}$$

**Method 2**

Since $\angle A$ is acute, $\sin A > 0$.

Therefore, $\sin A = \sqrt{1 - \cos^2 A} = \sqrt{1 - \left(\frac{7}{25}\right)^2} = \sqrt{\frac{576}{625}} = \frac{24}{25}$

Proceed as in Method 1 above.

**Example 2** Solve $\cos 2x + 1 > \cos x$ for $0 \le x < 2\pi$.

**Solution** **Method 1**

Graph $y = \cos 2x + 1$ and $y = \cos x$ on a graphing calculator. Use the zoom or rescale features to determine the points of intersection and note the interval(s) in which the first graph is above the second.

**Method 2**

Graph $y = \cos 2x + 1 - \cos x$ and note the interval(s) in which the graph is above the $x$-axis.

**Method 3**

Solve the related equation $\cos 2x + 1 = \cos x$.

$2 \cos^2 x - 1 + 1 = \cos x \quad \leftarrow \cos 2x = 2 \cos^2 x - 1$
$2 \cos^2 x - \cos x = 0$
$\cos x \,(2 \cos x - 1) = 0$

$\cos x = 0 \qquad \text{or} \qquad \cos x = 0.5$
$x = \frac{\pi}{2}, \frac{3\pi}{2} \qquad\qquad x = \frac{\pi}{3}, \frac{5\pi}{3}$

Do a sign analysis of $\cos 2x + 1 - \cos x$ in each interval defined by these four solutions: Each method gives the solution $0 \le x < \frac{\pi}{3}$, $\frac{\pi}{2} < x < \frac{3\pi}{2}$, or $\frac{5\pi}{3} < x \le 2\pi$.

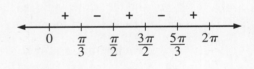

# Exercises

**17.** Suppose that $\sin \alpha = \frac{\sqrt{2}}{2}$ and $\cos \beta = \frac{15}{17}$, where $\frac{\pi}{2} < \alpha < \pi$ and $\frac{3\pi}{2} < \beta < 2\pi$. Find $\sin(\alpha - \beta)$ and $\cos(\alpha + \beta)$.

**18.** If $\cos A = \frac{7}{9}$ and $\angle A$ is acute, find $\cos 2A$ and $\sin \frac{A}{2}$.

**19.** *Writing* Suppose $\alpha$ and $\beta$ are first-quadrant angles and $\sin \alpha = \cos \beta$. Explain why $\sin 2\alpha = \sin 2\beta$.

**20.** *Application* Recall from Chapter 8 that you can determine the direction angle $\alpha$ of a rotated conic section by using the formula $\tan 2\alpha = \frac{B}{A - C}$ if the conic has equation $Ax^2 + Bxy + Cy^2 + Dx + Ey + F = 0$, $A \ne C$, and $0 < 2\alpha < \pi$. Find $\alpha$ to the nearest tenth of a degree for the conic section with equation $3x^2 + 4xy + y^2 - 4 = 0$.

# CHAPTER REVIEW

## Chapter 10: Trigonometric Addition Formulas

**Complete these exercises before trying the Practice Test for Chapter 10. If you have difficulty with a particular problem, review the indicated section.**

1. Simplify and evaluate $\cos 250° \cos 40° + \sin 250° \sin 40°$. *(Section 10-1)*

2. If $\tan \alpha = 2$ and $\tan \beta = \frac{1}{4}$, find **(a)** $\tan (\alpha + \beta)$ and **(b)** $\tan (\alpha - \beta)$. *(Section 10-2)*

3. Simplify. *(Section 10-3)*

   **a.** $2 \cos^2 \frac{x}{2} - 1$    **b.** $\dfrac{\sin 80°}{1 + \cos 80°}$    **c.** $\sin x \cos x$

4. Describe two ways to solve the equation $\tan 2x + \tan x = 0$ for $0° \leq x < 360°$. Then choose one of the methods and solve the equation. *(Section 10-4)*

**Simplify the given expression.**

1. $\sin \dfrac{\pi}{6} \cos \dfrac{5\pi}{6} + \sin \dfrac{5\pi}{6} \cos \dfrac{\pi}{6}$    2. $2 \sin \left( \dfrac{\pi}{4} - \dfrac{x}{2} \right) \cos \left( \dfrac{\pi}{4} - \dfrac{x}{2} \right)$

3. $(1 - \cos^2 x)(1 - \cot^2 x)$    4. $(\sin x - \cos x)^2 - 1$

**Evaluate each of the following if $\sin A = \dfrac{5}{13}$, $\cos B = \dfrac{15}{17}$, and $0 < B < \dfrac{\pi}{2} < A < \pi$.**

5. $\sin (A + B)$    6. $\cos (2A - B)$

7. $\tan (A - B)$    8. $\sin \dfrac{1}{2} B$

9. Find the acute angle, to the nearest tenth of a degree, formed by the intersection of the graphs of $3x + 4y = 12$ and $2x - y = -3$.

**Solve each equation for $0 \leq x < 2\pi$.**

10. $2 \sin^2 x = 2 + \cos 2x$    11. $\sin 2x + \sin x = 0$

12. $\csc x \sin 2x = 2 + \tan \dfrac{1}{2} x$    13. $\cot x \tan 2x = 3$

**Prove that the given equation is an identity.**

14. $\sin^2 \dfrac{x}{2} + \cos x = \cos^2 \dfrac{x}{2}$    15. $\cos^4 y - \sin^4 y - \cos 2y = 0$

**Consider the functions $f(x) = \sin 2x$ and $g(x) = \cos x$ on the interval $0 \leq x \leq 2\pi$.**

16. Solve $f(x) = g(x)$.    17. Solve $f(x) < g(x)$.

18. Sketch the graphs of $f$ and $g$ on the same set of axes.

Let $P(x) = 4 + 3x - 0.5x^3$.

1. Find each real zero of $P$ to the nearest hundredth.

2. Find the sum and product of the roots of $P(x) = 0$.

3. Use synthetic division to show that $x - 2$ is not a factor of $P(x)$.

**Simplify.**

4. $\log_3 54 - \log_3 2\sqrt{3}$

5. $\dfrac{2 \tan \dfrac{7\pi}{12}}{1 - \tan^2 \dfrac{7\pi}{12}}$

6. $e^{(-1/2)\ln 25}$

**Solve for x.**

7. $2x^2 + 4x + 3 = 0$

8. $x^3 + 4 = 15x$

9. $|x + 2.5| > 1.8$

10. A farmer wants to build four attached pens, as shown in the figure at the right, using a wall as one side and 200 m of fencing for the other sides. Express the total area of the pens in terms of $x$ and find the maximum possible area.

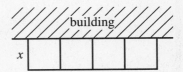

11. Give the domain, range, and zeros of the function
$f(x) = -\sqrt{4 - x^2}$. Does $f$ have an inverse?

12. Find an equation of the perpendicular bisector of the segment with endpoints $(-5, -2)$ and $(-3, 8)$. Then find the inclination of the line.

**Simplify.**

13. $\cos x\,(\tan x + \cot x)$

14. $\sin(-x) + \sin(\pi + x)$

15. Solve $2 \cot \theta = \csc^2 \theta$ for $0° \le x < 360°$.

16. Let $g(x) = -\dfrac{2}{x}$. Show that $g(g(x)) = x$ and explain what this indicates about $g$.

17. Describe the symmetries of the graph of (**a**) function $f$ in Exercise 11 and (**b**) function $g$ in Exercise 16.

18. Give the amplitude and period of the function $y - 2 = 3 \cos \dfrac{\pi}{2}(x + 1)$ and name three points on the graph of the function.

19. In $\triangle ABC$, $AB = 14$, $BC = 10$, and $\angle B = 108°$. Find $AC$ and the area of $\triangle ABC$, each to three significant digits.

20. If $\angle A$ is an obtuse angle and $\cos A = -\dfrac{15}{17}$, find the exact values of
$\cot A$, $\cos 2A$, and $\sin \dfrac{A}{2}$.

21. Suppose each person at a meeting shakes hands with every other person exactly once. Let $H(n)$ represent the number of handshakes when $n$ people attend the meeting.

| $n$ | 2 | 3 | 4 | 5 | 6 | ... | $n$ |
|---|---|---|---|---|---|---|---|
| $H(n)$ | 1 | 3 | 6 | 10 | 15 | ... | ? |

Find a model for $H(n)$ and use it to determine the number of handshakes if 20 people attend the meeting.

# Polar Coordinates and Complex Numbers

*Sections 11-1, 11-2, 11-3 and 11-4*

| OVERVIEW | Section 11-1 is concerned with polar coordinates and polar graphs. Sections 11-2, 11-3, and 11-4 focus on writing complex numbers and their powers and roots in polar and rectangular form. |
|---|---|

## KEY TERMS

**EXAMPLE/ILLUSTRATION**

**Polar coordinates of a point $P$** (p. 395)

the ordered pair $(r, \theta)$, where $r$ is the directed distance from a fixed point $O$ (called the *pole*) to $P$ and $\theta$ is the measure of an angle formed by ray $OP$ and a reference ray (called the *polar axis*)

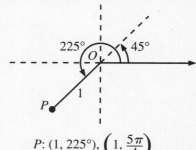

$P$: $(1, 225°)$, $\left(1, \dfrac{5\pi}{4}\right)$

$(-1, 45°)$, $\left(-1, \dfrac{\pi}{4}\right)$

A negative value for $r$ shows that $P$ is on the ray opposite the terminal side of $\theta$.

**Polar equation** (p. 396)

an equation written in terms of $r$, $\theta$, or both

$r = 3$, $\theta = \pi$, $r = 4 \sin \theta$

**Argand diagram** (p. 403)

a diagram that can be used to graph a complex number $a + bi$ using a point $(a, b)$ or an arrow from the origin to $(a, b)$

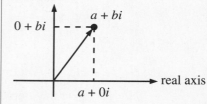

**Polar form of a complex number** (p. 403)

$z = r(\cos \theta + i \sin \theta)$, often written as $r$ cis $\theta$

$\sqrt{2}(\cos \theta + i \sin \theta)$, or $\sqrt{2}$ cis $\theta$

**Absolute value of a complex number $a + bi$** (p. 403)

$|a + bi| = \sqrt{a^2 + b^2}$

$\begin{aligned} |3 - 2i| &= \sqrt{3^2 + (-2)^2} \\ &= \sqrt{13} \end{aligned}$

## UNDERSTANDING THE MAIN IDEAS

### Polar and rectangular coordinates

- If you know the polar coordinates $(r, \theta)$ of a point, you can find the rectangular coordinates $(x, y)$ by using the equations $x = r \cos \theta$ and $y = r \sin \theta$. (See Example 2(b) on text page 397.)

- If you know the rectangular coordinates $(x, y)$ of a point, you can find the polar coordinates $(r, \theta)$ by using the equations $r = \pm\sqrt{x^2 + y^2}$ and $\tan \theta = \frac{y}{x}$. The rectangular coordinates will help you identify the quadrant in which the point lies; you need to know the quadrant to identify $\theta$. *Remember:* There are many polar coordinates for a given point. (See text page 395 and Example 2(a) on text page 397.)

### Complex numbers

- If you know the rectangular form, $a + bi$, of a complex number you can find the polar form, $r$ cis $\theta$, by using the equations $r = \sqrt{a^2 + b^2}$ $(r \geq 0)$ and $\tan \theta = \frac{b}{a}$. Be careful to identify the quadrant of $\theta$ before you find $\theta$.

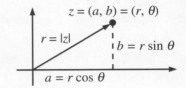

- If you know the polar form, $r$ cis $\theta$, of a complex number, you can find the rectangular form, $a + bi$, by using the formulas $a = r \cos \theta$ and $b = r \sin \theta$.

- To multiply two complex numbers in polar form, multiply their absolute values and add their polar angles, that is, if $z_1 = r(\cos \alpha + i \sin \alpha)$ and $z_2 = s(\cos \beta + i \sin \beta)$, then $z_1z_2 = rs(\cos (\alpha + \beta) + i \sin (\alpha + \beta))$. (See Example 3 on text page 405.)

- To find the *n*th power of a complex number $z = r(\cos \theta + i \sin \theta)$, use De Moivre's theorem: $z^n = r^n(\cos n\theta + i \sin n\theta)$, where $n$ is an integer. (See the Example on text page 408.)

- To find the *n*th roots of a complex number $z = r(\cos \theta + i \sin \theta)$, use the formula $\sqrt[n]{z} = \sqrt[n]{r} \left( \cos \frac{\theta + k \cdot 360°}{n} + i \sin \frac{\theta + k \cdot 360°}{n} \right)$ for $k = 0, 1, 2, \ldots, n - 1$, and positive integer $n$.

## CHECKING THE MAIN IDEAS

**Match each point or complex number with its polar form.**

1. $(1, -\sqrt{3})$     A. 2 cis 270°     B. $(2, \pi)$

2. $(-2, 0)$     C. $(-2, 120°)$     D. 2 cis 120°

3. $-1 + i\sqrt{3}$     E. $\left( 2, \frac{3\pi}{2} \right)$     F. 2 cis 300°

4. $-2i$

5. Plot point $P(3, 200°)$. Give two other pairs of polar coordinates for $P$.

6. Give polar coordinates $(r, \theta)$, where $\theta$ is in radians, for the point $(-4, 4)$.

7. Give rectangular coordinates for $\left( -2, \frac{7\pi}{6} \right)$.

8. Express $8 - 15i$ in polar form. Give the angle measure to the nearest degree.

9. Express 5 cis 70° in rectangular form.

10. Express the product $\left( 6 \text{ cis } \frac{7\pi}{12} \right) \left( \frac{1}{3} \text{ cis } \frac{\pi}{6} \right)$ in polar form and in rectangular form.

*ADVANCED MATHEMATICS Student Resource Guide* **101**

11. **Critical Thinking**  Manoj said, "If I want to express a complex number

$a + bi$ in polar form, I use the equations $\cos \theta = \dfrac{a}{\sqrt{a^2 + b^2}}$ and

$\sin \theta = \dfrac{b}{\sqrt{a^2 + b^2}}$ . Then I am sure to get the right value for $\theta$."

Comment on Manoj's method. Is it valid? Is he correct when he claims that it will give him the right value for $\theta$?

12.  **a.** Express $z = 1 + i$ in polar form.

  **b.** Find $z^{-1}$, $z^0$, and $z^2$.

  **c.** Simplify $(1 + i)(1 + i)$ and compare the product with your answer for $z^2$ in part (b).

13.  **a.** Find the four fourth roots of $-8 + 8i\sqrt{3}$. Give the answers in rectangular form.

  **b.** Find the fourth power of the first-quadrant root in simplest form. Did you get the answer you expected?

## USING THE MAIN IDEAS

**Example 1**  Sketch the polar graph of $r = 1 - \cos \theta$. Then give a rectangular equation of the graph.

**Solution**  Use the following symmetry tests. A curve has symmetry:
- with respect to the horizontal axis ($\theta = 0°$) if replacing $\theta$ by $-\theta$ gives an equivalent equation.
- with respect to the origin (the pole) if replacing $r$ by $-r$, or replacing $\theta$ by $(\theta + 180°)$, gives an equivalent equation.
- with respect to the line $\theta = 90°$ if replacing $r$ by $-r$ and $\theta$ by $-\theta$, or replacing $\theta$ by $(180° - \theta)$, gives an equivalent equation.

For the given equation, since $1 - \cos(-\theta) = 1 - \cos \theta$, the graph is symmetric with respect to the line $\theta = 0°$. Since $\cos \theta \neq \cos(\theta + 180°)$ and $\cos \theta \neq \cos(180° - \theta)$, the graph is *not* symmetric with respect to the origin or the line $\theta = 90°$, respectively.

Make a table of values:

| $\theta$ | 0° | 45° | 60° | 90° | 120° | 135° | 180° |
|---|---|---|---|---|---|---|---|
| $r$ | 0 | $\approx 0.3$ | 0.5 | 1 | 1.5 | $\approx 1.7$ | 2 |

| $\theta$ | 225° | 240° | 270° | 300° | 315° | 360° |
|---|---|---|---|---|---|---|
| $r$ | $\approx 1.7$ | 1.5 | 1 | 0.5 | $\approx 0.3$ | 0 |

Plot the points and draw a smooth curve, as shown at the right. The graph is a *cardioid*.

$$r = 1 - \cos \theta$$
$$r^2 = r - r \cos \theta$$
$$x^2 + y^2 = \pm\sqrt{x^2 + y^2} - x$$
$$x^2 + y^2 + x = \pm \sqrt{x^2 + y^2}$$
$$(x^2 + y^2 + x)^2 = x^2 + y^2 \quad \leftarrow \text{rectangular form}$$

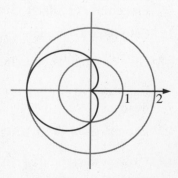

**Example 2** Let $z_1 = 1 - 2i$ and $z_2 = -4 - 3i$.

    **a.** Find $z_1z_2$ in rectangular form by multiplying $z_1$ and $z_2$.

    **b.** Find $z_1$, $z_2$, and $z_1z_2$ in polar form. Show that $z_1z_2$ in polar form agrees with $z_1z_2$ in rectangular form (from part (a)).

**Solution**  **a.** $z_1z_2 = (1 - 2i)(-4 - 3i)$

$$= -4 + 8i - 3i + 6i^2$$
$$= (-4 - 6) + 5i$$
$$= -10 + 5i$$

    **b.** $r = |z_1| = \sqrt{1^2 + (-2)^2} = \sqrt{5}$

$\tan \theta = \dfrac{-2}{1} = -2$ and $270° < \theta < 360°$, so $\theta \approx 297°$

Therefore, $z_1 = \sqrt{5}$ cis $297°$

$|z_2| = \sqrt{(-4)^2 + (-3)^2} = 5$

$\tan \theta = \dfrac{-3}{-4} = 0.75$ and $180° < \theta < 270°$, so $\theta \approx 217°$

Therefore, $z_2 = 5$ cis $217°$

$z_1z_2 = 5\sqrt{5}$ cis $(297° + 217°)$

$\quad\quad = 5\sqrt{5}$ cis $514°$, or $5\sqrt{5}$ cis $154°$   $\leftarrow$ $514° - 360° = 154°$

Since $5\sqrt{5}(\cos 154° + i \sin 154°) \approx -10 + 5i$, the rectangular and polar forms agree.

## Exercises

**14.** Find polar coordinates for point $(9, 4)$. Give $r$ to the nearest tenth and $\theta$ to the nearest tenth of a radian.

**15.** Sketch the polar graph of $r = 2 + \sin \theta$ (a limaçon) and give a rectangular equation of the graph.

**16.** Let $z_1 = -2 - 2i$ and $z_2 = -12 + 5i$.

    **a.** Find $z_1z_2$ in rectangular form by multiplying $z_1$ and $z_2$.

    **b.** Find $z_1$, $z_2$, and $z_1z_2$ in polar form. Show that $z_1z_2$ in polar form agrees with $z_1z_2$ in rectangular form.

    **c.** Show $z_1$, $z_2$, and $z_1z_2$ in an Argand diagram.

**17.** *Writing* Suppose a complex number $z$ is multiplied by $i$. Explain how this affects the graph of $z$.

**18.** *Application* Explain the connection between the polar coordinate system and radar.

**19.** Find the reciprocal of $z = -1 + i$ in rectangular form: **(a)** by simplifying $\dfrac{1}{z}$ and **(b)** by using De Moivre's theorem.

**20.** Find the fifth roots of $-i$ in polar form.

     *ADVANCED MATHEMATICS Student Resource Guide* **103**

# CHAPTER REVIEW

## Chapter 11: Polar Coordinates and Complex Numbers

Complete these exercises before trying the Practice Test for Chapter 11. If you have difficulty with a particular problem, review the indicated section.

1. Give the rectangular coordinates for the point $(8, -45°)$. *(Section 11-1)*

2. Describe how to sketch the polar graph of a polar equation without using a graphing calculator. *(Section 11-1)*

3. Express $-2i$ and $-1 - i$ in polar form. Find their product in rectangular form and in polar form. *(Section 11-2)*

4. If $z = -1 - i$, find $z^4$ in polar form and in rectangular form. Then show $z$ and $z^4$ in an Argand diagram. *(Section 11-3)*

5. Express $4i$ in polar form. Then find the two square roots of $4i$. Square each one to verify that it is a square root of $4i$. *(Section 11-6)*

---

Give polar coordinates $(r, \theta)$, where $\theta$ is in degrees, for each point.

1. $(0, 2)$     2. $(6, -6)$     3. $(-4, 3)$

Give the rectangular coordinates for each point.

4. $(4, 135°)$     5. $\left(8, \dfrac{7\pi}{6}\right)$     6. $(10, 310°)$

Sketch the polar graph of each equation. Use radian measure for $\theta$.

7. $r = 1 + 2 \sin \theta$     8. $r = 2 + \sin \theta$

9. $r = \cos 2\theta$     10. $r = \dfrac{\theta}{2}$

Let $z_1 = 1 + i$, $z_2 = -1 - i\sqrt{3}$, and $z_3 = \dfrac{3}{5} - \dfrac{4}{5}i$.

11. Find $|z_1|$.     12. Express $z_1$ in polar form.

13. Express $z_2$ in polar form.     14. Find $z_1 z_2$ in polar form.

15. Express $(z_3)^{10}$ in rectangular form.

16. Find the three cube roots of $z_3$.

Let $z_1 = \sqrt{3} - i$, $z_2 = 2i$, and $z_3 = -\sqrt{3} - i$.

17. Show $z_1, z_2,$ and $z_3$ in an Argand diagram.

18. On the Argand diagram for Exercise 17, include the graph of the circle $|z| = 2$.

19. Explain how $z_1, z_2,$ and $z_3$ are related to the circle $|z| = 2$.

---

Let $f(x) = -2x + 1$ and $g(x) = x^2 + 2x$.

1. Give the domain, range, and zeros of $g$.

2. Find $(g - f)(x)$, $(f \circ g)(x)$, and $(g \circ f)(x)$.

3. Find the inverse of $f$ and then graph $f$ and $f^{-1}$.

4. Find each point $(x, y)$ that is on the graph of both functions.

---

5. Find the cube roots of $-\sqrt{3} + i$ in polar form.

6. Evaluate $\tan(\text{Cos}^{-1} 0)$ and $\cos(\text{Tan}^{-1} 0)$. Explain your answers.

**Graph each system of equations. Label the points of intersection.**

7. $y^2 - x^2 = 4$
   $y = 2x - 2$

8. $x^2 + y^2 = 4$
   $y = \frac{5}{2}x^2 - 2$

9. Approximate (**a**) the measures of the angles of a triangle with sides of lengths 4 cm, 5 cm, and 8 cm and (**b**) the area of the triangle.

10. If $\sqrt{5}$ is a root of $x^4 + 2x^3 + 5x^2 - 10x - 50 = 0$, find the other three roots.

**Simplify.**

11. $3 \log 2 + 6 \log \sqrt{5}$

12. $(27^{5/3} - 81^{1/2})3^{-2}$

13. $\ln \sqrt[6]{e}$

14. $\sin\left(x + \frac{3\pi}{2}\right)$

15. $(1 + i)^{20}$

16. $1 - \sin x \cos x \tan x$

17. Describe the symmetry of the graph of $9x^2 + 6xy + 4y^2 - 36 = 0$. Then identify the graph.

18. Find the exact values of $\sin 157.5°$ and $\tan 157.5°$.

19. Suppose lines are drawn in a circle to form as many regions of the circle as possible. (See the figure at the right.) Let $R(n)$ be the number of regions formed by $n$ lines: $R(1) = 2$, $R(2) = 4$, $R(3) = 7$, and $R(4) = 11$. Find a formula for $R(n)$. For what value of $n$ is $R(n) = 37$?

3 lines
7 regions

20. If $\sin x = -\frac{12}{13}$ and $\frac{\pi}{2} < x < \frac{3\pi}{2}$, find $\cos x$, $\sin 2x$, and $\cos(\pi - x)$.

**Solve for $x$. If necessary, round to the nearest hundredth.**

21. $5^x = 350$

22. $2x^3 = x^2 + 16x + 15$

**Solve for $0° \leq \theta < 360°$. If necessary, give answers to the nearest tenth of a degree.**

23. $\sec^2 \theta = 2(\tan \theta + 2)$

24. $\sin 2\theta > -\sin \theta$

25. Find a rectangular equation for $r + 4 \cos \theta = 0$ and describe the graph of the equation.

26. $\triangle ABC$ has vertices $A(2, 0)$, $B(-1, 6)$, and $C(-3, -1)$. Find an equation of the line that contains the altitude to $\overline{AB}$. At what point does this line intersect $\overline{AB}$?

27. Find an equation for a cosine curve with amplitude $\frac{5}{2}$, period $\pi$, and which contains the origin.

28. The sum of the lengths of two sides of a triangle is 8 m and the measure of the included angle is 150°. If $x$ represents the shorter side, express the area of the triangle in terms of $x$ and find the maximum area.

# Introduction to Vectors

## Sections 12-1 and 12-2

**OVERVIEW** — Sections 12-1 and 12-2 introduce the basic vector operations through the use of vector diagrams and a coordinate approach.

## KEY TERMS

| | EXAMPLE/ILLUSTRATION |
|---|---|
| **Vector** (p. 419)<br>a quantity, such as force or velocity, that has both direction and magnitude (size or length) | 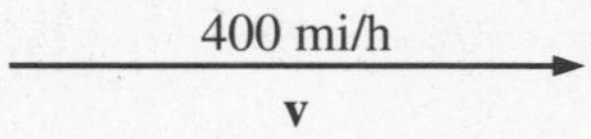 400 mi/h →<br>**v**<br>The velocity vector **v** has magnitude 400 mi/h and direction east. |
| **Negative of a vector** (p. 421)<br>a vector of the same length as the given vector but the opposite direction | 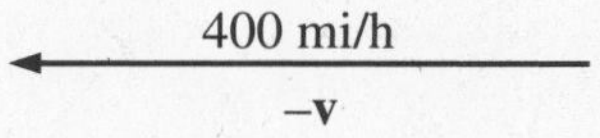 ← 400 mi/h<br>**−v** |
| **Zero vector** (p. 421)<br>a point, or a vector with length 0 | ●<br>**0** |
| **Component form of a vector** (p. 426)<br>specifying a vector in terms of an ordered pair | $\overrightarrow{AB} = (x_2 - x_1,\ y_2 - y_1)$ |

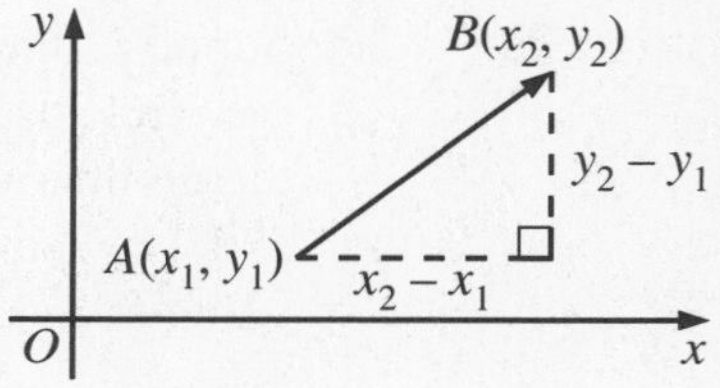

## UNDERSTANDING THE MAIN IDEAS

### Geometric representation of vectors

- To add two vectors, **u** and **v**:

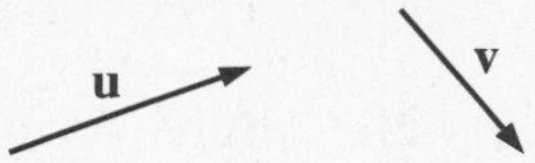

    1. place the "tail" of **v** at the "tip" of **u**; **u + v** is the vector extending from the "tail" of **u** to the "tip" of **v**, as shown, or

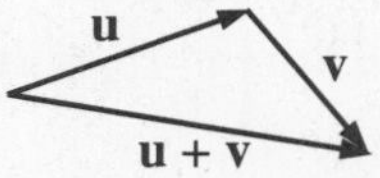

    2. form a parallelogram with **u** and **v** "tail-to-tail", as shown; **u + v** extends from the "tails" of **u** and **v** to the opposite vertex of the parallelogram.

- To find the difference, **u − v**, of vectors **u** and **v**, add the negative of **v** to **u**:

    **u − v = u + (−v)**

    Note that **u − u = u + (−u) = 0**.

- To multiply a vector **u** by a real number $k$ (called *scalar multiplication*), multiply the length of **u** by $|k|$ and reverse the direction if $k < 0$. You can think of scalar multiplication as repeated addition.

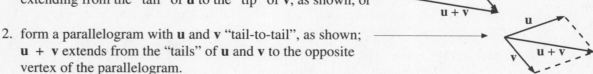

The distributive laws of scalar multiplication are
$k(\mathbf{v} + \mathbf{w}) = k\mathbf{v} + k\mathbf{w}$ and $(k + m)\mathbf{v} = k\mathbf{v} + m\mathbf{v}$ and the
associative law of scalar multiplication is $k(m\mathbf{v}) = (km)\mathbf{v}$.

### Algebraic representation of vectors

- To write $\overrightarrow{AB}$ in component form, subtract the coordinates of $A$ from those of $B$:
  If $A = (x_1, y_1)$ and $B = (x_2, y_2)$, $\overrightarrow{AB} = (x_2 - x_1, y_2 - y_1)$.
- To find the magnitude of $\mathbf{v} = (a, b)$, use the formula $|\mathbf{v}| = \sqrt{a^2 + b^2}$. Note that $|\overrightarrow{AB}| = \sqrt{(x_2 - x_1)^2 + (y_2 - y_1)^2}$.
- To add two vectors in component form, add their corresponding coordinates: $(a, b) + (c, d) = (a + c, b + d)$.
- To subtract two vectors in component form, subtract their corresponding components: $(a, b) - (c, d) = (a - c, b - d)$.
- To multiply a vector by a scalar, multiply each coordinate by the scalar: $k(a, b) = (ka, kb)$.

## CHECKING THE MAIN IDEAS

**ABCD is a rhombus with a 60° angle and sides of length 4. Match each vector or value with an equal vector or value. (*Hint:* Two vectors with the same length and direction are equal.)**

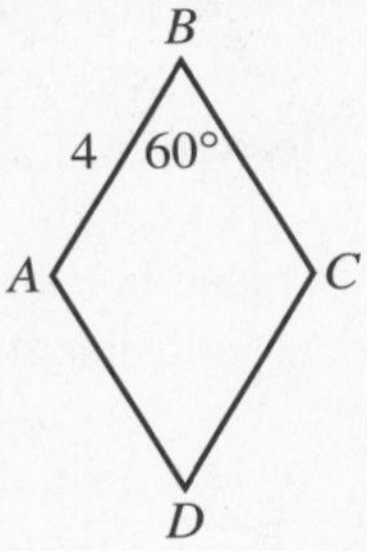

1. $\overrightarrow{AD}$
2. $|\overrightarrow{BC}|$
3. $\overrightarrow{AD} + \overrightarrow{DC}$
4. $\overrightarrow{BC} + \overrightarrow{CA}$
5. $|\overrightarrow{BC} + \overrightarrow{CD}|$
6. $\overrightarrow{AD} + \overrightarrow{CB}$

**A.** $\overrightarrow{AB} + \overrightarrow{BC}$
**B.** $4\sqrt{3}$
**C.** $-\overrightarrow{DC}$
**D.** $\mathbf{0}$
**E.** $\overrightarrow{BC}$
**F.** $4$

7. Copy vectors $\mathbf{u}$ and $\mathbf{v}$ shown at the right. Then sketch $2\mathbf{u} + \mathbf{v}$ and $\mathbf{u} - 2\mathbf{v}$.

8. Draw a vector diagram to show that $\frac{1}{2}(4\mathbf{w}) = \left(\frac{1}{2} \cdot 4\right)\mathbf{w}$.

9. Let $A = (5, 1)$ and $B = (1, -5)$. Find $\overrightarrow{AB}$ and $|\overrightarrow{AB}|$.

10. Let $\mathbf{u} = (-3, -4)$ and $\mathbf{v} = (5, 2)$. Calculate each expression.

   **a.** $\mathbf{u} - \mathbf{v}$   **b.** $|\mathbf{u} - \mathbf{v}|$   **c.** $4\mathbf{u} + \frac{1}{2}\mathbf{v}$   **d.** $\dfrac{\mathbf{u}}{|\mathbf{u}|}$

11. *Critical Thinking* If $\mathbf{v}$ is a vector, explain why $|\mathbf{v}|$ represents a real number and not a vector. What kind of real number is $|\mathbf{v}|$? Why?

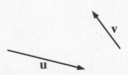

## USING THE MAIN IDEAS

**Example 1** $\mathbf{F}_1$ is a force of 3 N (newtons) pulling an object north and $\mathbf{F}_2$ is a force of 1.5 N pulling the object west. Use a vector diagram to estimate the magnitude and direction of the resultant force vector $\mathbf{F}_1 + \mathbf{F}_2$. Then use trigonometry to find the magnitude, to the nearest hundredth, and the direction, to the nearest tenth of a degree, of $\mathbf{F}_1 + \mathbf{F}_2$.

**Solution** Draw a scale diagram, measuring angles and lengths carefully. By measuring with a ruler and protractor, you see that $|\mathbf{F}_1 + \mathbf{F}_2| \approx 3.5$ N and that $\alpha \approx 27°$, so the compass direction of $\mathbf{F}_1 + \mathbf{F}_2$ is about $360° - 27° = 333°$.

$|\mathbf{F}_1 + \mathbf{F}_2| = \sqrt{3^2 + (1.5)^2} \approx 3.35$

$\tan \alpha = \dfrac{1.5}{3} = 0.5$, so $\alpha \approx 26.6°$

$\mathbf{F}_1 + \mathbf{F}_2$ has magnitude 3.35 N and compass direction $360° - 26.6°$, or $333.4°$.

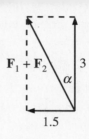

**Example 2** An airplane is flying due north at 320 km/h. The plane encounters a wind of 40 km/h from the southwest. Make a vector diagram. Then calculate the resultant speed and direction of the airplane (measured clockwise from north).

**Solution** Make a scale drawing like the one shown at the right. Measuring $\overrightarrow{AC}$ and $\angle A$, $|\overrightarrow{AC}| \approx 360$ and $\angle A \approx 5°$. This gives an estimated resultant speed of about 360 km/h and compass direction 5°.

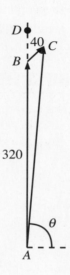

**Method 1**

Note that $\angle DBC = 45°$, so $\angle CBA = 135°$. Use the law of cosines and then the law of sines as in Example 1 on text pages 420–421.

**Method 2**

Write each vector in component form. Then $\overrightarrow{AB} = (0, 320)$ since the plane is heading due north at 320 km/h. The wind is from the southwest at 40 km/h, so the components of the wind vector are:

$x = 40 \cos 45° \approx 28.28$
$y = 40 \sin 45° \approx 28.28$

Thus, $\overrightarrow{BC} \approx (28.28, 28.28)$.

$\overrightarrow{AC} = \overrightarrow{AB} + \overrightarrow{BC} \approx (0, 320) + (28.28, 28.28) = (28.28, 348.28)$

$|\overrightarrow{AC}| \approx \sqrt{(28.28)^2 + (348.28)^2} \approx 349.4$

Let $\theta$ be the direction angle of $\overrightarrow{AC}$.

$\tan \theta \approx \dfrac{348.28}{28.28} \approx 12.32$, so $\theta \approx 85.4°$

The airplane's resultant direction is $90° - \theta$, or $90° - 85.4° = 4.6°$ and its resultant speed is about 349.4 km/h.

## Exercises

12. $\mathbf{F}_1$ is a force of 0.9 N pulling an object south and $\mathbf{F}_2$ is a force of 4 N pulling the object west. Use a diagram to estimate the magnitude and direction of the resultant force vector $\mathbf{F}_1 + \mathbf{F}_2$. Then use trigonometry to find the magnitude and compass direction of $\mathbf{F}_1 + \mathbf{F}_2$.

13. *Writing* Explain whether vector subtraction is commutative or not. Include a vector diagram in your explanation.

14. If $A = (2, 0)$ and $B = (10, -4)$, find the coordinates of the point $P$ that is $\dfrac{3}{4}$ of the way from $A$ to $B$. (See Example 4 on text page 428.)

15. A ship is heading on a course of 210° at a speed of 20 knots. Find the north-south component and east-west component of its velocity vector.

# Vectors and Parametric Equations and the Dot Product

## Sections 12-3 and 12-4

| OVERVIEW | In Section 12-3, vector and parametric equations are used to describe motion in the plane. In Section 12-4, the dot product is used to identify perpendicular vectors and to find the angle between two vectors. The section is also concerned with parallel vectors and properties of the dot product. |
|---|---|

## KEY TERMS

| | |
|---|---|
| **Vector equation of line AB** (p. 432)<br><br>$\overrightarrow{OP} = \overrightarrow{OA} + t\overrightarrow{AB}$, where $P$ is any point on line $AB$, $t$ is a real number, and $\overrightarrow{AB}$ is a *direction vector* of the line | 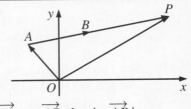<br><br>$\overrightarrow{AP} = t\overrightarrow{AB}$, that is, $\overrightarrow{AP}$ is a scalar multiple of $\overrightarrow{AB}$. |
| **Parametric equations** (p. 433)<br><br>equations in one variable, such as $x = f(t)$ and $y = g(t)$, or $x = f(\theta)$ and $y = g(\theta)$, that give the coordinates of points on a line or curve in the plane; $t$ is called the *parameter*. | 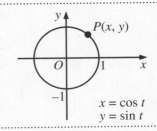<br><br>$x = \cos t$<br>$y = \sin t$ |
| **Dot product of two vectors** (p. 441)<br><br>$\mathbf{u} \cdot \mathbf{v} = ac + bd$, where $\mathbf{u} = (a, b)$ and $\mathbf{v} = (c, d)$ | $(6, 4) \cdot (-6, 9) = 6(-6) + 4(9)$<br>$= 0$ |

## UNDERSTANDING THE MAIN IDEAS

### Vector equations

- You can find a vector equation $\overrightarrow{OP} = \overrightarrow{OA} + t\overrightarrow{AB}$ if you know:
    1. a point $A$ on the line and a direction vector $\overrightarrow{AB}$, or
    2. two points on the line. (See Example 1 on text page 432.)
    A line has infinitely many vector equations.
- If you know a vector equation for a line, you can find the coordinates of points on the line by substituting values for $t$.
- A vector equation for an object moving with constant velocity is
$$(x, y) = (x_0, y_0) + t(a, b)$$
where $t$ represents units of time, $(x, y)$ is the position of the object at time $t$, $(x_0, y_0)$ is the position of the object at time $t = 0$ (called the *initial position*), $(a, b)$ is the constant velocity vector, and $\sqrt{a^2 + b^2}$ is the speed of the object.

### Parametric equations

- Another form of the vector equation for a line is $(x, y) = (x_0, y_0) + t(a, b)$ where $(x, y)$ is any point on the line, $(x_0, y_0)$ is a given or fixed point on the line, and $(a, b)$ is a direction vector of the line. From this form of the vector equation, we can get the parametric equations for the line:

$$x = x_0 + at \quad \text{and} \quad y = y_0 + bt$$

These equations give the $x$- and $y$-coordinates of points on the line in terms of the parameter $t$ ($x_0$, $y_0$, $a$, and $b$ are constants). A line has infinitely many pairs of parametric equations.

- To write a pair of parametric equations as a single rectangular (or Cartesian) equation in $x$ and $y$, eliminate $t$.

### Dot product

- The dot product of two vectors is a real number, *not* a vector.
- Vectors $\mathbf{u}$ and $\mathbf{v}$ are perpendicular if and only if $\mathbf{u} \cdot \mathbf{v} = 0$.
- Vectors $\mathbf{u}$ and $\mathbf{v}$ are parallel if and only if $\mathbf{u} = k\mathbf{v}$ ($k$ is a scalar).
- The dot product has these properties:
  Commutative: $\mathbf{u} \cdot \mathbf{v} = \mathbf{v} \cdot \mathbf{u}$
  Associative: $k(\mathbf{u} \cdot \mathbf{v}) = (k\mathbf{u}) \cdot \mathbf{v}$, where $k$ is a scalar
  Distributive: $\mathbf{u} \cdot (\mathbf{v} + \mathbf{w}) = \mathbf{u} \cdot \mathbf{v} + \mathbf{u} \cdot \mathbf{w}$
  Magnitude property: $\mathbf{u} \cdot \mathbf{u} = |\mathbf{u}|^2$
  (Recall that $|\mathbf{u}| = \sqrt{a^2 + b^2}$ if $\mathbf{u} = (a, b)$.)
- You can find the measure of the angle between two vectors $\mathbf{u}$ and $\mathbf{v}$ by using the formula $\cos \theta = \dfrac{\mathbf{u} \cdot \mathbf{v}}{|\mathbf{u}||\mathbf{v}|}$, where $0° \leq \theta \leq 180°$.
  Remember that if $\cos \theta$ is negative, then $90° < \theta \leq 180°$.

## CHECKING THE MAIN IDEAS

1. Which point lies on the line $(x, y) = (-5, 4) + t(2, -1)$?
   **A.** $(1, 1)$      **B.** $(-7, 3)$      **C.** $(-3, 5)$      **D.** $(-1, 6)$

2. If $\mathbf{u} \cdot \mathbf{v} = -8$, $|\mathbf{u}| = 4$, and $|\mathbf{v}| = 2\sqrt{2}$, find the angle between $\mathbf{u}$ and $\mathbf{v}$.
   **A.** $45°$      **B.** $120°$      **C.** $150°$      **D.** $135°$

3. Find a vector equation and corresponding parametric equations for the line containing $(5, 3)$ and $(-2, -4)$. (See Example 1 on text page 432.)

4. Find a vector equation and corresponding parametric equations of the moving object with constant velocity $(4, -3)$ and initial position $(-6, 2)$.

5. A line has vector equation $(x, y) = (3, 5) + t(2, -2)$. Find a pair of parametric equations and a Cartesian equation of the line.

6. *Critical Thinking* Describe the curve having parametric equations $x = 3 + t$ and $y = 4 - 2t$, where $-3 \leq t \leq 2$. Explain your answer.

7. Let $\mathbf{u} = (8, -6)$ and $\mathbf{v} = (-4, 3)$. Show that (a) $\mathbf{u} \cdot \mathbf{v} = \mathbf{v} \cdot \mathbf{u}$, (b) $\mathbf{v} \cdot \mathbf{v} = |\mathbf{v}|^2$, and (c) $\mathbf{u}$ and $\mathbf{v}$ are parallel vectors.

8. Find the measure of the angle between the vectors $(-1, 7)$ and $(-1, -1)$ to the nearest tenth of a degree. (See Example 2 on text page 443.)

# USING THE MAIN IDEAS

**Example 1** Find a vector equation of the line through $(-3, -1)$ and parallel to the line $(x, y) = (2, 7) + t(1, 5)$.

 The line $(x, y) = (2, 7) + t(1, 5)$ has parametric equations $x = 2 + t$ and $y = 7 + 5t$. Solving these equations for $t$, $t = x - 2$ and $t = \frac{y-7}{5}$, respectively. Thus, $x - 2 = \frac{y-7}{5}$, or $y = 5x - 3$.

The line $y = 5x - 3$ has slope 5. Notice that the direction vector $(1, 5)$ gives the same result: slope $= \frac{\text{change in } y}{\text{change in } x} = \frac{5}{1} = 5$.

Since any parallel line will also have direction vector $(1, 5)$, $(x, y) = (-3, -1) + t(1, 5)$ is the required vector equation.

**Example 2** In $\triangle PQR$, $P(2, 1)$, $Q(4, 7)$, and $R(-2, 4)$. Find the measure of $\angle P$ to the nearest tenth of a degree.

**Solution** **Method 1**
See Example 3 on text page 443.

**Method 2**
Use the fact, presented on text page 296, that the slope of a line is equal to $\tan \alpha$, where $\alpha$ is the angle measured counterclockwise from the positive $x$-axis to the line and $0° \le \alpha < 180°$.

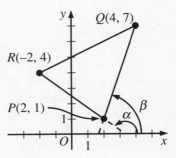

Line $PR$ has slope $\frac{4-1}{-2-2} = -\frac{3}{4}$.

Thus, $\tan \alpha = -\frac{3}{4}$ and $\alpha \approx 143.13°$ $\leftarrow$ Since $\tan \alpha < 0$, $90° < \alpha < 180°$

Line $PQ$ has slope $\frac{7-1}{4-2} = 3$.

Thus, $\tan \beta = 3$ and $\beta \approx 71.57°$ $\leftarrow$ Since $\tan \alpha > 0$, $0° < \alpha < 90°$

From the diagram, $\angle P = \alpha - \beta \approx 143.13° - 71.57° \approx 71.6°$.
(Notice that this agrees with the answer on text page 443.)

## Exercises

9. Find a vector equation of the line through $(9, 4)$ and parallel to the line $(x, y) = (-3, -2) + t(1, -4)$.

10. At time $t$, the position of an object moving with constant velocity is given by the parametric equations $x = 1 + 5t$ and $y = -1 + 12t$.
   **a.** Find the velocity and speed of the object.
   **b.** When and where does the object cross the line $2x - y = 5$?
      (See Example 3 on text page 434.)

11. Given $A(0, 0)$, $B(4, 3)$, and $C(1, -2)$, find the approximate measure of $\angle BAC$ **(a)** by using the dot product and **(b)** by using angles of inclination, as described in Example 2 above.

12. *Application* Find the work done by the force $\mathbf{F} = (3, 4)$ in moving a particle from $A(1, 2)$ to $B(5, 8)$. (*Hint:* From part (b) of Exercise 25 on text page 445, Work $= \mathbf{F} \cdot \mathbf{s}$, where $\mathbf{s} = \overrightarrow{AB}$.)

---

# Vectors in Three Dimensions

## Sections 12-5 and 12-6

| OVERVIEW | In these sections, work with vectors is extended to three dimensions. Section 12-5 presents properties of three-dimensional vectors and equations of a sphere and a line in three-dimensional space. Section 12-6 focuses on equations and graphs of planes. |
|---|---|

## KEY TERMS

**Three-dimensional coordinate system** (p. 446)
a system for locating points by using three perpendicular axes
($x$-axis, $y$-axis, and $z$-axis) that intersect at the origin

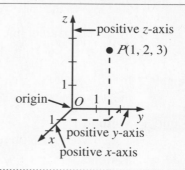

**$xy$-coordinate plane** (p. 446)
the plane containing the $x$- and $y$-axes

**$xz$-coordinate plane** (p. 446)
the plane containing the $x$- and $z$-axes

**$yz$-coordinate plane** (p. 446)
the plane containing the $y$- and $z$-axes

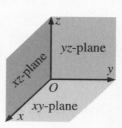

## UNDERSTANDING THE MAIN IDEAS

### Vectors in three dimensions

- A vector in three-dimensional space is an ordered triple $(a, b, c)$
  of numbers. If $A = (x_1, y_1, z_1)$ and $B = (x_2, y_2, z_2)$, then:

  1. vector $\overrightarrow{AB} = (x_2 - x_1, y_2 - y_1, z_2 - z_1)$

  2. length of vector $\overrightarrow{AB} = \sqrt{(x_2 - x_1)^2 + (y_2 - y_1)^2 + (z_2 - z_1)^2}$

  3. midpoint of segment $AB = \left( \dfrac{x_1 + x_2}{2}, \dfrac{y_1 + y_2}{2}, \dfrac{z_1 + z_2}{2} \right)$

- Three-dimensional vectors have the following properties:

  1. Addition: $(x_1, y_1, z_1) + (x_2, y_2, z_2) = (x_1 + x_2, y_1 + y_2, z_1 + z_2)$

  2. Scalar multiplication: $k(x_1, y_1, z_1) = (kx_1, ky_1, kz_1)$

  3. Magnitude: $|(x_1, y_1, z_1)| = \sqrt{x_1^2 + y_1^2 + z_1^2}$

  4. Dot product: $(x_1, y_1, z_1) \cdot (x_2, y_2, z_2) = x_1 x_2 + y_1 y_2 + z_1 z_2$

  5. The angle $\theta$ between two vectors $\mathbf{u} = (x_1, y_1, z_1)$ and
     $\mathbf{v} = (x_2, y_2, z_2)$: $\cos \theta = \dfrac{\mathbf{u} \cdot \mathbf{v}}{|\mathbf{u}||\mathbf{v}|}$.

6. Vectors **u** and **v** are perpendicular if and only if $\mathbf{u} \cdot \mathbf{v} = 0$.

7. Vectors **u** and **v** are parallel if and only if $\mathbf{u} = k\mathbf{v}$.

### Equation of a sphere

A sphere with radius $r$ and center $(x_0, y_0, z_0)$ has equation
$(x - x_0)^2 + (y - y_0)^2 + (z - z_0)^2 = r^2$.

### Equation of a line

If a line contains $(x_0, y_0, z_0)$ and has direction vector $(a, b, c)$, then:
1. its vector equation is $(x, y, z) = (x_0, y_0, z_0) + t(a, b, c)$, and
2. its parametric equations are $x = x_0 + at$, $y = y_0 + bt$, and
   $z = z_0 + ct$.

### Equation of a plane

The equation of a plane is $ax + by + cz = d$, where $a$, $b$, and $c$ are not all equal to 0. Also $d = ax_0 + by_0 + cz_0$, where $(a, b, c)$ is a nonzero vector perpendicular to the plane at the point $(x_0, y_0, z_0)$.

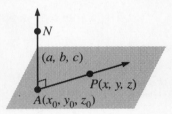

## CHECKING THE MAIN IDEAS

Let $A = (4, -3, -1)$ and $B = (0, 1, -1)$.

1. $|\overrightarrow{AB}| = \underline{\quad ? \quad}$
   **A.** $(2, -1, -1)$    **B.** $(-4, 4, 0)$    **C.** $4\sqrt{2}$    **D.** $2\sqrt{5}$

2. If $\mathbf{u} = (5, 7, -3)$ and $\mathbf{v} = (-2, -4, 0)$, find $\mathbf{u} \cdot \mathbf{v}$.
   **A.** $(3, 3, -3)$    **B.** $(-7, -11, 3)$    **C.** $-38$    **D.** $\sqrt{179}$

3. Which vector is perpendicular to the vector $(7, 2, -4)$?
   **A.** $(2, 3, 5)$    **B.** $(14, 4, -8)$    **C.** $(4, 0, -7)$    **D.** $(1, 1, 2)$

4. Which vector is perpendicular to the plane with equation
   $5y - 2z = -10$?
   **A.** $(0, 2, 5)$    **B.** $(0, 5, -2)$    **C.** $(0, 0, 5)$    **D.** $(-10, 0, 0)$

5. Find the midpoint of $\overline{AB}$ if $A = (8, -5, 7)$ and $B = (3, 4, 9)$.

6. If $\mathbf{u} = (7, -3, 0)$ and $\mathbf{v} = (1, 0, -4)$, simplify each expression.
   **a.** $\mathbf{u} - 3\mathbf{v}$        **b.** $|\mathbf{u} - 3\mathbf{v}|$        **c.** $\mathbf{u} \cdot \mathbf{v}$

7. Find an equation of the sphere with radius 11 and center $(1, -1, 4)$.
   Show that the point $(3, 5, -5)$ is on the sphere.

8. Find vector and parametric equations for the line containing $A(5, -2, 4)$
   and $B(6, 1, 1)$. (See Example 2 on text page 448.)

9. Sketch the plane with equation $3x + 4y = 12$. Which coordinate axis
   is parallel to this plane? (See text page 452.)

10. The vector $(1, 5, -2)$ is perpendicular to a plane that contains the point
    $P(7, 1, 4)$. Find an equation of the plane.

11. *Critical Thinking* Write a few sentences about the relationship
    between the graph of the plane $ax + by + cz = s$ and the graph of the
    plane $ax + by + cz = t$.

     *ADVANCED MATHEMATICS Student Resource Guide*

## USING THE MAIN IDEAS

**Example 1** Find the center and the radius of the sphere with equation
$$x^2 + y^2 + z^2 + 4x - 10y = 52.$$

**Solution** Group the terms so you can complete the square in $x$, $y$, and $z$:
$$(x^2 + 4x + \underline{\ ?\ }) + (y^2 - 10y + \underline{\ ?\ }) + z^2 = 52$$
$$\left(\frac{4}{2}\right)^2 \qquad\qquad \left(-\frac{10}{2}\right)^2$$
$$\downarrow \qquad\qquad\qquad \downarrow$$
$$(x^2 + 4x + \mathbf{4}) + (y^2 - 10y + \mathbf{25}) + z^2 = 52 + \mathbf{4} + \mathbf{25}$$
$$(x + 2)^2 + (y - 5)^2 + z^2 = 81$$

The center is $(-2, 5, 0)$ and the radius is $\sqrt{81} = 9$.

**Example 2** Find an equation of the plane tangent to the sphere
$$(x - 2)^2 + (y - 1)^2 + z^2 = 25 \text{ at the point } P(2, -2, -4).$$

**Solution** The sphere has center $C(2, 1, 0)$. Vector $\overrightarrow{CP}$ is perpendicular to the tangent plane, as shown at the right.

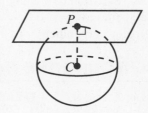

$\overrightarrow{CP} = (2 - 2, -2 - 1, -4 - 0) = (0, -3, -4)$

Therefore, the equation of the plane has the form $0x - 3y - 4z = d$, or $-3y - 4z = d$. Since $P$ lies on the plane, substitute the coordinates of $P$: $-3(-2) - 4(-4) = d$, or $d = 22$.

An equation of the tangent plane is $-3y - 4z = 22$, or $3y + 4z = -22$.

## Exercises

12. Find the center and radius of the sphere with equation
$$x^2 + y^2 + z^2 + 6x + 8y - 16z = 32.$$

13. Find the measure of the angle between the vectors $(1, 0, -1)$ and $(3, -5, -4)$ to the nearest tenth of a degree. (See Example 4 on text page 449.)

14. Line $L$ has vector equation $(x, y, z) = (3, -4, -2) + t(-1, 1, 5)$.
    **a.** Name two points on $L$.
    **b.** Write a vector equation of the line through $(-1, 0, 7)$ and parallel to $L$. (See Example 3 on text page 449.)

15. *Application* Show that $A(0, 0, 0)$, $B(1, 2, 4)$, $C(4, 0, -2)$, and $D(3, -2, -6)$ are the vertices of a parallelogram.

16. Consider the points $A(1, 5, 6)$ and $B(-1, -3, 0)$.
    **a.** Find a Cartesian equation of the plane that is perpendicular to $\overline{AB}$ at its midpoint $M$. (See Example 2 on text page 454.)
    **b.** Show that $P(6, 1, 1)$ lies on the plane described in part (a).
    **c.** Show that $P$ is equidistant from $A$ and $B$.

17. Find an equation of the plane tangent to the sphere
$$(x - 3)^2 + y^2 + (z + 1)^2 = 121 \text{ at the point } P(-6, 2, 5).$$

18. *Writing* Suppose you are given two vectors $\mathbf{u} = (a, b, c)$ and $\mathbf{v} = (d, e, f)$. Discuss two methods you could use to show that the vectors are parallel. Which method would you choose and why?

# Determinants and Their Applications

## Sections 12-7, 12-8, and 12-9

**OVERVIEW**

In Section 12-7, determinants are introduced. In Section 12-8, determinants are used to solve systems of equations and to find areas and volumes. Section 12-9 focuses on the cross product, its properties, and applications such as finding areas and equations of planes.

## KEY TERMS

**EXAMPLE/ILLUSTRATION**

**Determinant** (p. 458)
a value associated with a square array of numbers (called *elements*)

$$\begin{vmatrix} a & b \\ c & d \end{vmatrix} = ad - bc$$

**Minor of an element** (p. 458)
the determinant that remains when you cross out the row and column containing that element

$$\begin{vmatrix} a & b & c \\ d & e & f \\ g & h & i \end{vmatrix} \rightarrow$$ The minor of element $e$ is $\begin{vmatrix} a & c \\ g & i \end{vmatrix}$.

**Cross product of two vectors** (p. 465)
a vector associated with two vectors

If $\mathbf{u} = (a, b, c)$ and $\mathbf{v} = (d, e, f)$, then $\mathbf{u} \times \mathbf{v} =$
$$\left( \begin{vmatrix} b & c \\ e & f \end{vmatrix}, -\begin{vmatrix} a & c \\ d & f \end{vmatrix}, \begin{vmatrix} a & b \\ d & e \end{vmatrix} \right)$$

## UNDERSTANDING THE MAIN IDEAS

### Evaluating a determinant

- To find the value of a 2 × 2 determinant (2 rows, 2 columns), multiply the first and fourth numbers, multiply the second and third numbers, and subtract, as shown.

$$\begin{vmatrix} 2 & 3 \\ 5 & 1 \end{vmatrix} = 2(-1) - 5(3) = -17$$

- To find the value of a 3 × 3 determinant (3 rows, 3 columns), expand the determinant by the minors of a row or column (see text pages 458 –459), or:

  1. copy the first two columns to the right of the determinant;

  2. multiply each element in the first row of the original determinant by the other two elements on the left-to-right *downward* diagonal and add these products;

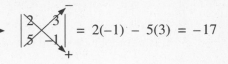

$$+ \quad + \quad + \qquad aei + bfg + cdh$$

  3. multiply each element in the last row of the original determinant by the other two elements on the left-to-right *upward* diagonal; subtract these products from the sum in step 2.

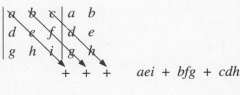

$$-gec - hfa - idb$$

$$\begin{vmatrix} a & b & c \\ d & e & f \\ g & h & i \end{vmatrix} = \begin{matrix} aei + bfg + cdh \\ - gec - hfa - idb \end{matrix}$$

## Solving systems of equations

- If you want to solve a system of equations in the form

$$ax + by = c$$
$$dx + ey = f$$

then by Cramer's Rule,

$$x = \frac{\begin{vmatrix} c & b \\ f & e \end{vmatrix}}{\begin{vmatrix} a & b \\ d & e \end{vmatrix}} \quad \text{and} \quad y = \frac{\begin{vmatrix} a & c \\ d & f \end{vmatrix}}{\begin{vmatrix} a & b \\ d & e \end{vmatrix}} \quad \text{provided that} \quad \begin{vmatrix} a & b \\ d & e \end{vmatrix} \neq 0.$$

- If $\begin{vmatrix} a & b \\ d & e \end{vmatrix} = 0$, then the system $\begin{array}{l} ax + by = c \\ dx + ey = f \end{array}$

has either no solution or an infinite number of solutions.

## Cross products

- To find the cross product of two vectors $\mathbf{u} = (a, b, c)$ and $\mathbf{v} = (d, e, f)$, write $\mathbf{u}$ and $\mathbf{v}$ in the form $\mathbf{u} = a\mathbf{i} + b\mathbf{j} + c\mathbf{k}$ and $\mathbf{v} = d\mathbf{i} + e\mathbf{j} + f\mathbf{k}$, where $\mathbf{i} = (1, 0, 0)$, $\mathbf{j} = (0, 1, 0)$, and $\mathbf{k} = (0, 0, 1)$. Then:

$$\mathbf{u} \times \mathbf{v} = \begin{vmatrix} \mathbf{i} & \mathbf{j} & \mathbf{k} \\ a & b & c \\ d & e & f \end{vmatrix} = \begin{vmatrix} b & c \\ e & f \end{vmatrix} \mathbf{i} - \begin{vmatrix} a & c \\ d & f \end{vmatrix} \mathbf{j} + \begin{vmatrix} a & b \\ d & e \end{vmatrix} \mathbf{k}$$

- The cross product of $\mathbf{u}$ and $\mathbf{v}$ has the following properties:
  1. $\mathbf{u} \times \mathbf{v}$ is perpendicular to both $\mathbf{u}$ and $\mathbf{v}$.
  2. $\mathbf{v} \times \mathbf{u} = -(\mathbf{u} \times \mathbf{v})$, that is, $\mathbf{v} \times \mathbf{u}$ and $\mathbf{u} \times \mathbf{v}$ have opposite directions.
  3. $|\mathbf{u} \times \mathbf{v}| = |\mathbf{u}||\mathbf{v}| \sin \theta$, where $\theta$ is the angle between $\mathbf{u}$ and $\mathbf{v}$
  4. $\mathbf{u} \times (\mathbf{v} + \mathbf{w}) = (\mathbf{u} \times \mathbf{v}) + (\mathbf{u} \times \mathbf{w})$ (distributive property)
  5. $\mathbf{u}$ is parallel to $\mathbf{v}$ if and only if $\mathbf{u} \times \mathbf{v} = \mathbf{0}$.

## Areas

- The area of the parallelogram with sides determined by $\mathbf{u} = (a, b)$ and $\mathbf{v} = (c, d)$ is equal to:

  1. the absolute value of the determinant $\begin{vmatrix} a & b \\ c & d \end{vmatrix}$.

  2. the magnitude of $\mathbf{u} \times \mathbf{v}$.
     (If $\mathbf{u} \times \mathbf{v} = (x, y, z)$, then the area is $\sqrt{x^2 + y^2 + z^2}$.)

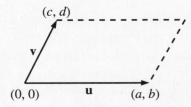

## Planes

To find a Cartesian equation of the plane containing three points $P$, $Q$, and $R$:
  1. find $\overrightarrow{PQ} \times \overrightarrow{PR}$, since the cross-product vector is perpendicular to the required plane; and
  2. substitute the coordinates of $P$, $Q$, or $R$ in the general form of the equation to find the Cartesian equation. (See the Example on text page 466.)

# CHECKING THE MAIN IDEAS

1. Let $\mathbf{u} = (1, 5, -3)$ and $\mathbf{v} = (2, -1, 0)$. Find $\mathbf{u} \times \mathbf{v}$.

   **A.** $-3$      **B.** $\sqrt{46}$      **C.** $(-3, 6, -11)$   **D.** $(-3, -6, -11)$

2. Evaluate $\begin{vmatrix} 1 & -1 & 0 \\ 0 & 3 & 2 \\ 4 & 0 & -1 \end{vmatrix}$.

   **A.** $-11$      **B.** $5$      **C.** $-5$      **D.** $11$

3. Which system can be solved using Cramer's rule?

   **A.** $4x - 6y = 9$          **B.** $6x + 9y = 0$
   $\quad -2x + 3y = 18$             $2x + 3y = -6$

   **C.** $6x + 8y = 15$       **D.** $4x + 6y = 9$

   $\quad 3x - 4y = 12$             $6x + 9y = 0$

4. **Critical Thinking** Suppose that each element of a row of a determinant is 0. What can you conclude about the determinant? Explain.

5. **a.** Show that $\begin{vmatrix} 3 & 4 & -1 \\ 2 & -2 & 5 \\ 1 & 0 & -3 \end{vmatrix} = - \begin{vmatrix} 2 & -2 & 5 \\ 3 & 4 & -1 \\ 1 & 0 & -3 \end{vmatrix}$.

   **b.** Write in words the property illustrated by part (a).

6. Solve by using Cramer's rule: $\begin{array}{l} 5x + 3y = 8 \\ 2x + y = 4 \end{array}$ .

7. Find the area of the triangle with vertices $P(5, 1)$, $Q(8, -3)$, and $R(7, -5)$. (See Example 3 on text page 463.)

**Let $\mathbf{u} = (3, 2, -1)$ and $\mathbf{v} = (-2, 0, 5)$.**

8. Find $\mathbf{u} \times \mathbf{v}$ and $\mathbf{v} \times \mathbf{u}$.

9. Show that $\mathbf{u} \times \mathbf{v}$ is perpendicular to $\mathbf{u}$.

10. Find the area of the triangle determined by $\mathbf{u} \times \mathbf{v}$.

# USING THE MAIN IDEAS

**Example 1** Solve by using Cramer's rule: $\begin{array}{l} ax + by = a + b \\ abx + aby = a^2 + b^2 \end{array}$ .

**Solution**

$$x = \frac{\begin{vmatrix} a+b & b \\ a^2+b^2 & ab \end{vmatrix}}{\begin{vmatrix} a & b \\ ab & ab \end{vmatrix}} = \frac{a^2b + ab^2 - (a^2b + b^3)}{a^2b - ab^2}$$

$$= \frac{ab^2 - b^3}{a^2b - ab^2}$$

$$= \frac{b^2(a-b)}{ab(a-b)}$$

$$= \frac{b}{a}$$

     *ADVANCED MATHEMATICS Student Resource Guide* **117**

$$y = \frac{\begin{vmatrix} a & a+b \\ ab & a^2+b^2 \end{vmatrix}}{\begin{vmatrix} a & b \\ ab & ab \end{vmatrix}} = \frac{a^3 + ab^2 - (a^2b + ab^2)}{a^2b - ab^2}$$

$$= \frac{a^3 - a^2b}{a^2b - ab^2}$$

$$= \frac{a^2(a-b)}{ab(a-b)}$$

$$= \frac{a}{b}$$

The solution is $\left(\dfrac{b}{a}, \dfrac{a}{b}\right)$.

**Example 2** Let $\theta$ be the angle between $\mathbf{u} = (1, 1, -1)$ and $\mathbf{v} = (1, -5, -1)$.
    **a.** Find $\sin\theta$ by using a cross product.
    **b.** Find $\cos\theta$ by using a dot product.
    **c.** Verify that $\sin^2\theta + \cos^2\theta = 1$.

**Solution** **a.** $\sin\theta = \dfrac{|\mathbf{u}\times\mathbf{v}|}{|\mathbf{u}||\mathbf{v}|} = \dfrac{|(-6, 0, -6)|}{\sqrt{3}\sqrt{27}}$

$$= \frac{6\sqrt{2}}{9} = \frac{2\sqrt{2}}{3}$$

    **b.** $\cos\theta = \dfrac{\mathbf{u}\cdot\mathbf{v}}{|\mathbf{u}||\mathbf{v}|} = \dfrac{-3}{\sqrt{3}\sqrt{27}} = \dfrac{-3}{9} = -\dfrac{1}{3}$

    **c.** $\left(\dfrac{2\sqrt{2}}{3}\right)^2 + \left(-\dfrac{1}{3}\right)^2 = \dfrac{8}{9} + \dfrac{1}{9} = 1$

## Exercises

**11.** Show that $\begin{vmatrix} a & b \\ c & d \end{vmatrix} = \begin{vmatrix} a - kb & b \\ c - kd & d \end{vmatrix}$.

**Solve each system of equations by using Cramer's rule.**

**12.** $ax - 2by = 4ab$
    $2ax + 3by = ab$

**13.** $\quad x + y + z = 0$
    $2x + y + 3z = 2$
    $3x + 5y - 4z = 1$

**14.** *Writing*  You know from Section 12-6 that an equation of the form $ax + by + cz = d$, where $a$, $b$, $c$, and $d$ are real numbers, is a plane. Suppose you are given a system of three equations in three variables like the system in Exercise 13. Describe the possible number of solutions of the system and relate each possibility to the possible intersections of three different planes.

**15.** Find an equation of the plane containing the points $P(1, 0, 3)$, $Q(-1, 1, 2)$, and $R(5, 4, -1)$. (See the Example on text page 466.)

**16.** *Application*  Given points $A(0, 0, 0)$, $B(0, 2, 4)$, and $C(0, 3, 0)$, find the area of $\triangle ABC$ by using **(a)** a cross product and **(b)** a dot product.

## Chapter 12: Vectors and Determinants

**QUICK CHECK**

*Chapter 12*

**Complete these exercises before trying the Practice Test for Chapter 12. If you have difficulty with a particular problem, review the indicated section.**

1. Draw two vectors **u** and **v**. Sketch $2\mathbf{u} + \mathbf{v}$ and $3\mathbf{v} - \mathbf{u}$. *(Section 12-1)*

2. If $P$ has polar coordinates $(3, 110°)$ and $O$ is the origin, find the component form of $\overrightarrow{OP}$. *(Section 12-2)*

3. At time $t$, the position of an object moving with constant velocity is given by the vector equation $(x, y) = (0, 4) + t(12, 5)$. Find the velocity and speed of the object and find the parametric equations that describe the path of the object. *(Section 12-3)*

4. If $\mathbf{u} = (9, -3)$, $\mathbf{v} = (2, 6)$, and $\mathbf{w} = (-3, -9)$, show that **u** and **v** are perpendicular and that **v** and **w** are parallel. *(Section 12-4)*

5. If $\mathbf{u} = (3, 5)$ and $\mathbf{v} = (2, -8)$, find the measure of the angle between **u** and **v**. *(Section 12-4)*

6. Name three points on the line with vector equation $(x, y, z) = (2, -1, 0) + t(1, 1, -1)$. *(Section 12-5)*

7. The vector $(1, 2, 3)$ is perpendicular to a plane that contains the point $(3, -2, -1)$. Find a Cartesian equation of the plane and then find its intercepts. *(Section 12-6)*

8. Evaluate (**a**) $\begin{vmatrix} 5 & 2 \\ 2 & -1 \end{vmatrix}$ and (**b**) $\begin{vmatrix} 1 & 3 & -2 \\ -1 & 0 & 1 \\ 0 & 4 & 2 \end{vmatrix}$. *(Section 12-7)*

9. Solve the system $\begin{array}{l} 4x + 7y = 28 \\ 3x + 5y = 15 \end{array}$ by using Cramer's rule. *(Section 12-8)*

10. Let $\mathbf{u} = (1, 0, -1)$ and $\mathbf{v} = (5, 1, 0)$. Show that the vector $\mathbf{u} \times \mathbf{v}$ is perpendicular to both **u** and **v**, and then find the area of the parallelogram determined by **u** and **v**. *(Section 12-9)*

**PRACTICE TEST**

*Chapter 12*

**Point $A$ divides $\overline{XY}$ into two segments with lengths in the ratio 3:5. Let $\mathbf{v} = \overrightarrow{XA}$. Express each vector in terms of v.**

1. $\overrightarrow{AY}$                    2. $\overrightarrow{XY}$

**Let $\mathbf{u} = (3, 4)$ and $\mathbf{v} = (-5, 2)$. Calculate each expression.**

3. $|\mathbf{u} - 2\mathbf{v}|$          4. $\mathbf{u} \cdot \mathbf{v}$          5. $\dfrac{\mathbf{u}}{|\mathbf{u}|}$

**A ship is on a course of 200° at a speed of 20 knots.**

6. Find the north-south component of its velocity vector.

7. Find the east-west component of its velocity vector.

**At time $t$, the position of an object moving with constant velocity is given by the parametric equations $x = 2 + 3t$ and $y = -3 + t$.**

8. What is the velocity and speed of the object?

9. Write a vector equation that describes the path of the object.

Let $A = (4, 1, 3)$, $B = (0, 3, -1)$, and $C = (6, 3, 2)$.

**10.** Find the length of $\overline{AB}$.      **11.** Find the midpoint of $\overline{AB}$.

**12.** Show that $\overrightarrow{AB}$ is perpendicular to $\overrightarrow{AC}$.    **13.** Find the area of $\triangle ABC$.

**Find each of the following for the plane whose equation is**
$3x + 4y - 2z = 5$.

**14.** a vector perpendicular to the plane     **15.** a point in the plane

**16.** Solve the system $\begin{array}{r} x + 2y - 3z = -6 \\ 2x - y + z = -5 \\ x - 2z = -5 \end{array}$ by using Cramer's rule.

**Let $u = (1, 0, -2)$, $v = (-3, 1, 5)$, and $w = (-2, 4, 3)$. Calculate each expression.**

**17.** $u \times v$        **18.** $v \times w$        **19.** $v \cdot (w \times u)$

---

## MIXED REVIEW
Chapters 1–12

**1.** If $\frac{1}{4} + \frac{\sqrt{7}}{4}i$ is a root of $4x^3 + x + 1 = 0$, find the other two roots.

**Simplify.**

**2.** $2 \sin 345° \cos 345°$    **3.** $\log 0.1 + e^{\ln 8}$     **4.** $(1 + i\sqrt{3})^9$

**Solve for $x$.**

**5.** $x^4 + 3 < 4x^2$    **6.** $|4x - 7| = 3$     **7.** $6x^3 + 4x + 3 = 13x^2$

**8.** $9^{x+1} = 27\sqrt{3}$    **9.** $\cos x = \frac{1}{\sec x}$     **10.** $\sin x = 0.6$

**11.** The equation $h(t) = -4.9t^2 + 7t + 20$ gives the height of a ball in meters $t$ seconds after it is thrown in the air. Find the maximum height reached by the ball and the number of seconds it is in the air until it first hits the ground.

**12.** Find the inclination of any line perpendicular to the line $2x - 5y = 10$.

**13.** When the equation of a periodic function is changed from $y = f(x)$ to $y = f(3x)$, is its amplitude or period changed, and how is it changed?

**14.** Find the cube roots of $-8i$.

**15.** If $A = (-3, 4, 1)$, $B = (1, 0, -1)$, and $C = (2, 1, 0)$, find a vector equation for $\overleftrightarrow{AB}$, and an equation for the plane that contains $A$, $B$, and $C$.

**16.** Solve $2 \cos^2 \theta = \sin \theta - 1$ for $0° \le \theta < 360°$.

**17.** If $\alpha$ and $\beta$ are first-quadrant angles, $\sin \alpha = \frac{21}{29}$, and $\cos \beta = \frac{15}{17}$, find $\cos(\alpha + \beta)$, $\tan 2\beta$, and $\tan \frac{\alpha}{2}$.

**18.** Solve $\triangle ABC$ if $\angle A = 20°$, $a = 36$, and $b = 40$.

**19.** If $u = (-1, 3)$ and $v = (4, 8)$, show that the vector $2u - v$ is perpendicular to $u$.

**20.** If $f$ is a linear function such that $f(2) = 8$ and $f(5) = 2$, find the zero of $f$ and find $f(2.5)$.

---

# Finite Sequences and Series

| OVERVIEW | In Sections 13-1 and 13-2, the concept of a sequence is introduced. The focus is on defining arithmetic and geometric series explicitly and recursively. In Section 13-3, the concept of a series is introduced. Also, formulas for sums of arithmetic and geometric series are developed and applied. |
|---|---|

## KEY TERMS

**EXAMPLE/ILLUSTRATION**

| KEY TERMS | EXAMPLE/ILLUSTRATION |
|---|---|
| **Sequence** (pp. 473, 474)<br>a set of numbers, called *terms*, arranged in a particular order; a function whose domain is the set of positive integers | $t_1 = 2$<br>$t_2 = 3$<br>$t_3 = 4$<br>$t_n = n + 1$<br>← $n$th term |
| **Arithmetic sequence** (p. 473)<br>a sequence in which the difference of any two consecutive terms is a constant, called the *common difference* | $7, 12, 17, 22, 27, \ldots$<br><br>$12 - 7 = 5, 17 - 12 = 5, \ldots$<br>common difference $d = 5$ |
| **Geometric sequence** (p. 473)<br>a sequence in which the ratio of any two consecutive terms is a constant, called the *common ratio* | $1000, 500, 250, 125, \ldots$<br><br>$\dfrac{500}{1000} = \dfrac{250}{500} = \dfrac{125}{250} = \dfrac{1}{2}$<br><br>common ratio $r = \dfrac{1}{2}$ |
| **Recursive definition of a sequence** (p. 479)<br>an initial condition, such as specifying the first term, and a formula or rule that tells how to compute a term from the previous term(s) | $7, 12, 17, 22, \ldots$<br><br>explicit definition: $t_n = 5n + 2$<br>recursive definition: $t_1 = 7$ and<br>$t_n = t_{n-1} + 5$ |
| **Series** (p. 486)<br>an expression formed by adding the terms of a sequence | $7 + 12 + 17 + 22 + \ldots$ |

## UNDERSTANDING THE MAIN IDEAS

### Sequences

- If you have an explicit formula for $t_n$, the $n$th term of a sequence, you can find the terms of the sequence by substituting $n = 1, 2, 3, 4, \ldots$, into the formula in turn.

- To find a formula for $t_n$, try to find the pattern in the terms of the sequence. If you get the same number when you *subtract* consecutive terms, the sequence is arithmetic. Substitute the value of $t_1$ (the first term) and $d$ (the common difference) in the formula $t_n = t_1 + (n - 1)d$ to get a formula for $t_n$.

(See part (a) of Example 1 on text pages 474–475.) If you get the same number when you *divide* a term by the preceding term, the sequence is geometric. Substitute the value of $t_1$ and the value of the common ratio $r$ in the formula $t_n = t_1 \cdot r^{(n-1)}$ to get a formula for $t_n$. (See part (a) of Example 2 on text page 475.) If the sequence is neither arithmetic nor geometric, try to relate the terms of the sequence to their positions in the list, that is, to write $t_n$ in terms of $n$.

- If you know a recursive definition for a sequence, you can find the terms of the sequence by first substituting $t_1$ into the formula for $t_n$ to get $t_2$; then substituting $t_2$ to get $t_3$, and so on. (See text page 479.)
- To find a recursion equation for a given sequence, try to find a relationship between each term and the term preceding it.

### Series

- Like a sequence, you can define a series either explicitly or recursively. If $S_n$ represents the sum of the first $n$ terms, then the explicit definition is $S_n = t_1 + t_2 + t_3 + \ldots + t_n$ and the recursive definition is $S_0 = 0, S_n = S_{n-1} + t_n$ for $n \geq 1$. The formula $S_n = S_{n-1} + t_n$ shows that to get the sum of the first $n$ terms, add the $n$th term to the sum of the first $(n-1)$ terms.
- If an arithmetic series has $n$ terms, then the sum of the series can be found using the formula $S_n = \dfrac{n(t_1 + t_n)}{2}$.
- If a geometric series has $n$ terms, then the sum of the series can be found using the formula $S_n = \dfrac{t_1(1 - r^n)}{1 - r}$, where $r$ is the common ratio and $r \neq 1$.

## CHECKING THE MAIN IDEAS

1. The formula $t_n = 3^{2n}$ is best described as:
   A. an explicit definition of a geometric series.
   B. a recursive definition of a geometric series.
   C. an explicit definition of a geometric sequence.
   D. a recursive definition of a geometric sequence.

2. If $S_1 = 1, S_2 = 3, S_3 = 7$, and $S_4 = 15$, then $t_5 = $ __?__ .
   A. 23      B. 31      C. 32      D. 16

3. Find the first four terms of the sequence $t_n = n - \dfrac{1}{n}$ and state whether the sequence is arithmetic, geometric, or neither.

4. Find a formula for the $n$th term of the sequence $3, -9, 27, -81, \ldots$. Then sketch the graph of the sequence.

5. State whether the sequence $100, 90, 80, 70, \ldots$ is arithmetic, geometric, or neither. Find a formula for $t_n$.

6. Give the first five terms of the sequence defined recursively by $t_1 = 10$ and $t_n = 2t_{n-1} - 5$.

7. Define the sequence $1, \dfrac{1}{1 \cdot 2}, \dfrac{1}{1 \cdot 2 \cdot 3}, \dfrac{1}{1 \cdot 2 \cdot 3 \cdot 4}, \ldots$ recursively.

8. Find the sum of the first 80 positive odd integers. (See Example 2 on text page 487.)

9. Show that $(-1)^1 + (-1)^2 + (-1)^3 + \ldots + (-1)^n = \dfrac{(-1)^n - 1}{2}$.

10. **Critical Thinking** How is the idea of a sequence related to the idea of an ordered pair?

## USING THE MAIN IDEAS

**Example 1** In an arithmetic sequence, $t_3 = 12$ and $t_6 = 96$. Find $t_{11}$.

**Solution** Substitute the known values in the formula $t_n = t_1 + (n - 1)d$.

$$t_3 = 12 = t_1 + (3 - 1)d \quad \rightarrow \quad t_1 + 2d = 12$$
$$t_6 = 96 = t_1 + (6 - 1)d \quad \rightarrow \quad t_1 + 5d = 96$$
$$\text{Subtract:} \quad \overline{\phantom{t_1 + } -3d = -84}$$
$$d = 28$$

Substitute 28 for $d$: $\quad t_1 + 2(28) = 12$, so $t_1 = -44$

Thus, $t_n = -44 + (n - 1)(28)$

Therefore, $t_{11} = -44 + (11 - 1)(28) = 236$

**Example 2** Find a formula for the $n$th term of the series $1 + 3 + 5 + 7 + \ldots$. Give a recursive definition for $S_n$, the sum of the first $n$ terms of the series.

**Solution** The series is arithmetic with $d = 2$, so $t_n = t_1 + (n - 1)d = 1 + (n - 1)2 = 2n - 1$. Thus, a recursive definition is $S_0 = 0$ and $S_n = S_{n-1} + (2n - 1)$.

## Exercises

11. Find $t_{60}$ of an arithmetic sequence for which $t_4 = 8$ and $t_7 = 23$.

12. Find $t_6$ of a geometric sequence for which $t_1 = 16$ and $t_4 = 54$.

13. **Writing** If an arithmetic sequence has first term $t_1$ and common difference $d$, give a recursive definition for the sequence. Write an explanation of the recursive definition.

14. **Application** Suppose the value of a car depreciates 20% per year. Express this fact with a recursive definition. About how long will it take for the car's value to be half its original value?

15. Show that $a + (a + 1) + (a + 2) + \ldots + a^2 = \dfrac{a(a^3 + 1)}{2}$. (*Hint:* Express $n$ in terms of $a$.)

16. Find a formula for the $n$th term of the series $64 + 32 + 16 + \ldots$. Give a recursive definition for $S_n$, the sum of the first $n$ terms of the series.

17. Write a computer program, like the one given in Example 1 on text pages 486–487, that will print the sum of the series $1^1 + 2^2 + 3^2 + 4^2 + 5^2 + \ldots + 50^2$. Then run the program to find the sum.

# Infinite Sequences and Series

_Sections 13-4, 13-5, 13-6, and 13-7_

| OVERVIEW | Section 13-4 presents methods for determining whether an infinite sequence has a limit and if so, how to find the limit. Section 13-5 focuses on finding the sum of an infinite geometric series. In Section 13-6, series are represented in a compact form using sigma notation. Section 13-7 discusses mathematical induction. |
|---|---|

## KEY TERMS

EXAMPLE/ILLUSTRATION

**Infinite sequence or series** (p. 493)
a sequence or series that has no last term

infinite sequence: 1, 3, 9, 27, ...
infinite series:
$$1 + 3 + 9 + 27 + ...$$

**Limit of an infinite sequence** (p. 493)
a "target value" that the terms of a sequence approach more and more closely as $n$ gets larger and larger (that is, as $n$ "goes to infinity")

The larger $n$ gets, the closer $\left(1 + \frac{1}{n}\right)^n$ gets to its limit, $e$.

**$n$th partial sum** (p. 500)
the sum, $S_n$, of the first $n$ terms of a series:
$$S_n = t_1 + t_2 + t_3 + ... + t_n$$

$t_1 = 0.9, t_2 = 0.09,$
$t_3 = 0.009, ...$
$S_n = 0.9 + 0.09 + 0.009 + ...$
$+ 9 \cdot 10^{-n}$

**Sequence of partial sums** (p. 500)
the sequence $S_1, S_2, S_3, ...$

$S_1 = 0.9$
$S_2 = 0.9 + 0.09 = 0.99$
$S_3 = 0.999$
$S_4 = 0.9999$
$\vdots$

**Sum of an infinite series** (p. 500)
If the sequence of partial sums has a finite limit $S$, that is, if the sum of the series approaches a certain value more and more closely as you add more terms, then that value is the sum of the infinite series.

Notice that the sequence of partial sums above is approaching 1 more and more closely as $n$ increases.

**Convergent series** (p. 500)
an infinite series that has a sum

The series $0.9 + 0.09 + 0.009 + ...$ is convergent; it converges to 1.

**Divergent series** (p. 500)
a series whose partial sums approach infinity or have no finite limit

The series $9 + 90 + 900 + ...$ diverges, because the partial sums get larger and larger rather than approaching a fixed value.

| Sigma notation (p. 506)<br>   a way to represent a series or its sum in compact form by using the Greek letter sigma | $\sum\limits_{k=1}^{4} 2k = (2 \cdot 1) + (2 \cdot 2) + (2 \cdot 3) + (2 \cdot 4)$ |
| --- | --- |
| Mathematical induction (p. 510)<br>   a method of proof used to show that a statement is true for all positive integers | (See the box on text page 511.) |

## UNDERSTANDING THE MAIN IDEAS

### Limit of an infinite sequence

- If you are given a formula for the $n$th term of a sequence, you can check whether the sequence has a limit by using a calculator to evaluate $t_n$ for larger and larger values of $n$. (See Example 1 on text page 494.) This method will also work in Example 2 on text pages 494–495.

- For an infinite *geometric* sequence, the limit of the sequence $t_n = r^n$ is 0 if $|r| < 1$.

- A sequence has *no* limit (or *diverges*) if:

  1. the terms increase without bound. $\longrightarrow$ 2, 4, 8, 16, 32, …
  2. the terms decrease without bound. $\longrightarrow$ $-5, -10, -15, -20, …$
  3. the terms do not "target in" on *one* specific value. $\longrightarrow$ $1, -1, 1, -1, 1, -1, …$

  For sequence (1), we write $\lim\limits_{n \to \infty} 2^n = \infty$.

  For sequence (2), we write $\lim\limits_{n \to \infty} (-5n) = -\infty$.

  For sequence (3), we say the limit does not exist.

### Sum of an infinite geometric series

- If an infinite geometric series has common ratio $r$ with $|r| < 1$, then $t_1 + t_1 r + t_1 r^2 + \ldots$ converges to the sum $S = \dfrac{t_1}{1 - r}$.

- If $|r| \geq 1$ and $t_1 \neq 0$, then the series diverges.

### Sigma notation

- In an expression such as $\sum\limits_{k=1}^{4} 2k$, $2k$ is the *summand*, 1 and 4 are the *limits of summation*, and $k$ is the *index*. Any letter can be used for the index, but the same letter should be used in both the index and the summand. The limits of summation are the first and last values of the index. If the series is infinite, then the infinity symbol, $\infty$, is used for the upper limit. If a series is represented in sigma form, you can write it in expanded form by substituting the values of the index and writing out the sum of the terms:

$$\sum\limits_{k=1}^{4} 2k = 2 \cdot 1 + 2 \cdot 2 + 2 \cdot 3 + 2 \cdot 4.$$

- To write a series using sigma notation, try to find a formula in terms of $k$ for each term, $t_k$, in the series. (See the Example on text page 507.)
- Properties of finite sums:

  1. $\displaystyle\sum_{i=1}^{n} (a_i + b_i) = \sum_{i=1}^{n} a_i + \sum_{i=1}^{n} b_i$

  2. $\displaystyle\sum_{i=1}^{n} ca_i = c \sum_{i=1}^{n} a_i$

**Mathematical induction**

To prove a statement by mathematical induction, first show that the statement is true for $n = 1$, and then prove that the statement is true for $n = k + 1$, assuming that the statement is true for $n = k$. (See Examples 1 and 2 on text pages 511–512.)

## CHECKING THE MAIN IDEAS

1. Find $\displaystyle\lim_{n \to \infty} \frac{n-1}{n+1}$.

   **A.** $\infty$       **B.** 1       **C.** 0       **D.** does not exist

2. Which geometric series converges?

   **A.** $2 - 2^{-1} + 2^{-2} + 2^{-3} - \dots$      **B.** $2 - 2^2 + 2^3 - 2^4 + \dots$

   **C.** $2 - 2 + 2 - 2 + \dots$             **D.** $2 + 2 + 2 + 2 + \dots$

3. Find $\displaystyle\lim_{n \to \infty} \left(1 - \frac{1}{n}\right)$.

4. Find the limit of the sequence $1, -\frac{1}{2}, \frac{1}{3}, -\frac{1}{4}, \dots$ or state that the limit does not exist.

5. *Critical Thinking* Suppose the common ratio of a convergent infinite geometric series is a negative number. Is the sum *always*, *sometimes*, or *never* a negative number? Explain.

6. Find the sum of the series $8 + 2 + \frac{1}{2} + \dots$ .

7. Find the first three terms of an infinite geometric series with sum 10 and first term 2.

8. Write $\displaystyle\sum_{k=1}^{5} (-k)^{k+1}$ in expanded form.

9. Write the series in Exercise 6 using sigma notation.

10. Consider the statement $\dfrac{1}{2} - \dfrac{1}{4} - \dfrac{1}{8} - \dots - \dfrac{1}{2^n} = \dfrac{1}{2^n}$.

    **a.** Show that the statement is true when $n = 1$.

    **b.** Let $n = k$. Write the resulting statement.

    **c.** Assume that the statement written in part (b) is true. Prove that the general statement must be true for $n = k + 1$.

## USING THE MAIN IDEAS

**Example 1** Find the value of $x$ if the series $1 + \dfrac{x}{2} + \dfrac{x^2}{4} + \dfrac{x^3}{8} + \cdots$
converges to 4.

**Solution** This is an infinite series with common ratio $r = \dfrac{x}{2}$.

The sum is $\dfrac{t_1}{1-r} = \dfrac{1}{1-\dfrac{x}{2}}$, so $\dfrac{1}{1-\dfrac{x}{2}} = 4$.

Thus, $1 = 4\left(1 - \dfrac{x}{2}\right) = 4 - 2x$ and $2x = 3$, or $x = 1.5$.

(Notice that $|x| > 1$, but $r = \dfrac{x}{2} = \dfrac{1.5}{2} = 0.75$, so $|r| < 1$.)

**Example 2** Evaluate $\displaystyle\sum_{n=1}^{50} i^n$.

**Solution** We have $i + i^2 + i^3 + i^4 + i^5 + i^6 + \cdots + i^{49} + i^{50} =$
$i + (-1) + (-i) + 1 + i + (-1) + (-i) + \cdots + i + (-1)$.
Notice that the sum of every set of four consecutive terms is 0.
Therefore, the sum of the first 48 terms is 0 and the sum of the 50
terms is simply the sum of the 49th and 50th terms, which is
$i + (-1)$, or $i - 1$.

(*Note:* It is also possible to evaluate the sum as

$$S_n = \frac{t_1(1-r^n)}{1-r} = \frac{i(1-i^{50})}{1-i} = \frac{i(1-(-1))}{1-i}$$

$$= \frac{2i}{1-i} \cdot \frac{1+i}{1+i} = \frac{2i(1+i)}{2} = i + i^2 = i - 1.)$$

## Exercises

**Evaluate the given limit or state that the limit does not exist. If the
sequence approaches $\infty$ or $-\infty$, so state.**

11. $\displaystyle\lim_{n \to \infty} \log\left(\frac{20n+1}{2n-1}\right)$ 

12. $\displaystyle\lim_{n \to \infty} n \cos(n\pi)$

13. Find the value of $x$ if the series $1 - 2x + 4x^2 - 8x^3 + \cdots$ converges to 5.

14. For the series in Exercise 13, find the interval of convergence and the sum, expressed in terms of $x$. (See Example 2 on text page 501.)

15. *Writing* Explain how a repeating decimal is related to an infinite series.

16. *Application* Suppose a spider travels 2 m in 1 minute, 1 m the next minute, 0.5 m the next minute, and so on. If the spider could continue this way forever, how far would it travel altogether?

17. Evaluate $\displaystyle\sum_{n=1}^{100} (2n)(-1)^{n+1}$.

18. Use mathematical induction to show that $(1 \cdot 2) +$
$(2 \cdot 3) + (3 \cdot 4) + \cdots + n(n+1) = \dfrac{n(n+1)(n+2)}{3}$.

# CHAPTER REVIEW

## Chapter 13: Sequences and Series

Complete these exercises before trying the Practice Test for Chapter 13. If you have difficulty with a particular problem, review the indicated section.

1. Find the first four terms of the sequence $t_n = (\sqrt{2})^n$. Then tell whether the sequence is arithmetic, geometric, or neither. *(Section 13-1)*

2. State whether the sequence $3, 5, 7, 9, \ldots$ is arithmetic, geometric, or neither. Find a formula for $t_n$, the $n$th term of the sequence. *(Section 13-1)*

3. Give the first four terms of the sequence $t_1 = 2, t_n = (t_{n-1})^2$. *(Section 13-2)*

4. Find $S_{20}$ for the arithmetic series $4 + 8 + 12 + \ldots$. *(Section 13-3)*

5. Find $S_6$ for the geometric series $1000 + 100 + 10 + \ldots$. *(Section 13-3)*

6. Find **(a)** $\lim\limits_{n \to \infty} \dfrac{n^2}{n+8}$ and **(b)** $\lim\limits_{n \to \infty} \cos n\pi$. *(Section 13-4)*

7. Find the sum of the infinite geometric series $1 + \dfrac{1}{8} + \dfrac{1}{64} + \dfrac{1}{512} + \ldots$. *(Section 13-5)*

8. Express the series in Exercise 7 using sigma notation. *(Section 13-6)*

9. Explain how you could use mathematical induction to prove that $4 + 8 + 12 + \ldots + 4n = 2n(n + 1)$. *(Section 13-7)*

State whether the given sequence is arithmetic, geometric, or neither. Find a formula for $t_n$, the $n$th term of each sequence.

1. $24, 12, 6, 3, \ldots$ 　　2. $5, 8, 13, 20, 29, \ldots$ 　　3. $84, 80, 76, 72, \ldots$

Give a recursive definition for each sequence.

4. $3, 6, 12, 24, \ldots$ 　　　　　　5. $3, 4, 7, 11, 18, 29, \ldots$

Find the indicated term of the indicated sequence.

6. Arithmetic: $t_5 = 40, t_8 = 61, t_{20} = ?$

7. Geometric: $t_3 = 10, t_7 = 160, t_{12} = ?$

Find the specified sum of the given series.

8. $S_{200}$: $5 + 8 + 11 + \ldots$ 　　　　9. $S_{12}$: $24 + 12 + 6 + \ldots$

Find the given limit. If the limit does not exist, so state.

10. $\lim\limits_{n \to \infty} \left(1 - \dfrac{1}{n}\right)$ 　　　　11. $\lim\limits_{n \to \infty} (-1)^n \left(\dfrac{2n+1}{2n-1}\right)$

Let $f(x) = 1 - (x - 2) + (x - 2)^2 - (x - 2)^3 + \ldots$.

12. Find the interval of convergence for $f(x)$.

13. Find the sum of the terms of $f(x)$.

14. Express $f(x)$ using sigma notation.

15. Find the difference between the sum of all terms of $f(2.1)$ and the sum of the first 5 terms.

16. Prove $1 + 4 + 7 + \ldots + (3n - 2) = \dfrac{n(3n-1)}{2}$ by mathematical induction.

1. Find a rectangular equation and a pair of parametric equations for the line through points $(5, 6)$ and $(1, 8)$.

2. If $-\dfrac{\pi}{2} < \alpha < 0$, $\cos \alpha = \dfrac{35}{37}$, and $\beta = 210°$, find $\csc \alpha$, $\sin(\alpha - \beta)$, $\cos 2\alpha$, and $\sin \dfrac{\alpha}{2}$.

**Evaluate each expression.**

3. $\dfrac{1}{2} \log_8 4 + 4 \log_8 \dfrac{1}{2}$

4. $\sin\left(\text{Cos}^{-1}\left(-\dfrac{1}{3}\right)\right)$

5. $(1 - i)^{18}$

6. $9 - \dfrac{9}{2} + \dfrac{9}{4} - \dfrac{9}{8} + \ldots$

7. $10^{2 - \log 8}$

8. $\displaystyle\sum_{k=1}^{100} \cos \dfrac{k\pi}{4}$

9. Show that $\tan(\pi + \theta) = \tan \theta$ and explain what this shows about $\tan \theta$.

10. A financial report on an airport shuttle bus projects that the daily revenue collected will vary according to the fare charged, as shown by the table below. Predict the maximum possible daily revenue.

| Fare | $10 | $11 | $12 |
|---|---|---|---|
| Daily Revenue | $3000 | $3135 | $3240 |

**Solve each system.**

11. $9x - 5y = 18$
    $5x - 3y = 15$

12. $5x - y = 10$
    $y = x^2 - 7x + 10$

13. $16x^2 - y^2 = 64$
    $4x^2 + 4y^2 = 169$

14. A rectangle has sides of lengths 12 cm and 5 cm. Find the measures of the angles formed by the diagonals.

15. Find all four roots of $4x^4 + 16x^3 + 23x^2 - 4x - 6 = 0$.

**Solve for $x$.**

16. $4x^3 + 4x^2 > 9x + 9$

17. $8x^{-1.5} = 125$

18. $5^x = 100$

19. Let $f(x) = x^2$ and $g(x) = \dfrac{1}{2}x - 6$. Find $f(g(x))$ and $g(f(x))$.

20. Solve $\sin 2\theta = \cos^2 \theta$ for $0° \leq \theta < 360°$.

21. Find the sum of (**a**) the first 20 terms and (**b**) the first $n$ terms of the series $2 + 6 + 10 + 14 + 18 + \ldots$ .

22. In $\triangle XYZ$, $\angle X = 50°$, $\angle Y = 105°$, and $x = 30$. Find $z$ and the area of $\triangle XYZ$.

23. If $\mathbf{u} = (6, -8, -24)$ and $\mathbf{v} = (2, -1, 2)$, find $\mathbf{u} \cdot \mathbf{v}$, $\mathbf{u} \times \mathbf{v}$, and the angle between $\mathbf{u}$ and $\mathbf{v}$.

24. Describe the symmetries of the graph of the equation $x^2 + 4xy + y^2 = 4$. Then tell whether the graph is a circle, a parabola, or a hyperbola.

25. Simplify $\dfrac{1}{1 + \tan^2 x} + \dfrac{1}{1 + \cot^2 x}$.

26. Use mathematical induction to prove your answer to part (b) of Exercise 21.

# Matrix Operations

| OVERVIEW | Sections 14-1 and 14-2 introduce matrices and the basic matrix operations – addition, subtraction, scalar multiplication, and matrix multiplication. In Section 14-3, these skills are applied to finding the inverse of a matrix and to solving systems of linear equations using matrices. |
|---|---|

## KEY TERMS

### EXAMPLE/ILLUSTRATION

**Matrix** (p. 517)
a rectangular arrangement of numbers, called *elements*, enclosed by brackets

$$A = \begin{bmatrix} 3 & 1 & 0 \\ -1 & 4 & 6 \end{bmatrix} \rightarrow \text{rows}$$

columns

---

**Dimensions of a matrix** (p. 517)
the number of rows (horizontal lines) and the number of columns (vertical lines) of a matrix

The symbol $A_{2\times3}$ shows that $A$ is a 2 by 3 matrix. The number of rows is listed first.

---

**Transpose of a matrix** (p. 517)
the matrix obtained by interchanging the rows and columns of a matrix

The columns of $A^t$ are the same as the rows of $A$ shown above.

$$A^t = \begin{bmatrix} 3 & -1 \\ 1 & 4 \\ 0 & 6 \end{bmatrix}$$

---

**Zero matrix** (p. 530)
a matrix whose elements are all zero

$$O_{1\times3} = \begin{bmatrix} 0 & 0 & 0 \end{bmatrix}$$

---

**Square matrix** (p. 530)
a matrix with the same number of rows as columns

$$\begin{bmatrix} 0 & -1 \\ 3 & 5 \end{bmatrix}, \begin{bmatrix} 1 & 2 & 3 \\ 4 & 5 & 6 \\ 7 & 8 & 9 \end{bmatrix}$$

---

**Identity matrix** (p. 530)
an $n$ by $n$ (square) matrix, written as $I_{n\times n}$, whose elements along the diagonal from the upper left to lower right are 1 and whose other elements are 0

$$I_{3\times3} = \begin{bmatrix} 1 & 0 & 0 \\ 0 & 1 & 0 \\ 0 & 0 & 1 \end{bmatrix}$$

---

**Additive inverse of a matrix** (p. 531)
the matrix obtained by multiplying each element of a matrix by –1

$$-A = \begin{bmatrix} -3 & -1 & 0 \\ 1 & -4 & -6 \end{bmatrix}$$

---

**Multiplicative inverse of a matrix** (p. 531)
an $n \times n$ (square) matrix whose product with another $n \times n$ matrix is $I_{n\times n}$

$$\begin{bmatrix} 2 & 3 \\ 1 & 2 \end{bmatrix}\begin{bmatrix} 2 & -3 \\ -1 & 2 \end{bmatrix} = \begin{bmatrix} 1 & 0 \\ 0 & 1 \end{bmatrix}$$

# UNDERSTANDING THE MAIN IDEAS

## Comparing and combining matrices

- Two matrices are *equal* if and only if they have the same dimensions and the elements in all corresponding positions are equal. $\longrightarrow$

$$[1 \quad 0] \neq \begin{bmatrix} 1 \\ 0 \end{bmatrix}$$

$$[1 \quad 0] \neq [0 \quad 1]$$

- To add two matrices *with the same dimensions*, add the corresponding elements of the matrices. To subtract a matrix $B$ from another matrix $A$ *with the same dimensions* as $B$, subtract the elements of $B$ from the corresponding elements of $A$. Alternatively, add the additive inverse of $B$ to $A$: $A + B = A + (-B)$. (See Example 2 on text page 519.)

◄▰▰▰ **Caution:** If two matrices have different dimensions, they *cannot* be added or subtracted.

- To multiply two matrices, $A_{m \times n}$ and $B_{n \times p}$, form the product matrix whose dimensions are $m \times p$ and whose element in the $a$th row and $b$th column is obtained by multiplying the elements of the $a$th row of $A$ by the elements of the $b$th column of $B$ and adding these products. (See Example 1 on text page 524.)

◄▰▰▰ **Caution:** (1) The product $AB$ of two matrices is defined only if the number of columns of $A$ equals the number of rows of $B$; the product matrix $AB$ has the same number of rows as $A$ and the same number of columns as $B$. (2) In general, if $A$ and $B$ are square matrices, $AB \neq BA$. The order in which two matrices are multiplied is important.

- To find the multiplicative inverse, $A^{-1}$, of a $2 \times 2$ matrix $A$, use the following formula:

◄▰▰▰ **Caution:** $A^{-1}$ exists only if $|A| = ad - bc \neq 0$.

$$\text{If } A = \begin{bmatrix} a & b \\ c & d \end{bmatrix}, \text{ then } A^{-1} = \begin{bmatrix} \dfrac{d}{ad-bc} & \dfrac{-b}{ad-bc} \\ \dfrac{-c}{ad-bc} & \dfrac{a}{ad-bc} \end{bmatrix}.$$

## Properties of matrices

Let $A$, $B$, and $C$ be $m \times n$ matrices; let $O_{m \times n}$ be the $m \times n$ zero matrix.

- Addition of matrices is commutative. $\longrightarrow$ $A + B = B + A$
- Addition of matrices is associative. $\longrightarrow$ $(A + B) + C = A + (B + C)$
- $O$ is the identity matrix for addition. $\longrightarrow$ $A + O = O + A = A$
- The sum of a matrix and its additive inverse is $O$. $\longrightarrow$ $A + (-A) = -A + A = O$

Let $A$, $B$, and $C$ be $n \times n$ matrices; let $O_{n \times n}$ and $I_{n \times n}$ be the zero and identity matrices, respectively.

- Multiplication of matrices is *not* commutative. $\longrightarrow$ $AB \neq BC$
- Multiplication of matrices *is* associative. $\longrightarrow$ $(AB)C = A(BC)$
- $I$ is the identity matrix for multiplication. $\longrightarrow$ $A \cdot I = I \cdot A = A$
- The product of a matrix and its multiplicative inverse is $I$. $\longrightarrow$ $A \cdot A^{-1} = A^{-1} \cdot A = I$ if $A^{-1}$ exists.
- The product of a matrix and the zero matrix is the zero matrix. $\longrightarrow$ $A \cdot O = O \cdot A = O$
- Matrix multiplication is distributive. $\longrightarrow$ $A(B + C) = AB + AC$

$$(B + C)A = BA + CA$$

### Solving systems of linear equations

To solve the system $\begin{array}{l} ax + by = e \\ cx + dy = f \end{array}$, write the matrix equation

$\begin{bmatrix} a & b \\ c & d \end{bmatrix} \begin{bmatrix} x \\ y \end{bmatrix} = \begin{bmatrix} e \\ f \end{bmatrix}$. Then find the inverse of the coefficient

matrix, $\begin{bmatrix} a & b \\ c & d \end{bmatrix}$, and "left multiply" both sides of the matrix equation

by this inverse. (See Example 3 on text page 533.)

## CHECKING THE MAIN IDEAS

Let $A = \begin{bmatrix} 5 & 0 & 1 \\ 7 & -3 & -4 \end{bmatrix}$, $B = \begin{bmatrix} 1 & 0 \\ 5 & -2 \\ -1 & 0 \end{bmatrix}$, and $C = \begin{bmatrix} -1 & -3 \\ 0 & 5 \\ 2 & -2 \end{bmatrix}$.

1. Which sum is defined, $A + B$ or $B + C$? Find this sum.

2. Which product is defined, $AB$ or $BC$? Find this product.

3. Find $A^t$ and then find $A^t - B$.      4. Find $2B$ and then find $2B - C$.

5. Let $D = \begin{bmatrix} 3 & -2 \\ -1 & 2 \end{bmatrix}$. Find the additive inverse of $D$, the

   multiplicative inverse of $D$, and $D \cdot D$.

6. Solve $\begin{array}{l} 5x + 4y = 20 \\ 6x + 5y = 30 \end{array}$ by using matrices.

7. *Critical Thinking*  Describe two conditions that would result in a matrix having no inverse.

## USING THE MAIN IDEAS

**Example 1**  Let $A = \begin{bmatrix} 1 & -1 \\ 0 & 2 \end{bmatrix}$ and $B = \begin{bmatrix} 3 & 1 \\ 0 & -1 \end{bmatrix}$. Is $(AB)^t = B^t A^t$?

**Solution**  $AB = \begin{bmatrix} 1 & -1 \\ 0 & 2 \end{bmatrix} \begin{bmatrix} 3 & 1 \\ 0 & -1 \end{bmatrix} = \begin{bmatrix} 3 & 2 \\ 0 & -2 \end{bmatrix}$

$(AB)^t = \begin{bmatrix} 3 & 0 \\ 2 & -2 \end{bmatrix}$

$B^t \cdot A^t = \begin{bmatrix} 3 & 0 \\ 1 & -1 \end{bmatrix} \begin{bmatrix} 1 & 0 \\ -1 & 2 \end{bmatrix} = \begin{bmatrix} 3 & 0 \\ 2 & -2 \end{bmatrix}$

Thus, $(AB)^t = B^t \cdot A^t$.

**Example 2**  Solve each matrix equation for $X$.

a. $2X + \begin{bmatrix} 7 & 3 \\ -1 & -4 \end{bmatrix} = \begin{bmatrix} 9 & -7 \\ 3 & 6 \end{bmatrix}$

b. $\begin{bmatrix} 2 & 3 \\ -1 & -1 \end{bmatrix} X - \begin{bmatrix} 1 & 0 \\ -1 & -1 \end{bmatrix} = \begin{bmatrix} 4 & -2 \\ 1 & 0 \end{bmatrix}$

**Solution**  **a.**

$$2X + \begin{bmatrix} 7 & 3 \\ -1 & -4 \end{bmatrix} = \begin{bmatrix} 9 & -7 \\ 3 & 6 \end{bmatrix}$$

$$2X + \begin{bmatrix} 7 & 3 \\ -1 & -4 \end{bmatrix} + \begin{bmatrix} -7 & -3 \\ 1 & 4 \end{bmatrix} = \begin{bmatrix} 9 & -7 \\ 3 & 6 \end{bmatrix} + \begin{bmatrix} -7 & -3 \\ 1 & 4 \end{bmatrix}$$

$$2X + O = \begin{bmatrix} 2 & -10 \\ 4 & 10 \end{bmatrix}$$

$$X = \frac{1}{2}\begin{bmatrix} 2 & -10 \\ 4 & 10 \end{bmatrix} = \begin{bmatrix} 1 & -5 \\ 2 & 5 \end{bmatrix}$$

**b.**
$$\begin{bmatrix} 2 & 3 \\ -1 & -1 \end{bmatrix} X - \begin{bmatrix} 1 & 0 \\ -1 & -1 \end{bmatrix} = \begin{bmatrix} 4 & -2 \\ 1 & 0 \end{bmatrix}$$

$$\begin{bmatrix} 2 & 3 \\ -1 & -1 \end{bmatrix} X = \begin{bmatrix} 4 & -2 \\ 1 & 0 \end{bmatrix} + \begin{bmatrix} 1 & 0 \\ -1 & -1 \end{bmatrix}$$

$$\begin{bmatrix} 2 & 3 \\ -1 & -1 \end{bmatrix} X = \begin{bmatrix} 5 & -2 \\ 0 & -1 \end{bmatrix}$$

Now left-multiply each side by the inverse of $\begin{bmatrix} 2 & 3 \\ -1 & -1 \end{bmatrix}$.

The inverse of $\begin{bmatrix} 2 & 3 \\ -1 & -1 \end{bmatrix}$ is $\begin{bmatrix} \dfrac{-1}{-2+3} & \dfrac{-3}{-2+3} \\ \dfrac{1}{-2+3} & \dfrac{2}{-2+3} \end{bmatrix} = \begin{bmatrix} -1 & -3 \\ 1 & 2 \end{bmatrix}$.

$$\begin{bmatrix} -1 & -3 \\ 1 & 2 \end{bmatrix}\begin{bmatrix} 2 & 3 \\ -1 & -1 \end{bmatrix} X = \begin{bmatrix} -1 & -3 \\ 1 & 2 \end{bmatrix}\begin{bmatrix} 5 & -2 \\ 0 & -1 \end{bmatrix}$$

$$\begin{bmatrix} 1 & 0 \\ 0 & 1 \end{bmatrix} X = \begin{bmatrix} -5 & 5 \\ 5 & -4 \end{bmatrix}$$

$$X = \begin{bmatrix} -5 & 5 \\ 5 & -4 \end{bmatrix}$$

## Exercises

**8.** *Application*  The matrix $A$ at the right gives the number of calories, number of grams of fat, and number of milligrams of sodium in a serving of low-fat granola. The matrix $B$ gives similar information for $\frac{1}{2}$ cup of skim milk. Calculate $A + B$ and $A - B$, and give the real-world significance of each matrix.

$$A = \begin{bmatrix} 110 & 2 & 15 \end{bmatrix}$$

$$B = \begin{bmatrix} 40 & 0 & 60 \end{bmatrix}$$

**9.** Let $A = \begin{bmatrix} 1 & -1 \\ 0 & 2 \end{bmatrix}$ and $B = \begin{bmatrix} 3 & 1 \\ 0 & -1 \end{bmatrix}$.

  **a.** Is $(A + B)^2 = A^2 + B^2$? (*Note:* $A^2 = A \cdot A$)

  **b.** Is $(A + B)(A - B) = A^2 - B^2$?

**Solve each matrix equation for $X$.**

**10.** $3X - \begin{bmatrix} 5 & -1 \\ -4 & 0 \end{bmatrix} = \begin{bmatrix} 4 & -2 \\ 0 & 6 \end{bmatrix}$     **11.** $\begin{bmatrix} 2 & -8 \\ -4 & -1 \end{bmatrix} + \begin{bmatrix} 3 & 7 \\ 2 & 4 \end{bmatrix} X = \begin{bmatrix} 5 & -3 \\ 0 & 4 \end{bmatrix}$

   

# Applications of Matrices

## Sections 14-4, 14-5, and 14-6

| OVERVIEW | These three sections present several important applications of matrices. In Section 14-4, communication matrices are used to solve communication network problems. In Section 14-5, transition matrices are used to make predictions about the future states of various systems. In Section 14-6, transformation matrices are used to analyze transformations of the coordinate plane. |
|----------|---|

## KEY TERMS

**EXAMPLE/ILLUSTRATION**

**Communication matrix** (p. 538)

a matrix that represents routes by which information can be sent and received

$$\begin{array}{c} \phantom{From} \\ \text{From} \end{array} \begin{array}{c} \phantom{A} \\ A \\ B \\ C \end{array} \begin{array}{c} \text{To} \\ \begin{array}{ccc} A & B & C \end{array} \\ \begin{bmatrix} 0 & 1 & 0 \\ 0 & 0 & 1 \\ 1 & 1 & 0 \end{bmatrix} \end{array}$$

The 1 in the first row shows that a message can be sent from A to B. The second 0 in the first row shows that a message *cannot* be sent from A to C.

**Transition matrix** (p. 543)

a matrix that specifies a transition from one observation or prediction to the next

See Example 1 on text page 543.

**Steady-state matrix** (p. 545)

a matrix that a sequence of matrices $M_0, M_1, M_2, \ldots, M_n, \ldots$ approaches as $n$ gets larger and larger

See text pages 544–545.

**Transformation of the plane** (p. 551)

a function whose domain and range are sets of points

$T: (x, y) \rightarrow (x - y, x + y)$
$T: (1, 0) \rightarrow (1, 1)$
$T: (0, 1) \rightarrow (-1, 1)$

**Transformation matrix** (p. 552)

a matrix that specifies a transformation of the plane

$$\begin{array}{c} \overbrace{x \rightarrow x - y}^{} \\ T = \begin{bmatrix} 1 & -1 \\ 1 & 1 \end{bmatrix} \\ \underbrace{\phantom{xxxxx}}_{y \rightarrow x + y} \end{array}$$

# UNDERSTANDING THE MAIN IDEAS

## Communication matrices

- In a communication matrix, a "1" in any row indicates that direct communication from one point to another is possible. A "0" indicates that direct communication is not possible. We assume that no message is sent directly from a point back to that point.

- $M^2$ gives the number of two-step communication paths, (for example, $A \rightarrow B \rightarrow C$), $M^3$ gives the number of three-step paths, and so on. Also, $M + M^2$ gives the number of communication paths using either 1 or 2 steps, $M + M^2 + M^3$ gives the number of paths that are at most three steps, and so on.

## Transition matrices

- To solve a transition problem, you need a transition matrix $T$ and an initial or current observation matrix $M_0$. The next observation or prediction, $M_1$, equals $M_0T$. Similarly, $M_2 = M_1T$, $M_3 = M_2T$, and so on.

$$T = \begin{array}{c} \\ H \\ N \end{array} \begin{array}{cc} H & N \\ \left[ \begin{array}{cc} 0.7 & 0.3 \\ 0.1 & 0.9 \end{array} \right] \end{array}$$

$$M_0 = [\, 0.15 \quad 0.85 \,]$$

$$M_1 = [\, 0.15 \quad 0.85 \,] \left[ \begin{array}{cc} 0.7 & 0.3 \\ 0.1 & 0.9 \end{array} \right]$$

$$= [\, 0.19 \quad 0.81 \,]$$

- If a matrix $S$ has the property that $ST = S$, then $S$ is the steady-state matrix for $T$.

$$[\, 0.25 \quad 0.75 \,] \left[ \begin{array}{cc} 0.7 & 0.3 \\ 0.1 & 0.9 \end{array} \right] = [\, 0.25 \quad 0.75 \,]$$

## Transformation matrices

- Given a linear transformation $T: (x, y) \rightarrow (ax + by, cx + dy)$ that maps a region $R$ to an image region $R'$:

  1. the transformation matrix $T$ is $\left[ \begin{array}{cc} a & b \\ c & d \end{array} \right]$.

  2. $T$ maps $(1, 0)$ to $(a, c)$ and $(0, 1)$ to $(b, d)$.

  3. $\dfrac{\text{area of } R'}{\text{area of } R} = |ad - bc|$.

  4. $R$ and $R'$ have the same orientation (clockwise or counterclockwise) if $ad - bc > 0$ and have opposite orientations if $ad - bc < 0$.

  5. if $ad - bc = 0$, then $R$ is mapped to a line through the origin.

  6. the image point $(x', y')$ of a point $(x, y)$ is given by the product
  $$\left[ \begin{array}{cc} a & b \\ c & d \end{array} \right] \left[ \begin{array}{c} x \\ y \end{array} \right]; x' = ax + by \text{ and } y' = cx + dy.$$

- If $S$ and $T$ are transformations with transformation matrices
  $$S = \left[ \begin{array}{cc} a & b \\ c & d \end{array} \right] \text{ and } T = \left[ \begin{array}{cc} e & f \\ g & h \end{array} \right], \text{ then the matrix representing}$$
  the composite $S \circ T$ is the matrix $ST = \left[ \begin{array}{cc} a & b \\ c & d \end{array} \right] \left[ \begin{array}{cc} e & f \\ g & h \end{array} \right]$.

◀▰▰▰▰ **Caution:** Be sure the transformation matrix is on the left.

## CHECKING THE MAIN IDEAS

Match each matrix with an appropriate name. Each name should
be used only once.

1. $\begin{bmatrix} 2 & -5 \\ 1 & 0 \end{bmatrix}$

2. $\begin{bmatrix} 1 & 0 \\ 0 & 1 \end{bmatrix}$

3. $\begin{bmatrix} \dfrac{1}{2} & \dfrac{1}{2} \\ \dfrac{3}{4} & \dfrac{1}{4} \end{bmatrix}$

**A.** communication matrix

**B.** transformation matrix

**C.** transition matrix

**D.** dominance matrix

4. Write a communication matrix for the communication network shown
   at the right.

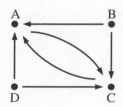

5. Suppose Computer *A* can send data to and receive data from Computers
   *B*, *C*, and *D*. Computer *B* can send data to Computer *A* and receive data
   from Computers *A* and *C*. Computer *C* can send data to Computers *A*,
   *B*, and *D*, and receive data from Computer *A*. Computer *D* can send
   data to Computer *A* and receive data from Computers *A* and *C*.

   **a.** Draw a communication network.

   **b.** Write the communication matrix *M*.

   **c.** Find the matrix that represents the number of ways that data can be
   sent using a path of *at most* two steps. Show that the entries in the
   first row are correct.

6. In a certain region, it is expected that of land that is currently being
   used for agricultural purposes, 85% will be used for agricultural
   purposes next year and 15% will not. Of land that is currently being
   used for nonagricultural purposes, 10% will be used for agricultural
   purposes next year and 90% will not.

   **a.** Find the transition matrix *T*.

   **b.** Suppose that 50,000 acres are currently being used for agricultural
   purposes and 10,000 for nonagricultural purposes. Give a $1 \times 2$
   matrix $M_0$ which represents this information.

   **c.** Find the number of acres of land used for agricultural purposes
   one year from now and two years from now. Describe the trend.

   **d.** Show that the matrix $[\, 24{,}000 \quad 36{,}000 \,]$ is the steady-state matrix
   for *T*. Explain what the matrix means.

7. *Critical Thinking* Describe the transformation
   $T: (x, y) \rightarrow (2x + 3y, 4x + 6y)$ and explain your reasoning.

8. Consider the transformation $T: (x, y) \rightarrow (-y, -x)$.

   **a.** Plot points $A(1, 2)$, $B(0, 4)$, and $C(-2, 3)$, and draw $\triangle ABC$. Then
   plot their images $A'$, $B'$, and $C'$, and draw $\triangle A'B'C'$.

   **b.** Describe the transformation *T*.

   **c.** Write the transformation matrix and find its determinant.

   **d.** Find the ratio of the area of $\triangle ABC$ and the area of $\triangle A'B'C'$. Then
   compare their orientation.

# USING THE MAIN IDEAS

**Example 1** Find the steady-state matrix $S$ for the transition matrix

$$T = \begin{bmatrix} 0.8 & 0.2 \\ 0.4 & 0.6 \end{bmatrix}.$$

**Solution** Matrix $S$ must satisfy $ST = S$. Also, the elements must add up to 1.
Therefore, let $S = \begin{bmatrix} x & 1 - x \end{bmatrix}$.

$$\begin{bmatrix} x & 1 - x \end{bmatrix}\begin{bmatrix} 0.8 & 0.2 \\ 0.4 & 0.6 \end{bmatrix} = \begin{bmatrix} x & 1 - x \end{bmatrix}$$

$0.8x + 0.4(1 - x) = x$ and $0.2x + 0.6(1 - x) = 1 - x$

$$\begin{array}{ll} 0.4x + 0.4 = x & 0.6 - 0.4x = 1 - x \\ -0.6x = -0.4 & 0.6x = 0.4 \\ x = \dfrac{2}{3} & x = \dfrac{2}{3} \end{array}$$

The steady-state matrix is $\begin{bmatrix} \dfrac{2}{3} & \dfrac{1}{3} \end{bmatrix}$.

**Example 2** Suppose a translation $G$ maps a point $P(2, 4)$ to the point $P'(0, 7)$.
Describe the motion of the translation and write a matrix equation for $G$.

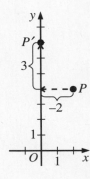

**Solution** Notice that to go from $P$ to $P'$ you move 2 units left and 3 units up.
Thus, $x - 2 = x'$ and $y + 3 = y'$. In the form of a matrix equation,

$$\begin{bmatrix} x \\ y \end{bmatrix} + \begin{bmatrix} -2 \\ 3 \end{bmatrix} = \begin{bmatrix} x' \\ y' \end{bmatrix}.$$

# Exercises

9. Let $M$ be a communication matrix with no ones. What can you conclude?

10. Let $M$ be a communication matrix. Write an expression to represent all the possible paths using at most $n$ steps.

11. **Application** Suppose the transition matrices below give the market shares for competing products $A$, $B$, and $C$. Describe the real-world situation illustrated by each matrix.

a.
$$\begin{array}{c} \\ A \\ B \\ C \end{array} \begin{array}{ccc} A & B & C \\ \begin{bmatrix} 1 & 0 & 0 \\ 0 & 1 & 0 \\ 0 & 0 & 1 \end{bmatrix} \end{array}$$

b.
$$\begin{array}{c} \\ A \\ B \\ C \end{array} \begin{array}{ccc} A & B & C \\ \begin{bmatrix} 1 & 0 & 0 \\ 1 & 0 & 0 \\ 1 & 0 & 0 \end{bmatrix} \end{array}$$

12. Find the steady-state matrix for the transition matrix $T = \begin{bmatrix} 0.7 & 0.3 \\ 0.2 & 0.8 \end{bmatrix}$.

13. Let $S = \begin{bmatrix} 1 & 2 \\ 2 & -3 \end{bmatrix}$ and $T = \begin{bmatrix} 3 & 0 \\ 1 & 4 \end{bmatrix}$.

a. Find the matrix for $S \circ T$ and the matrix for $T \circ S$.
b. **Writing** Explain why $(S \circ T)(P) \neq (T \circ S)(P)$, where $P$ is the point $(x, y)$.

14. Suppose a translation $G$ maps a point $P(-3, -1)$ to the point $P'(3, 1)$. Describe the motion of the translation and write a matrix equation for $G$.

*ADVANCED MATHEMATICS* Student Resource Guide **137**

## Chapter 14: Matrices

**QUICK CHECK**

**Chapter 14**

Complete these exercises before trying the Practice Test for Chapter 14. If you have difficulty with a particular problem, review the indicated section.

Let $A = \begin{bmatrix} 1 & 0 & 5 \\ -3 & -1 & 0 \end{bmatrix}$ and $B = \begin{bmatrix} 7 & -2 \\ 0 & 3 \\ -4 & -1 \end{bmatrix}$.

1. Explain why the sum $A + B$ is not defined. Then find $A^t - 2B$. *(Section 14-1)*

2. Find $AB$ and $BA$. Are the products equal? *(Section 14-2)*

3. Show that $-A + A = O_{2\times3}$. *(Section 14-3)*

4. Matrix $M$ at the right describes a communication network.
   a. Draw a diagram to represent the network.
   b. Explain how you could determine the number of ways a message can be sent using one relay.
   c. Find the total number of ways a message can be sent from X using one relay. *(Section 14-4)*

$$\begin{array}{c} & \text{To} \\ & \begin{array}{ccc} \text{X} & \text{Y} & \text{Z} \end{array} \\ \text{From} \begin{array}{c} \text{X} \\ \text{Y} \\ \text{Z} \end{array} & \begin{bmatrix} 0 & 1 & 1 \\ 0 & 0 & 1 \\ 1 & 1 & 0 \end{bmatrix} \end{array}$$

5. Suppose the matrix at the right is a transition matrix that gives the percents of customers who will switch from a national brand of applesauce to a store brand in consecutive purchases, and from a store brand to a national brand.

$$T = \begin{array}{c} \\ \text{national} \\ \text{store} \end{array} \begin{array}{c} \begin{array}{cc} \text{national} & \text{store} \end{array} \\ \begin{bmatrix} 0.8 & 0.2 \\ 0.3 & 0.7 \end{bmatrix} \end{array}$$

   a. If 70% of current customers bought a national brand of applesauce last time, what percent will buy a national brand next time?
   b. Show that $S = [\,0.6 \quad 0.4\,]$ is the steady-state matrix. Explain what this means. *(Section 14-5)*

6. Let $S = \begin{bmatrix} 1 & 0 \\ 0 & -2 \end{bmatrix}$ and $T = \begin{bmatrix} 1 & 0 \\ 0 & 1 \end{bmatrix}$.

   a. *Complete:* $S(x, y): \rightarrow (\underline{\ ?\ }, \underline{\ ?\ })$ and $T(x, y): \rightarrow (\underline{\ ?\ }, \underline{\ ?\ })$
   b. If $S$ maps $\triangle ABC$ to $\triangle A'B'C'$, compare their areas. *(Section 14-6)*

**PRACTICE TEST**

**Chapter 14**

For Exercises 1–6, use matrices $A = \begin{bmatrix} 2 & 3 \\ 4 & -1 \end{bmatrix}$, $B = \begin{bmatrix} 4 & -6 \\ 2 & -3 \end{bmatrix}$, and $C = \begin{bmatrix} 1 & 7 \\ 2 & -4 \\ 3 & 5 \end{bmatrix}$. Find each matrix, if possible.

1. $A - 2B$
2. $AC$
3. $CB$
4. $C^t$
5. $A^{-1}$
6. $B^{-1}$

Given the matrix equation $\begin{bmatrix} 2 & 3 \\ 4 & -1 \end{bmatrix} X = \begin{bmatrix} -6 \\ 16 \end{bmatrix}$.

7. What linear system does the equation represent?

8. Solve the matrix equation for $X$.

The diagram at the right shows the paths of communication between four computers in an office.

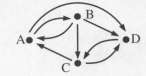

9. Write the matrix $T$ that models this network. Label the rows and columns in alphabetical order. Then find $T^2$.

10. Which computer has the greatest number of two-step paths to it?

**Each year 5% of the people that use UltraWhite toothpaste change to another brand and 10% of those using another brand switch to UltraWhite. Currently UltraWhite has 30% of the toothpaste market.**

11. Write a transition matrix $T$.

12. Write a $1 \times 2$ matrix $M_0$ describing the current market share.

13. What will UltraWhite's market share be in 2 years?

**Consider the transformation $T: (x, y) \rightarrow (2x + y, x)$.**

14. Write the transformation matrix.

15. Find the images of $A(0, 0)$, $B(0, 6)$, and $C(8, 0)$.

---

**MIXED REVIEW**

*Chapters 1–14*

1. Solve the system $\begin{matrix} 8x - 7y = 56 \\ 5x - 4y = 20 \end{matrix}$ by using a matrix equation.

2. Consider the series $5 + 10 + \dots$ . Find the sum of the first 20 terms if the series is (**a**) arithmetic and (**b**) geometric.

3. Find the measure of $\angle A$ and the area of $\triangle ABC$, given $A(0, 0, 0)$, $B(1, 1, 1)$, and $C(1, -5, 1)$.

4. If $z = 2\sqrt{3} - 2i$, find $z^8$ and the two square roots of $z$.

5. The diagram at the right illustrates a communication network. Find the matrix that represents the number of ways messages can be sent using *at most* one relay.

**Evaluate.**

6. $\displaystyle\sum_{n=1}^{\infty} 4 \cdot \left(-\frac{2}{3}\right)^n$     7. $\tan(\text{Cos}^{-1} 0.8)$     8. $\ln \sqrt[4]{e^5} + \log_4 32$

9. The transformation $T: (x, y) \rightarrow (6x - 2y, -3x + y)$ maps every point of the plane onto a line. Find the slope and an equation of the line. Then find the transformation matrix.

10. An airplane heading southwest at 500 knots encounters a wind of 50 knots blowing toward the east. Find the resultant speed and direction of the plane.

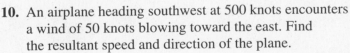

11. Let $\alpha$ and $\beta$ be acute angles with $\sin \alpha = \frac{3}{5}$ and $\cos \beta = \frac{15}{17}$. Find $\sin(\alpha - \beta)$, $\cos 2\alpha$, and $\sin \frac{1}{2}\beta$.

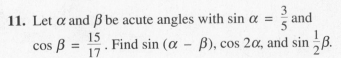

12. Find $\displaystyle\lim_{n \to \infty} \frac{\sin n\pi}{n}$ and $\displaystyle\lim_{n \to \infty} \frac{n}{\sin n\pi}$ .

13. Find the first 6 terms of the sequence defined recursively by $t_1 = 2$, $t_n = t_{n-1} + (2n - 1)$. Find an explicit definition for the sequence.

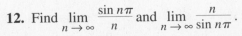

---

| OVERVIEW | These sections introduce some basic methods of organized counting (combinatorics). In Section 15-1, Venn diagrams help you visualize sets which may have common elements. In Section 15-2, the multiplication, addition, and complement principles are applied to counting problems. |

## KEY TERMS

**EXAMPLE/ILLUSTRATION**

**Venn diagram** (p. 565)
a method of representing sets as overlapping circles in a rectangular "universe"

$T$ = families who own trucks
$C$ = families who own cars
$U$ = all families

**Intersection of sets ($A \cap B$)** (p. 565)
the set of elements belonging to both set $A$ and set $B$

The intersection of sets $T$ and $C$ below is the set of families having both a car and a truck.

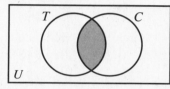

**Union of sets ($A \cup B$)** (p. 565)
the set of elements belonging to set $A$ or set $B$ or both

$T \cup C$ in the example below is the set of all families having a car, or a truck, or both.

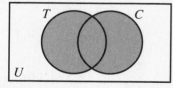

**Complement of a set ($\overline{A}$)** (p. 566)
the set of all elements which do *not* belong to $A$

$\overline{T}$ in the example below is the set of all families *not* having a truck.

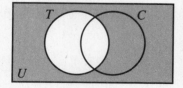

**Mutually exclusive actions or events** (p. 573)
actions or events which cannot occur at the same time

A professional baseball player can play in the American League or the National League, but cannot play in both at the same time.

## UNDERSTANDING THE MAIN IDEAS

### Inclusion-exclusion principle

For any sets $A$ and $B$, $n(A \cup B) = n(A) + n(B) - n(A \cap B)$. Since the elements in the set $A \cap B$ are in both sets $A$ and $B$, this number of elements must be subtracted since these elements are counted once in $n(A)$ and then again in $n(B)$. If sets $A$ and $B$ are *mutually exclusive*, then $n(A \cap B) = 0$ and $n(A \cup B) = n(A) + n(B)$.

### Multiplication principle

- To determine the number of ways a two-step action can be performed, multiply the number of ways the first step can be done by the number of ways the second step can be performed.
- If there are more than two steps, the principle is the same: find the product of the number of choices at each step.

### Addition principle

If sets $A$ and $B$ are *mutually exclusive*, $n(A \cup B) = n(A) + n(B)$. That is, the number of elements in $A$ *or* $B$ is the number of elements in $A$ *plus* the number of elements in $B$.

## CHECKING THE MAIN IDEAS

1. Thirty-six of 50 employees signed up for the health plan. Eighteen joined the pension plan. Only 5 took neither benefit. How many enrolled in both the health plan and the pension plan?

**There are five choices of ice cream and three choices of cookies.**

2. How many different desserts are there if you have one scoop of ice cream *and* one cookie?

3. How many different desserts are there if you have either one scoop of ice cream *or* a cookie?

4. *Critical Thinking* Why are the answers to Exercises 2 and 3 different? Which principles do they illustrate?

## USING THE MAIN IDEAS

**Example 1** A club has 15 members, 5 seniors and 10 underclassmen. How many ways can the offices of president, vice-president, secretary, and treasurer be filled if the president must be a senior and the vice-president must be an underclassman?

**Solution** This is a 4-step action, so use the multiplication principle.

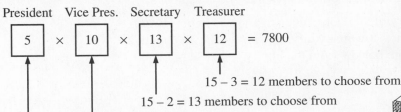

President   Vice Pres.   Secretary   Treasurer

$$5 \times 10 \times 13 \times 12 = 7800$$

15 − 3 = 12 members to choose from

15 − 2 = 13 members to choose from

10 underclassmen to choose from

5 seniors to choose from

**Example 2** Of 150 students who are taking at least one science class, 45 are taking Biology, 75 are taking Chemistry, and 75 are taking Physics. Fifteen students are taking both Chemistry and Physics, 35 are taking only Chemistry, 25 are taking Biology and Chemistry, and no one is taking all three classes. How many students are taking only Biology?

**Solution** Draw a Venn Diagram showing the given information. Let $B$ = biology students, $C$ = chemistry students, and $P$ = physics students.

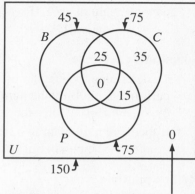

All 150 students are taking at least one of the science classes.

• Use the inclusion-exclusion and complement principles to determine the numbers in the remaining regions. Since 35 students are taking only Chemistry, $150 - 35 = 115$ are taking Biology or Physics (or both). This means $n(B \cup P) = 115$.

$$n(B \cup P) = n(B) + n(P) - n(B \cap P)$$
$$115 = 45 + 75 - n(B \cap P)$$
$$5 = n(B \cap P)$$

Since 45 students are taking Biology, and of these there are 25 taking Biology *and* Chemistry, 5 taking Biology *and* Physics, and none taking all three classes, then $45 - (25 + 5 + 0) = 45 - 30$, or 15 students are taking only Biology.

• Finally, check that the sum of the regions equals $n(U) = 150$. ✔ Notice that the number of students taking only Physics is $75 - (15 + 0 + 5)$, or 55.

## Exercises

5. *Application* Half of 120 patients are given an experimental treatment for their disease, while the other half were left untreated. Seventy patients improved, including $\frac{1}{3}$ of the patients who had no treatment. What fraction of the treated patients improved?

6. Copy the diagram at the right and shade the region which represents $\overline{A} \cap (B \cap C)$.

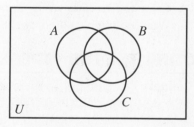

7. Which of these events are mutually exclusive?
   **A.** Wearing a plaid shirt and wearing a striped tie
   **B.** Failing French and failing Chemistry
   **C.** Being under 25 years old and being president of the United States
   **D.** Getting a walk and hitting a home run in the same baseball game

8. How many three letter "words" can be formed if the middle letter must be a vowel (a, e, i, o, or u) and the other letters must be consonants?

9. Your teacher has a library of 20 books. You are allowed to borrow up to three books at one time. How many different ways can you choose a three-book combination to borrow from the library?

# Permutations, Combinations, and the Binomial Theorem

## *Sections 15-3, 15-4, and 15-5*

**OVERVIEW** **Sections 15-3 and 15-4 describe methods of solving problems involving permutations and combinations. Section 15-5 describes Pascal's triangle and the Binomial Theorem, an important application of combinatorics to algebra.**

| KEY TERMS | EXAMPLE/ILLUSTRATION |
|---|---|
| **Permutation** (p. 578)<br>an arrangement of objects in which the order *is* important | 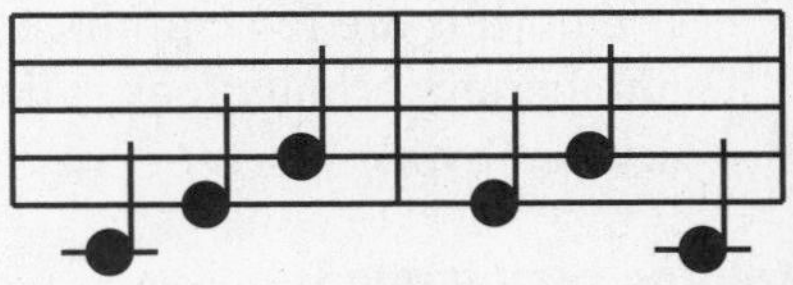DO MI SOL MI SOL DO<br>Two permutations of three notes. (Order is important.) |
| **Combination** (p. 578)<br>an arrangement of objects in which the order is *not* important | 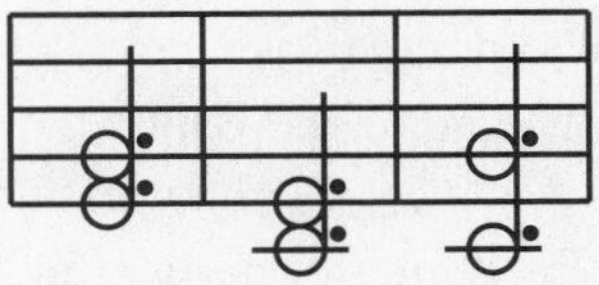SOL MI / MI DO / SOL DO<br>These chords are combinations of 2 notes. Since they are played together, order is not important. |
| **Binomial coefficients** (p. 590)<br>coefficients of the terms of the expansion of $(a + b)^n$ | $(a + b)^3 = 1a^3 + 3a^2b + 3ab^2 + 1b^3$<br>The numbers 1, 3, 3, 1 above are the binomial coefficients. |
| **Pascal's triangle** (p. 590)<br>a triangular arrangement of binomial coefficients | See text page 590. |

## UNDERSTANDING THE MAIN IDEAS

### Counting permutations

- To find the number of permutations of $r$ things chosen from a set of $n$ things, multiply

$$\underbrace{n \times (n-1) \times (n-2) \times \dots \times (n-r+1)}_{r \text{ factors}}$$

This number is denoted by the symbol ${}_nP_r$, and its value is given by the formula $\frac{n!}{(n-r)!}$.

- To find the number of permutations of a set in which some elements are the same, such as the letters in the word "CINCINNATI":

  count the elements $\longrightarrow$ $n = 10$

  count the repeated elements separately $\longrightarrow$ 2 C's, 3 I's, 3 N's

  Use the formula $\dfrac{n!}{n_1!n_2! \dots n_k!}$, where $n_1, n_2, \dots, n_k$ are

  the number of times each repeated element appears $\longrightarrow$ $\dfrac{10!}{2!3!3!} = 50{,}400$

- If $n$ objects are to be arranged in a circle rather than a line, the number of permutations is the number of linear permutations divided by $n$.

### Counting combinations

- The number of ways $r$ things can be selected from a set of $n$ things is $_nC_r = \dfrac{n \times (n-1) \times \dots (n-r+1)}{r!} = \dfrac{_nP_r}{r!}$.

### Pascal's triangle

- The numbers in the $n$th row of Pascal's triangle are $_nC_0, _nC_1, _nC_2, \dots, _nC_n$. The first and last of these are always 1.
- Each interior number in Pascal's triangle is the sum of the numbers diagonally above it (to the right and left).

### Binomial theorem

To expand $(a + b)^n$, follow this pattern for $(a + b)^5$:

$$(a + b)^5 = \square a^5b^0 + \square a^4b^1 + \square a^3b^2 + \square a^2b^3 + \square a^1b^4 + \square a^0b^5$$

Notice that the powers of $a$ are decreasing and the powers of $b$ are increasing from left to right, and that the sum of the powers for each term is always 5.

Now fill in the coefficients from the *fifth* row of Pascal's triangle (see page 590): 1   5   10   10   5   1. Thus,

$$(a + b)^5 = 1a^5b^0 + 5a^4b^1 + 10a^3b^2 + 10a^2b^3 + 5a^1b^4 + 1a^0b^5$$
$$= a^5 + 5a^4b + 10a^3b^2 + 10a^2b^3 + 5ab^4 + b^5$$

## CHECKING THE MAIN IDEAS

1. How many different 3-letter "words" can be formed from the letters in the word CANOE?

2. How many different ways can five children arrange themselves for a game of ring-around-the-rosie?

3. How many different ways can a teacher assign 10 homework problems from a set of 25?

4. What is the coefficient of $a^4b^9$ in the expansion of $(a + b)^{13}$?

5. How many different arrangements are there of the digits 166555?

## USING THE MAIN IDEAS

**Example 1** From a club of 20 members, 15 are needed to work at their giant yard sale — 5 members to set up, 7 others to work during the sale, and 3 more to clean up. How many different ways can the crews be assigned?

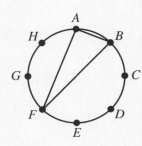

**Solution** Use the multiplication principle:

| ways to choose set-up crew | $\times$ | ways to choose sale crew | $\times$ | ways to choose clean-up crew | $=$ | number of different work arrangements |
|---|---|---|---|---|---|---|

The order in which people are chosen for each crew does not matter, so this is a combination rather than a permutation.

- The number of ways to choose 5 set-up workers from 20 members is

(five factors) $\rightarrow$ $\dfrac{20 \times 19 \times 18 \times 17 \times 16}{5!} = 15{,}504 = {}_{20}C_5$.

- Once the set-up crew is chosen, there are 15 remaining members from which to choose the 7 salespeople.

$${}_{15}C_7 = \frac{15 \cdot 14 \cdot 13 \cdot 12 \cdot 11 \cdot 10 \cdot 9}{7!} = 6435 \text{ ways}$$

- There are now 8 people left from whom to choose the 3-member clean-up crew.

$${}_8C_3 = \frac{8 \cdot 7 \cdot 6}{3!} = 56 \text{ ways}$$

Therefore, the number of different work assignments is $15{,}504 \times 6435 \times 56 = 5{,}587{,}021{,}440$.

**Example 2** In the expansion of $(x - 2y)^{15}$, find the term containing $x^6$.

**Solution** $(x - 2y)^{15}$ is like $(a + b)^{15}$, with $a = x$ and $b = -2y$.
In the expansion of $(a + b)^{15}$, the term with $a^6$ is ${}_{15}C_9\, a^6 b^9$.

The coefficient of the term $a^6 b^9$ is ${}_{15}C_9 = \dfrac{15!}{9!6!} = 5005$.

Substituting $x = a$ and $-2y = b$, we get
$5005a^6 b^9 = 5005x^6(-2y)^9 = 5005x^6(-512y^9) = -2{,}562{,}560x^6 y^9$.

## Exercises

6. A child has 10 identically-shaped blocks — 4 red, 3 green, 2 yellow, and 1 blue. How many different stacks of all 10 blocks are possible?

7. How many different triangles can be drawn with vertices on the eight points shown in the figure at the right?

8. How many ways can ten people be seated around a circular table if the host and hostess cannot be seated together?

9. A committee of 4 is to be chosen from a club with 10 male and 12 female members. The committee is to have at least two women. In addition, members Mr. and Mrs. Rodriguez refuse to serve together on the same committee. How many ways can the committee be chosen?

10. Expand $\left(2x - \dfrac{1}{x^2}\right)^7$ using the binomial theorem.

# CHAPTER REVIEW

## Chapter 15: Combinatorics

### QUICK CHECK

**Chapter 15**

Complete these exercises before trying the Practice Test for Chapter 15. If you have difficulty with a particular problem, review the indicated section.

Let $P$ = set of voters who voted for the Republican presidential candidate. Let $C$ = set of voters who voted for the Republican congressional candidate.

1. Describe each of these sets in words. *(Section 15-1)*
   **a.** $P \cap C$     **b.** $\overline{P} \cup \overline{C}$     **c.** $\overline{P \cup C}$     **d.** $C \cap \overline{P}$

2. Of 2000 voters who cast ballots, 1200 voted for the Republican presidential candidate, 800 voted for the Republican congressional candidate, and 700 did not cast a vote for either Republican candidate. How many voters cast ballots for both Republican candidates? *(Section 15-1)*

3. David has twelve shows recorded on videotape, as well as shows on four different channels on TV. If David plans to watch a show on TV and then one on videotape, in how many different ways can he do this? *(Section 15-2)*

4. Find $_{10}P_6$ and $_{10}C_6$. In what ways are they used? *(Section 15-3)*

5. How many different 5-card hands can be dealt from a standard deck of cards if all the cards are to be of the same suit? *(Section 15-3)*

6. How many different ways can the letters of the word REVIVE be arranged? *(Section 15-4)*

7. In the expansion of $(a + 3)^{15}$, what is the coefficient of the term $a^9$? *(Section 15-5)*

### PRACTICE TEST

**Chapter 15**

Draw a Venn diagram and shade the region representing the given set of students. Let $F$ = the set of French minors, $S$ = the set of seniors, and $B$ = the set of Biology majors.

1. $\overline{F} \cap S$                           2. $F \cap S \cap \overline{B}$

3. Describe the sets given in Exercises 1 and 2.

**Last year, 50% of the senior class played soccer, 52% played basketball, 26% played tennis, 4% played all three sports, and 3% did not play any of these sports. Also, 15% played both soccer and basketball, 11% basketball and tennis, and 9% soccer and tennis.**

4. Draw a Venn Diagram to represent the data.

5. What percent played only one sport?

**Evaluate each expression.**

6. $6!$          7. $\dfrac{10!}{5!}$          8. $\dfrac{8!}{5! \cdot 3!}$

9. In how many ways can 5 different jobs be assigned to 5 people in a group of 12?

---

**Find the number of possible arrangements of each set of items.**

10. the letters in the word SEARCH

11. the letters in the word LEVELED

12. seven people sitting around a circular table

**A box contains 5 English, 4 mathematics, and 6 history textbooks, all different. They are to be put on a shelf.**

13. In how many ways can the books be arranged?

14. In how many ways can they be arranged if books of the same subject are to be kept together?

**Consider the expansion of $(a - 3)^{10}$.**

15. Find the 4th term of the expansion.

16. Find the term containing $a^8$.

---

**MIXED REVIEW**

*Chapters 1–15*

1. How many communication matrices are possible which describe a network of three stations (a station does not communicate with itself)?

**Consider the transformation $T: (x, y) \rightarrow (3x + y, -2x)$.**

2. Find the image under $T$ of $\triangle PQR$ with vertices $P(0, 0)$, $Q(3, 4)$, and $R(0, 3)$.

3. Find the matrix $T$ and the value of its determinant $|T|$. What information does $|T|$ provide?

4. Find the matrix $T^{-1}$ and the transformation $T^{-1}: (x, y) \rightarrow (?, ?)$.

**Find the sums. If the sum does not exist, so state.**

5. $\displaystyle\sum_{n=2}^{20} 3n$

6. $\displaystyle\sum_{n=1}^{\infty} 3\left(-\frac{1}{5}\right)^n$

7. $\displaystyle\sum_{n=1}^{\infty} (10 + n)$

**In Exercises 8–16, sketch the graph of the relation.**

8. $r = 1 + \sin 2\theta$

9. $xy = -2$

10. $4y^2 + 16x^2 = 64$

11. $y = \log_2 x$

12. $\dfrac{x}{10} - \dfrac{y}{6} = 1$

13. $y = |x - 5|$

14. $y = 3 \sin 2\left(x - \dfrac{\pi}{4}\right)$, where $-\pi \le x \le \pi$

15. $f(n) = \left(-\dfrac{2}{3}\right)^n$, where $n = 0, 1, 2, \ldots$

16. $(x, y) = (2t, 1 + t^2), t \ge 0$

17. Use the binomial theorem to expand $\left(\dfrac{1}{4} + \dfrac{3}{4}\right)^6$. What is the sum of the seven resulting terms? Why?

---

## Sections 16-1, 16-2, 16-3, and 16-4

| OVERVIEW | Sections 16-1 and 16-2 introduce the basic concepts of probability. In Section 16-3, the binomial probability theorem is applied to probability problems. Section 16-4 discusses the applications of combinatorics to probability. |
|---|---|

## KEY TERMS

|  | EXAMPLE/ILLUSTRATION |
|---|---|
| **Probability** (p. 598)<br> a number from 0 to 1 which gives the likelihood of a particular event occurring | For a coin flip,<br> $P(\text{Heads}) = \frac{1}{2}$, $P(\text{Tails}) = \frac{1}{2}$ |
| **Sample space** (p. 598)<br> the set of all possible outcomes of an experiment (the outcomes must be mutually exclusive) | For a coin flip,<br> sample space = {Heads, Tails} |
| **Event** (p. 598)<br> a particular outcome, or set of outcomes, of an experiment | {Heads} is one event. |
| **Conditional probability** (p. 607)<br> the probability that a second event will occur given that a particular first event has already occurred (written $P(A \mid B)$) | What is the probability I can buy a car *if* I get a job? |
| **Independent events** (p. 607)<br> two events for which the occurrence of one event does not affect the probability of the occurrence of the other event | Flipping heads on a penny and heads on a dime are independent. |

## UNDERSTANDING THE MAIN IDEAS

### Determining the probability of an event theoretically

- Determine the sample space for the experiment.
- If all the outcomes are equally likely, the probability is

$$\frac{\text{number of successful outcomes}}{\text{number of elements in the sample space}}.$$

### Probability of *A* or *B*

- The general rule is $P(A \text{ or } B) = P(A) + P(B) - P(A \text{ and } B)$.
- If *A* and *B* are mutually exclusive, then $P(A \text{ and } B) = 0$, and thus $P(A \text{ or } B) = P(A) + P(B)$.
- Since $P(A \text{ or not } A) = 1$, then $P(A) = 1 - P(\text{not } A)$ and $P(\text{not } A) = 1 - P(A)$

### Conditional probability

- The probability of *A* given *B* (written $P(A \mid B)$) is $\dfrac{P(A \text{ and } B)}{P(B)}$.
- If *A* and *B* are independent events, then $P(A \mid B) = P(A)$ and $P(B \mid A) = P(B)$.

### Using combinations to solve probability problems

- Combinatorics (Chapter 15) should be used whenever a sample space becomes too large to conveniently count. Many calculators have keys for $_nC_r$, $_nP_r$, and factorials.

### Binomial experiment

For a problem like "What is the probability of rolling four 3's in ten rolls of a die?," use the binomial probability theorem (see text page 614).

- Determine $A$ (success in a trial). $\longrightarrow$ $A$ = rolling 3

- Find $P(A)$. $\longrightarrow$ $P(A) = \dfrac{1}{6}$

- Use the formula $P(k\ A\text{'s}) = {}_nC_k\, p^k(1 - p)^{n-k}$. $\longrightarrow$ $P(\text{four 3's}) = {}_{10}C_4 \left(\dfrac{1}{6}\right)^4\left(\dfrac{5}{6}\right)^6$

$$\approx 0.054$$

## CHECKING THE MAIN IDEAS

**In Exercises 1 through 5, a die is rolled and then a coin is flipped.**

1. What is the sample space and how large is it?

2. What is the probability of getting a 2 and a tail?

3. What is the probability of getting an even number and a head?

4. What is the probability of flipping a tail if you roll a 6?

5. What is the probability of rolling a 5 or flipping a head?

6. If two differently-colored dice are rolled, what is the probability of having a total greater than 7 if one die is less than 3? (See the sample space on page 601.)

7. If you take a five-question true-false test by guessing at the answers, what is the probability you will get three or more questions right?

## USING THE MAIN IDEAS

**Example 1** A refrigerator has 110 cans of soft drinks, both diet and regular. The flavors are orange and cola. The Venn diagram shows the quantities of each type of soda. Suppose a can is taken at random from the refrigerator. What is the probability that the soda is:
**a.** diet? **b.** regular? **c.** regular cola?
**d.** orange or cola? **e.** cola or regular?

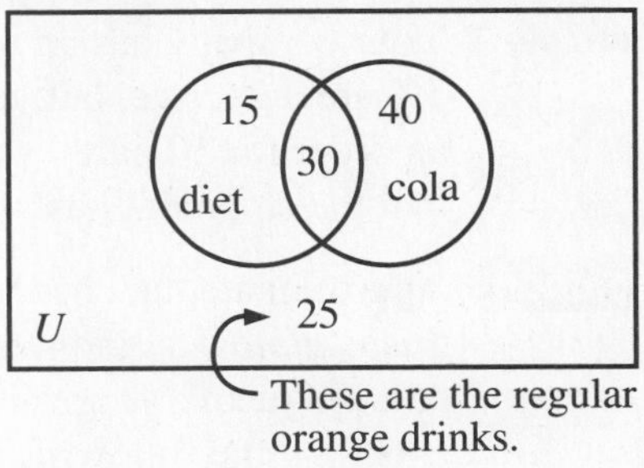

15
40
30
diet
cola
$U$
25

These are the regular orange drinks.

**Solution** Since the can is taken "at random," each can is as likely to be taken as any other.

  **a.** The probability of a diet soda being taken is

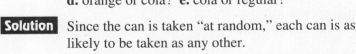

$\dfrac{15 + 30}{110} = \dfrac{45}{110} = \dfrac{9}{22}$.

  **b.** Since taking a regular drink is the same as not taking a diet drink, $P(\text{regular}) = 1 - P(\text{diet})$. Thus, using the result in part (a),

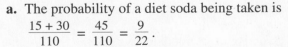

$P(\text{regular}) = 1 - \dfrac{9}{22} = \dfrac{13}{22}$.

**c.** The Venn diagram shows 40 regular colas, so the probability of drawing one is $\frac{40}{110} = \frac{4}{11}$.

**d.** *All* sodas are either orange or cola, so the probability of drawing one or the other is 1.

**e.** The Venn diagram at the right is shaded to show the sodas which are either colas or regular (or both). There are $30 + 40 + 25 = 95$ such cans, so the probability is $\frac{95}{110} = \frac{19}{22}$.

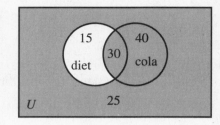

**Example 2 a.** In Example 1, what is the probability that if a cola is drawn it will be a diet cola?

**b.** Given that a diet drink is drawn, what is the probability that it will be a diet cola?

**Solution** Part (a) asks for $P(D \mid C)$, (the probability of drawing a diet drink given that it is a cola) while part (b) asks for $P(C \mid D)$ (the probability of drawing a cola given that it is a diet drink).

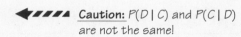

Caution: $P(D \mid C)$ and $P(C \mid D)$ are not the same!

**a.** The Venn diagram shows 70 colas, 30 of which are diet. This means that $P(D \mid C) = \frac{30}{70} = \frac{3}{7}$. Using the formula,

$$P(D \mid C) = \frac{P(D \text{ and } C)}{P(C)} = \frac{\frac{30}{110}}{\frac{70}{110}} = \frac{3}{7}.$$

**b.** The Venn diagram shows 45 diet drinks, 30 of which are colas. So $P(C \mid D) = \frac{30}{45} = \frac{2}{3}$. Using the formula,

$$P(C \mid D) = \frac{P(C \text{ and } D)}{P(D)} = \frac{\frac{30}{110}}{\frac{45}{110}} = \frac{2}{3}.$$

**Example 3** Your company has an inventory of 1000 electronic parts. You know 5% are defective, but you do not know which particular parts. To fill an order for 50 parts, you select 50 at random. What are the chances that the shipment has no more than one defective part?

**Solution** Approximate the chances with the binomial probability theorem, since this is *almost* a binomial experiment (see Example 2 on text page 614). Always justify your use of the binomial probability theorem by checking the following three points:

 *n* repeated trials … 50 selections of parts
 independent trials … No, (since there's no replacement) but close
      because 1000 is so large.
 exactly two outcomes (*A* or not *A*) … defective or not defective

To use the theorem, we need to know *p* and *k*.

$P(A) = P(\text{defect}) = 0.05$ because 5% are defective.

The values of $k$ are 0 and 1, since we want to know the chances of zero or one defect.

The probability of $k$ occurrences of $A$ is $_nC_k\,p^n(1-p)^{n-k}$, so the probability of sending zero defects is $_{50}C_0\,(0.05)^0(0.95)^{50}$ and the probability of one defect is $_{50}C_1\,(0.05)^1(0.95)^{49}$.

Now find the sum of these probabilities:
$_{50}C_0\,(0.05)^0(0.95)^{50} + \ _{50}C_1(0.05)^1(0.95)^{49} \approx$
$1(0.0769) + 50(0.05)(0.0810) \approx 0.2794 \leftarrow$ Use a calculator here.

The chances of shipping no more than one defective part is 27.9%. Notice that this means there is a 72.1% (100% − 27.9%) chance of shipping more than one bad part!

## Exercises

**8.** Let $E$ be the event that there was an earthquake and $B$ be the event that a particular building is demolished in the earthquake. Express these probabilities *in words*: $P(B \mid E)$, $P(E \mid B)$, $P(B \mid \text{not } E)$, and $P(\text{not } E \mid \text{not } B)$.

**9.** *True* or *false*: A citizen of the United States is to be chosen at random. Since there are 50 states, the probability that a Californian will be chosen is $\frac{1}{50}$. Justify your answer.

**10.** A card is chosen from a well-shuffled deck. Which of these pairs of events are independent?

   **A.** "red" and "black"

   **B.** "queen" and "face card"

   **C.** "queen of clubs" and "club"

   **D.** "diamond" and "two"

**11.** A multiple-choice test has 10 questions, each with 4 choices for the answer. If the answers are chosen randomly, without even looking at the question, what are the chances of choosing 8 or more correct answers?

**12.** *Writing* Discuss this statement: If a weather forecaster predicts a 30% chance of rain tomorrow, there is no way she can be proved wrong, whether it rains or not. How should you decide if a forecaster is good?

# Applications of Probability

## Sections 16-5 and 16-6

**OVERVIEW**  Section 16-5 discusses many practical applications of conditional probability, using tree diagrams to visualize the information. Section 16-6 shows how to find and use expected value to determine if a game is fair.

## KEY TERMS

| | EXAMPLE/ILLUSTRATION |
|---|---|
| **Expected value** (p. 630)<br>the average gain (or loss) by a player if a game were played many times |  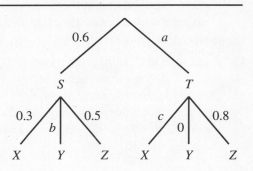<br>Expected value = $0 |
| **Fair game** (p. 630)<br>a game with an expected value of zero | The spinner above is a fair game. |

## UNDERSTANDING THE MAIN IDEAS

### Working with conditional probability

A tree diagram is a useful tool for organizing conditional probabilities. The sum of the probabilities branching from any given point must be 1. (See the tree diagrams on text page 624 and the top of text page 625.)

### Expected value

To calculate expected value, take each outcome and multiply its payoff by its probability. Then add these products. (See Example 1 on text page 631.)

## CHECKING THE MAIN IDEAS

**For Exercises 1–3, refer to the tree diagram at the right.**

1. Find the values of $a$, $b$, and $c$.

2. Find $P(X \mid S)$.　　　　3. Find $P(Z)$.

4. Two players, $A$ and $B$, play this game: A die is rolled. If the result is odd, $A$ pays $B$ the value of the die (in dollars). If the result is even, $B$ pays $A$ the value of the die. Calculate $A$'s and $B$'s expected values.

5. In Exercise 4, which of the two players would you rather be? Devise a way to change the game to make it fair.

## USING THE MAIN IDEAS

**Example 1** A blood test is devised to help detect a disease which occurs in 5% of the population. The test is 90% accurate, that is, it gives a positive result for 90% of the infected people and a negative result for 90% of the uninfected people. Illustrate this situation with a tree diagram. Then, suppose *you* are tested and the result is positive. How likely are you to have the disease?

**Solution** First choose symbols for the relevant situations.
Let $D$ = having the disease, $D'$ = not having the disease, $T$ = testing positive, and $T'$ = testing negative.
Then sketch the tree and fill in as many probabilities as possible:

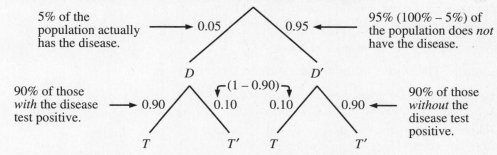

5% of the population actually has the disease. $\longrightarrow$ 0.05

0.95 $\longleftarrow$ 95% (100% – 5%) of the population does *not* have the disease.

$D$ $\qquad$ $D'$

90% of those *with* the disease test positive. $\longrightarrow$ 0.90

(1 – 0.90)

0.10 $\qquad$ 0.10

0.90 $\longleftarrow$ 90% of those *without* the disease test positive.

$T$ $\qquad$ $T'$ $\quad$ $T$ $\qquad$ $T'$

If *you* test positive, you want to know the probability of actually having the disease given a positive test, or $P(D \mid T)$.

$$P(D \mid T) = \frac{P(D \text{ and } T)}{P(T)}$$

$$= \frac{(0.05)(0.90)}{(0.05)(0.90) + (0.10)(0.95)}$$

$$= \frac{0.045}{0.14} \approx 0.321$$

Thus, you have only a 32% chance of actually having the disease, even though you tested positive!

**Example 2** A lottery offers a $2,000 first prize, ten "second prizes" of $500 each, and 100 "third prizes" worth $25 each. All 5000 tickets are expected to be sold. What is a fair price to pay for a ticket?

**Solution** "Fair" means that the expected value of the game is zero.
First, calculate the expected gain for playing this lottery:

| outcome | 1st prize | 2nd prize | 3rd prize |
|---|---|---|---|
| value | $2000 | $500 | $25 |
| probability | $\dfrac{1}{5000}$ | $\dfrac{10}{5000}$ | $\dfrac{100}{5000}$ |

$$\text{Average gain} = 2000 \cdot \frac{1}{5000} + 500 \cdot \frac{10}{5000} + 25 \cdot \frac{100}{5000}$$

$$= 0.40 + 1 + 0.50 = \$1.90$$

Since the expected value (which must be zero for the game to be fair) is the difference between the cost of the ticket and the average gain, then $1.90 is the fair price for the ticket. However, since lotteries are intended to make a profit, the price will be higher than $1.90.

## Exercises

**6.** As Example 1 illustrates, screening tests on a general population must be used with care. Often "high risk" groups, if they can be identified, are targeted for testing to increase the meaningfulness of the results. Repeat Example 1, this time assuming you are part of a high risk group with a 20% incidence of the disease. Again you test positive. Now what are your chances of having the disease?

**For Exercises 7–9, consider this game: A coin is flipped and a marble is chosen at random from bowl #1 if the coin lands heads or bowl #2 if the coin lands tails. See the bowls in the figure at the right.**

**Bowl #1**          **Bowl #2**

**7.** If a black marble is drawn, what is the chance it came from bowl #1?

**8.** Calculate the probability of drawing a black marble?

**9.** *Critical Thinking*  There are 5 black marbles out of a total of 15 marbles in the two bowls. Why is the answer to Exercise 8 not $\frac{5}{15} = \frac{1}{3}$?

**10.** To handle a lawsuit, a law firm has to decide whether to charge a client a fee of $4000, or to charge a "contingency" fee of $20,000 which will be paid only if they win the case. If the firm decides to charge the $4000 fee, what must they think of their chances of winning? (Your answer should state the probability of winning the case as a percent.)

**11.** For a woman entering a dress shop, there is a 50% probability she will buy nothing, a 35% probability she will buy one dress, a 10% probability she will buy two dresses, and a 5% chance she will buy three dresses. How many dresses can a woman visiting the shop be expected to buy?

# Chapter 16: Probability

Complete these exercises before trying the Practice Test for Chapter 16. If you have difficulty with a particular problem, review the indicated section.

In Exercises 1–2, two letters are selected randomly, without replacement, from the letters of the word "SPAIN."

1.  **a.** Write the sample space for this experiment.

    **b.** Find the probability that one of the letters drawn is A. *(Section 16-1)*

2. If one of the letters drawn is I, what is the probability that the other is A? *(Section 16-2)*

3. Find the probability of getting exactly 7 heads and 7 tails in 14 flips of a fair coin. *(Section 16-3)*

4. Five cards are dealt from a well-shuffled standard deck. Find the probability that the cards dealt are five consecutive cards of the same suit. *(Section 16-4)*

5. Jar A has 2 black marbles and 1 white marble. Jar B has 2 black and 3 white marbles. Without looking, a marble is drawn at random from Jar A and transferred to Jar B. Then a marble is drawn at random from Jar B. What is the probability that the second marble drawn is white? *(Section 16-5)*

6. In a game show, a contestant chooses prizes hidden behind three closed doors. One door hides a valueless "prize," another door hides a prize worth $500, and the remaining door hides a grand prize worth $10,000. What is the expected value (payoff) of this game? *(Section 16-6)*

Give an example for each of the following.

1. two mutually exclusive events    2. two independent events

Three cards are selected from a well-shuffled standard deck of 52 playing cards. Find the probability of drawing each of the following.

3. three black cards                 4. three face cards

5. at least one ace                  6. no pairs

A pair of fair dice are rolled. Find the probability of rolling each of the following.

7. a sum less than seven             8. a pair of numbers

9. a sum of seven or eleven          10. a pair of prime numbers

The probability that Jack makes a free throw is 0.6 and the probability that April makes a free throw is 0.7. Find the probability of each of the following.

11. Each of them makes their next free throw.

12. At least one of them makes their next free throw.

13. Jack makes 7 out of his next 10 free throws.

14. April makes at least 8 out of her next 10 free throws.

A committee of 6 is to be formed from 10 students and 8 faculty members. Find the probability of each of the following.

**15.** Exactly 2 students are on the committee.

**16.** At most 4 students are on the committee.

Jar A contains 4 red and 3 white marbles and Jar B contains 3 red and 2 white marbles. A die is rolled. If the number rolled is greater than 2, a marble is selected from Jar B, otherwise a marble is selected from Jar A.

**17.** Draw a tree diagram that displays this information.

**18.** Find the probability that the marble is red.

**19.** The marble selected is red. Find the probability that the marble came from Jar A.

For a particular game, the probability of winning $50 is 0.2 and the probability of winning $10 is 0.6.

**20.** How much would you expect to win in this game?

**21.** What would be a fair price to play this game?

---

**MIXED REVIEW**

*Chapters 1–16*

**1. a.** Copy and complete the table of values below for binomial probabilities with $n = 6$ and $p = 0.4$. Round your answers to the nearest hundredth.

| $r$ | 0 | 1 | 2 | 3 | 4 | 5 | 6 |
|-----|---|---|---|---|---|---|---|
| $p(r) = {}_6C_r(0.4)^r(0.6)^{6-r}$ | | | | | | | |

**b.** Graph the points $(r, p(r))$ and connect them with a smooth curve.

**2.** Repeat Exercise 1, this time with $n = 6$ and $p = 0.5$.

**3.** Compare and contrast the graphs from Exercises 1 and 2. Consider their slopes, symmetry, maxima, and minima.

**4.** Find the sum $\sum\limits_{i=1}^{10} \left(\frac{3i}{20}\right)^2$.

**Solve these trigonometric equations.**

**5.** $\sin x - \cos^2 x \sin x = -1$

**6.** $\sin 3x \cos 2x - \cos 3x \sin 2x = 0.5$

**7.** $2 \sin^2 \frac{1}{2}x = 2 + \cos x$

**8.** $\cos (45° - \theta) = 3 \cos \theta$

**Find the limit of the sequence or state that the limit does not exist.**

**9.** $t_n = {}_nC_3, n = 3, 4, 5, \ldots$

**10.** $t_1 = 100, t_n = \left(\frac{1}{2}t_{n-1} + 20\right)$

**11.** $t_n = \frac{18}{3^n}, n = 1, 2, 3, \ldots$

**12.** $t_n = \left(1 + \frac{0.06}{n}\right)^n, n = 1, 2, 3, \ldots$

**For Exercises 13 and 14, refer to $\triangle ABC$ at the right.**

**13.** Solve for $x$.

**14.** Find the measures of $\angle C$ and $\angle B$ to the nearest tenth of a degree.

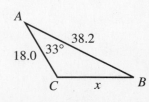

# Sections 17-1, 17-2, 17-3, and 17-4

**OVERVIEW**

**These sections show how to organize and summarize data using tables, graphs, and charts. They also show how to find and interpret averages and measures of dispersion. Section 17-4 discusses the most important distribution of data, the normal distribution.**

## KEY TERMS

**EXAMPLE/ILLUSTRATION**

**Data** (p. 639)
a list of numerical facts about some group

| | | | | |
|---|---|---|---|---|
| 21 | 17 | 28 | 21 | 20 |
| 15 | 22 | 26 | 20 | 34 |
| 26 | 24 | 25 | 22 | 22 |
| 30 | 27 | 29 | 24 | 19 |
| 25 | 22 | 23 | 30 | 19 |

**Stem-and-leaf plot** (p. 639)
a method for organizing and displaying data

| | |
|---|---|
| 1 | 5 7 9 9 |
| 2 | 0 0 1 1 2 2 2 2 3 4 4 5 5 6 6 7 8 9 |
| 3 | 0 0 4 |

3 | 0 means 30.

**Histogram** (p. 640)
a graph using horizontal or vertical bars to organize and display data

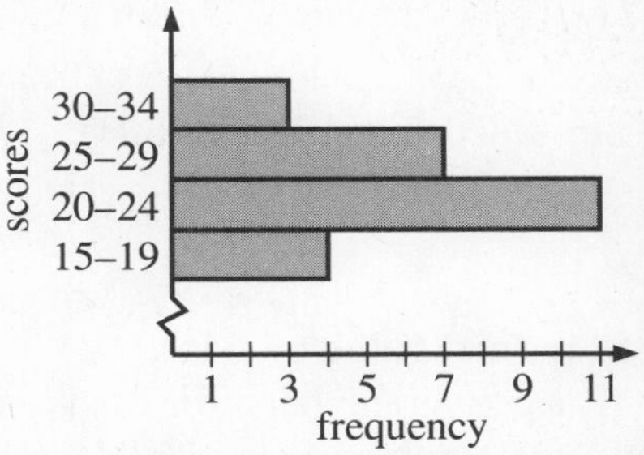


**Frequency polygon** (p. 641)
a line graph used to display data

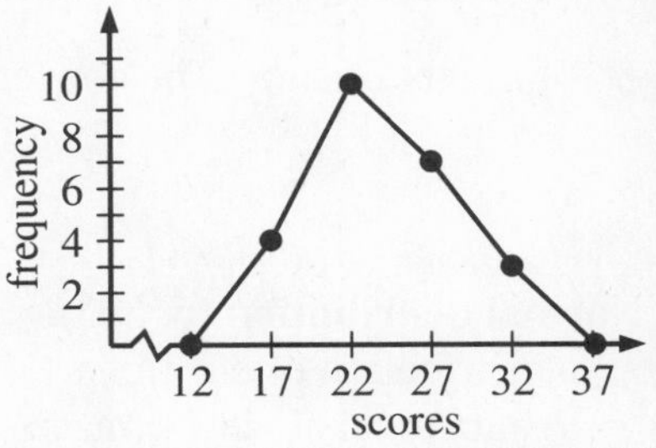


**Mean** (p. 641)
the sum of the items in a set of data divided by the number of items (written $\overline{x}$)

For the set of data above,

$$\overline{x} = \frac{21 + 17 + 28 + \ldots + 30 + 19}{25}$$

$$= \frac{591}{25} = 23.64$$

**Median** (p. 642)
the middle number of an ordered set of data with an odd number of items, or the mean of the two middle numbers if the ordered set of data has an even number of items

For the set of data above, the median is 23.

**Mode** (p. 642)

the item (or items) which occurs most often in a set of data (if no item is repeated or all the items occur the same number of times, then there is no mode)

For the set of data on page 157, the mode is 22.

---

**Box-and-whisker plot** (p. 649)

a method for displaying the median, quartiles, extremes, and any outliers of a set of data (The *lower quartile* is the median of the ordered numbers less than the median of the entire set of data; the *upper quartile* is the median of the ordered numbers greater than the median of the entire set of data; the *lower and upper extremes* are the greatest and least numbers (that are not outliers), respectively, of the set of data; an *outlier* is any data item whose distance from the nearer quartile is more than 1.5 times the interquartile range.)

lower extreme: 15
lower quartile: 20.5
median: 23
upper quartile: 26.5
upper extreme: 34
interquartile range: 6
outliers: none

---

**Variance** (p. 654)

the mean of the squares of the deviations from the mean of a set of data for each data item $\left(\text{written } s^2 = \dfrac{\sum\limits_{i=1}^{n} (x_i - \overline{x})^2}{n}\right)$

For the set of data on page 157,
$$s^2 = \frac{(21 - 23.64) + \ldots + (19 - 23.64)}{25}$$
$$= \frac{475.76}{25} = 19.0304$$

---

**Standard deviation** (p. 654)

the square root of the variance of a set of data (written $s$)

For the set of data on page 157,
$s = \sqrt{19.0304} \approx 4.36$

---

**Standard value** (p. 656)

a converted value for each item in a set of data which represents the number of standard deviations the data item is from the mean of the set of data $\left(\text{The conversion formula is } z_i = \dfrac{(x_i - \overline{x})}{s}.\right)$

For data item 34 in the set of data on page 157,
$$z = \frac{34 - 23.64}{4.36}$$
$$= \frac{10.36}{4.36} \approx 2.38$$

---

**Normal distribution** (p. 663)

a distribution of data for which the graph is bell-shaped and symmetrical about the mean (A *standard normal distribution* is the normal distribution having a mean of 0 and a standard deviation of 1.)

See the graph at the top of text page 663. (The general graph of the standard normal distribution is shown at the bottom of text page 663.)

---

**Percentiles** (p. 665)

the 99 points in an ordered set of data that divide the set into 100 equal parts

For the set of data on page 157, the 50th percentile is 23.

---

# UNDERSTANDING THE MAIN IDEAS

## Displaying data

There are several methods for displaying a set of data:

- **stem-and-leaf plot**—The stems are the first digits of the data items, each *different* digit used only once. (*Note:* If more than one line is needed to list all the leaves for a particular stem, a dot is used in the stem position on the next line rather than repeating the stem digit; see the plot on text page 644.) The leaves are the second digits of the data items. One leaf is entered in the plot for each data item, so several leaves may be the same digit. The leaves for each stem are arranged in order. The plot should include a key which defines the entries; the key informs a reader whether $3 \mid 2$ means 32 or whether it means 320.)

- **histogram**—Each bar on a histogram represents a group of values rather than a single value; this group is called a *class*, and all the classes for a set of data must be equal in size (see text page 640).

- **frequency polygon/cumulative frequency polygon**—These displays are line graphs constructed from *frequency tables/relative frequency tables* (see text page 640). The vertical axis on both types of display can be labeled with either integers or percents. The horizontal axis on both types is labeled with the midpoint of each class from the respective table.

- **box-and-whisker plot**—Begin by finding the median of the (ordered) set of data. Then find the median of the ordered data items below (less than) the median of the entire set of data; this number is the *lower quartile*. Find the median for the ordered data items above (greater than) the median of the entire set of data; this number is the *upper quartile*. Now identify the least number and greatest number in the set of data. These are called the *lower extreme* and *upper extreme*, respectively, unless either number qualifies as an *outlier* (see the discussion at the top of text page 650). After drawing a number line extending from the least number in the data set to the greatest number, construct the box from the lower to upper quartile (including a vertical line through the median) and the whiskers from the extremes to the quartiles (see the construction shown on text page 649). *Note:* The effect of any outliers on the construction of whiskers is shown at the top of text page 650.

## Finding the variance ($s^2$) and standard deviation ($s$) of a set of data

- There are three formulas to choose from for finding variance:

  (1) $s^2 = \dfrac{\sum\limits_{i=1}^{n}(x_i - \bar{x})^2}{n}$ : For each data item ($x_i$), find its distance from the mean ($x_1 - \bar{x}$) and square it. Find the sum of all the squares, and then divide this sum by the number of terms, $n$.

  (2) $s^2 = \dfrac{\sum\limits_{i=1}^{n} x_i^2}{n} - \bar{x}^2$: Square each data item ($x_i^2$), add the squares, and divide this sum by the number of terms, $n$. Then subtract the square of the mean of the set of data.

---

(3) (For data given in a frequency table)

$$s^2 = \dfrac{\sum\limits_{i=1}^{r} x_i^2 \cdot f_i}{n} - \bar{x}^2\text{:}$$ (See Example 2 on text page 656.)

- To find the standard deviation, take the square root of the variance.

 **Caution:** (For calculator users) There are two kinds of variance (and standard deviation), *sample* and *population*. The one you are learning (population) has *n* in the denominator of its formula. The other (sample) has *n* − 1. Find out which one your calculator gives!

### Normal distribution

The graph of a set of data which is normally distributed has the following characteristics:

- The graph is symmetrical about the mean of the set of data.
- The *x*-axis is a horizontal asymptote of the graph.
- About 68% of the data is within 1 standard deviation of the mean, about 95% is within 2 standard deviations, and about 99% is within 3 standard deviations. (See the graph at the top of text page 663.)

The *standard normal distribution* has mean 0 and standard deviation 1. Any normal distribution can be converted to a standard normal distribution by finding the *z*-score, or standard value, for each data item using the formula $z = \dfrac{x - \bar{x}}{s}$.

## CHECKING THE MAIN IDEAS

1. Find the mean, median, and mode of the set of data below.

   3   6   6   9   9   9   12   12   12   12

2. Find the variance and standard deviation of the set of data in Exercise 1.

3. The box-and-whisker plots at the right show students' test scores in two classes. Notice that both sets of data have the same median. Which set has the higher mean? the higher standard deviation?

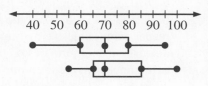

4. Find the standard values for the upper and lower quartiles of a standard normal distribution.

## USING THE MAIN IDEAS

**Example 1** Find the mean, median, and mode of the data shown in the histogram at the right.

**Solution** We see that 14 + 22 + 10 + 5 + 4 = 55 patients were studied. The histogram has grouped the data into classes so we do not know exactly how long each person waited. To estimate the mean, use 5, 15, 25, 35, and 45 minutes for the waiting times in each class, because these are the middle times of the classes.

$$\text{Mean} = \dfrac{\sum\limits_{i=1}^{r} x_i \cdot f_i}{n} = \dfrac{5 \cdot 14 + 15 \cdot 22 + 25 \cdot 10 + 35 \cdot 5 + 45 \cdot 4}{55} \approx 18.3$$

The median of 55 items is the 28th item (because there are 27 items above and 27 items below it). The 28th item is in the class 10–19, so the median is 15 minutes. Since the bar for the class 10–19 is the tallest, 15 minutes is the mode.

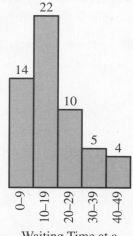

Waiting Time at a Doctor's Office (in min)

**Example 2** Find the standard deviation of the waiting times in Example 1.

**Solution** The waiting times are grouped into classes, so use the formula

$$s^2 = \frac{\sum\limits_{i=1}^{r} x_i^2 \cdot f_i}{n} - \bar{x}^2$$ to find the variance. (*Note:* $r = 5$ because

there are 5 groups; the mean, $\bar{x}$, is approximately 18.3, and the total
number of patients, $n$, is 55.)
The $x_i$'s are the middle times of the classes: 5, 15, 25, 35, and 45.
The $f_i$'s are the frequencies of the classes: 14, 22, 10, 5, and 4,
respectively.

$$s^2 = \frac{5^2 \cdot 14 + 15^2 \cdot 22 + 25^2 \cdot 10 + 35^2 \cdot 5 + 45^2 \cdot 4}{55} - 18.3^2 \approx 133.7$$

Therefore, the standard deviation, $s$, is $\sqrt{133.7} \approx 11.6$ minutes.

**Example 3** What percent of data in a standard normal distribution lies between
the standard values $z = -1.3$ and $z = 2.1$?

**Solution** The table on page 664 gives the proportion of data *to the left* of a
particular $z$-score. We will need to subtract $P(-1.3)$ from $P(2.1)$.

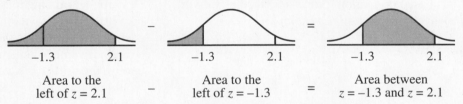

| Area to the left of $z = 2.1$ | $-$ | Area to the left of $z = -1.3$ | $=$ | Area between $z = -1.3$ and $z = 2.1$ |

From the table, $P(-1.3) = 0.0968$ and $P(2.1) = 0.9821$
Thus, $P(2.1) - P(-1.3) = 0.9821 - 0.0968 = 0.8853$
This means that 88.53% of the data lies between the standard values
$-1.3$ and 2.1.

## Exercises

5. A candy company makes bags of mints. The number of mints per bag
   is normally distributed with a mean of 65 and a standard deviation of
   1.8 mints. What percent of bags have fewer than 60 mints?

**In Exercises 6–7, decide which average (mean, median, or mode)
gives the best information about the data. Give your reasons.**

6. You ask 100 schoolmates for the call number (i.e. 98.9 FM) of their
   favorite radio station.

7. You collect data on the salaries of 100 workers: 90 factory workers,
   8 managers, the vice-president, and the president.

8. At Lotsa Pasta Restaurant, customers wait an average of 15 minutes
   for a table at lunch. Waiting times are normally distributed with $s =$
   4 minutes. What percent of customers are seated within 10 minutes?

9. Sketch a cumulative frequency polygon of the data in Example 1.

# Inferential Statistics

## Sections 17-5 and 17-6

| OVERVIEW | We apply inferential statistics when we use data about the past to predict the future and when we use data from a sample to tell us about an entire population. These sections discuss how to reduce error by employing good sampling techniques and how to estimate error through the use of confidence intervals. |
|---|---|

## KEY TERMS

### EXAMPLE/ILLUSTRATION

| | |
|---|---|
| **Population** (p. 669)<br>the whole set of persons or things of interest; the universal set | If we want to learn about the entire bacteria population in a lake, we might take several samples of water from different parts of the lake. |
| **Sample** (p. 669)<br>a subset of a population which is selected for study in order to gain information about the whole population | |
| **Probability sampling** (p. 670)<br>sampling methods based on random processes, such as *simple random sampling* and *stratified random sampling* | Drawing a grid on a map of the lake discussed above, numbering each grid section, and then using a random number generator to choose the grids for water testing would produce random water samples. |
| **Nonprobability sampling** (p. 670)<br>sampling methods which may lead to bias, such as *convenience sampling*, *judgment sampling*, and *sampling by questionnaire* | The results of a 900-number telephone call-in poll done by a late-night tabloid television show are *not* representative of an entire population since only a certain segment of the population can be expected to call. |

## UNDERSTANDING THE MAIN IDEAS

### Probability sampling

- In *simple random sampling*, a method of selection is devised so that each member of the population is equally likely to be selected. Examples are a random number generator (your calculator may have one), a lottery, a coin flip, and so on.
- In *stratified random sampling*, the entire population is divided into nonoverlapping subgroups (such as grade levels) called strata. Each stratum is randomly sampled, and the results are then weighted according to the proportional size of each subgroup.

### Nonprobability sampling

Guard against these sampling methods which lead to biased samples:

- convenience sampling—sampling only the most-accessible members of a population
- judgment sampling—relying on the subjective judgment of experts to select the sample
- sampling by questionnaire—sampling which reflects only the opinions of those who choose to, and are able to, respond

### Confidence intervals

Suppose a survey or poll shows that a certain proportion, denoted $\overline{p}$, of a sample population has a given characteristic.

- The value of $\overline{p}$ can be used to estimate the standard deviation:

$$s = \sqrt{\frac{\overline{p}(1-\overline{p})}{n}}$$

- A 95% confidence interval is $\overline{p} - 2s < p < \overline{p} + 2s$. (See Example 1 on text page 676.)

      ↑ sample proportion   ↑ population proportion   ↑ standard deviation

- A 99% confidence interval is $\overline{p} - 3s < p < \overline{p} + 3s$. (See Example 3 on text page 677.)

## CHECKING THE MAIN IDEAS

**In Exercises 1–2, state any errors you think might occur in the sampling situations.**

1. A phone poll on the last night of the 1992 Republican Convention showed a large boost in support for the Republican nominee, President George Bush. Another phone poll the next morning showed no boost.

2. A teacher who wants to see who is doing daily homework collects homework every other Friday.

3. Of the residents of a certain town, 30% are from lower income, 55% are from middle income, and 15% are from upper income families. A random sample of fifty members of each group is asked if they favor a property tax increase to build a new school. Thirty-three lower income, twenty-seven middle income, and eighteen upper income residents were in favor of the tax increase. Estimate the percent of the population in favor of the tax increase.

4. A random inspection of 100 bolts from a shipment of 10,000 showed that 8 were defective. Find a 95% confidence interval for the proportion $p$ of the total shipment which is defective.

## USING THE MAIN IDEAS

**Example**  You are the mayor of a town with a "911" service for emergency calls to police, fire, and ambulance. Unfortunately, many of the 911 calls are not true emergencies – they are pranks and routine calls. You feel that if fewer than 50% of the calls are emergencies, it will be worthwhile to launch a public education campaign to teach citizens about the proper use of 911. You order a study of calls. A random sample of 180 calls reveals that only 76 *were* emergencies. Should you institute the education campaign? (Use a 95% level of confidence.)

**Solution**  The sample had $\overline{p} = \dfrac{76}{180} \approx 0.4222$ as its proportion   **Caution:** The bar on $\overline{p}$ means that this proportion is from a sample (the population proportion is $p$). Do not confuse it with $\overline{x}$ or $\overline{z}$, which are means.

of true emergencies. This proportion is below 50%, but can you be 95% sure that fewer than 50% of *all* calls are emergencies? First, calculate the standard deviation using the sample proportion:

$$s = \sqrt{\frac{\overline{p}(1-\overline{p})}{n}}$$

$$= \sqrt{\frac{(0.4222)(0.5778)}{180}} \approx 0.0368$$

For 95% confidence, the margin of error is $2s = 2(0.0368) = 0.0736$. Therefore, the 95% confidence interval is
$0.4222 - 0.0736 < p < 0.4222 + 0.0736$, or
$0.3486 < p < 0.4958$.

This means that between 34.9% and 49.6% of all calls are emergencies and that the public education campaign would be worthwhile.

## Exercises

5. *Critical Thinking*  In the Example, suppose your town council insists on 99% confidence before spending money on the campaign. Does your data still prove the idea that fewer than 50% of all calls are emergencies? Does the data disprove it? What will you do?

6. *Writing*  Again refer to the situation described in the Example. Explain how you would randomly sample the 911 calls. Can you think of any problems obtaining a fair sample?

7. In a poll of 45 shoppers, 36 state that they are influenced by price in their choice of laundry detergent. Give a 95% confidence interval for the proportion of all shoppers who would say they are influenced by price.

8. Which of these changes in the design of a poll would reduce the margin of error by half?
   A. doubling the sample size        B. halving the sample size
   C. quadrupling the sample size     D. doubling the confidence interval

**In Exercises 9 and 10, you want to determine how much television your schoolmates watch by polling a sample.**

9. How could you do this using convenience sampling?

10. How could you do this using stratified random sampling?

## Chapter 17: Statistics

Complete these exercises before trying the Practice Test for Chapter 17. If you have difficulty with a particular problem, review the indicated section.

*Chapter 17*

For Exercises 1–5, refer to the following set of data listing the circumference (in meters) of trees in a section of forest.

| | | | | |
|---|---|---|---|---|
| 3.3 | 1.2 | 4.8 | 2.5 | 0.6 |
| 3.7 | 0.3 | 1.0 | 3.1 | 0.4 |
| 3.9 | 4.9 | 2.7 | 2.0 | 3.2 |
| 4.0 | 4.3 | 4.6 | 4.3 | 2.6 |

1. Make a stem-and-leaf plot for the data. *(Section 17-1)*

2. Find the mean and median of the set of data. *(Section 17-1)*

3. Draw a histogram for the data. *(Section 17-1)*

4. Make a box-and-whisker plot for the data. *(Section 17-2)*

5. Find the variance and standard deviation of the set of data. *(Section 17-3)*

6. A certain brand of light bulbs has a mean life expectancy of 1500 hours with a standard deviation of 200 hours. What percent of the bulbs burn longer than 2000 hours? *(Section 17-4)*

7. Give an example of a sample procedure that is (**a**) convenience sampling and (**b**) stratified random sampling. *(Section 17-5)*

8. A meat inspector tests 100 randomly-selected cuts of meat from a distributor and finds that 5 of the cuts are contaminated. Give a 99% confidence interval for the proportion of contaminated meat from this distributor. *(Section 17-6)*

The scores on a recent physics test were 91, 73, 74, 96, 84, 83, 72, 98, 73, 62, 68, 85, and 77. Use this data for Exercises 1–6.

*Chapter 17*

1. Make a stem-and-leaf plot for the data.

2. Make a box-and-whisker plot for the data.

3. Find the mean.                    4. Find the median.

5. Find the mode.                    6. Find the range.

7. Explain the difference between measures of central tendency and measures of variability.

8. Explain how standard values are used.

At the AJX company, 11 people earn $20,000, 3 people earn $40,000, and one person earns $200,000.

9. Make a frequency table.           10. Find the mean salary.

11. Find the variance.               12. Find the standard deviation.

On a recent calculus test, the results were normally distributed with $\bar{x} = 70$ and $s = 10$. Find each of the following.

13. the percent of scores greater than 90

14. the percent of scores between 80 and 90

15. the score which has 8% of the scores above it

If $\bar{p} = 0.3$ and $n = 100$, find each of the following.

16. the standard deviation of $\bar{p}$

17. a 95% confidence interval for $p$

18. a 99% confidence interval for $p$

19. A company employs 46 women and 28 men. In a survey of employees, 12 women and 8 men are randomly selected, with 9 of the women and 3 of the men favoring a proposed change in benefits. Estimate the percent of all employees favoring the change.

---

**MIXED REVIEW**

*Chapters 1–17*

1. Write an equation of the line through $(-1, 10)$ and $(17, 4)$.

2. What is the $y$-coordinate of the point on the line with $x$-coordinate 3 for the line described in Exercise 1?

3. Find an equation of the line through the origin and perpendicular to the line described in Exercise 1.

**Sketch the graph of each function.**

4. $y = -\frac{1}{4}x^4$

5. $y = 20x^{0.5}$

6. $y = 4(2.5)^x$

7. $y = 3\left(\frac{1}{2}\right)^x$

**Express $y$ in terms of $x$.**

8. $\log y = 3.4x + 1.3$

9. $\log y = \log 4 + 7 \log x$

**Find an expression for $\log y$.**

10. $y = 4(2.4)^x$

11. $y = \frac{2}{3}x^{-4}$

12. Using the graph at the right below, complete this definition of $f(x)$:

$$f(x) = \begin{cases} -x\, , & 0 \le x < 2 \\ \underline{\phantom{?}}\, , & 2 \le x < 4 \\ \underline{\phantom{?}}\, , & \underline{\phantom{?}} \\ \underline{\phantom{?}}\, , & \underline{\phantom{?}} \\ \underline{\phantom{?}}\, , & \underline{\phantom{?}} \end{cases}$$

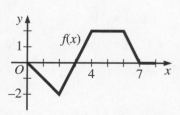

**For Exercises 13–15, refer to the graph in Exercise 12.**

13. Sketch the graph of $y = |f(x)|$.

14. Sketch the graph of $y = f(2x)$.

15. Sketch the graph of $y = f(x - 3) - 5$.

---

# Curve Fitting and Models

| **OVERVIEW** | This chapter is concerned with the analysis of paired data – observations collected together, such as rainfall and crop yield or age of wife and age of husband. The chapter discusses how to determine what (if any) mathematical function best describes the relationship between two variables, and how to use this model to make predictions. |
|---|---|

## KEY TERMS

**EXAMPLE/ILLUSTRATION**

**Scatter plot** (p. 683)
a graph of ordered pairs of data on a coordinate plane

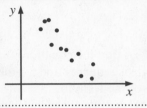

**Least-squares line** (p. 684)
the line which best fits the data points graphed in a scatter plot

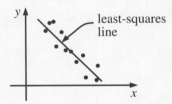

least-squares line

**Correlation coefficient *r*** (p. 685)
a number $r$, such that $-1 \leq r \leq 1$, which measures how closely the data points of a scatter plot fit the least-squares line

There is a strong negative correlation for the least-squares line in the graph above.

## UNDERSTANDING THE MAIN IDEAS

### Linear models $y = mx + b$

- Two variables are related linearly when a change in one variable is directly proportional to a change in the other. The points on the scatter plot above seem to cluster around a line.

- The best linear model is the least-squares line. For a set of data $\{(x_i, y_i)\}$ this is the line through $(\overline{x}, \overline{y})$ with slope $\dfrac{\overline{xy} - \overline{x} \cdot \overline{y}}{s_x^2}$.

- The correlation coefficient $r$ is $\dfrac{\overline{xy} - \overline{x} \cdot \overline{y}}{s_x s_y}$. It lies between $-1$ and $1$, inclusive.

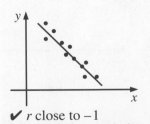

✔ $r$ close to $-1$
✔ Negative correlation
✔ Least-squares line has negative slope.

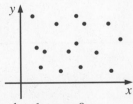

✔ $r$ close to $0$
✔ Very weak correlation
✔ No good linear model

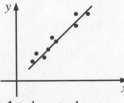

✔ $r$ close to $1$
✔ Positive correlation
✔ Least-squares line has positive slope.

## Other models

There are many other kinds of mathematical models, besides linear models. Two important nonlinear types are:

- Exponential curves of the form $y = ab^x$, where $b > 0, b \neq 1$

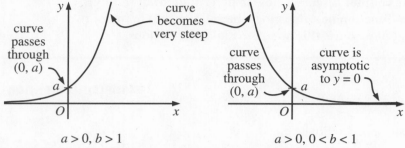

$$a > 0, b > 1 \qquad\qquad a > 0, 0 < b < 1$$

To find the equation for the best exponential model, find the least-squares line for the points $(x, \log y)$. This gives an equation of the form $\log y = mx + b$, which can be converted to $y = 10^b(10^m)^x$, or $y = ab^x$ $(a = 10^b$ and $b = 10^m)$.

- Power curves of the form $y = ax^m$

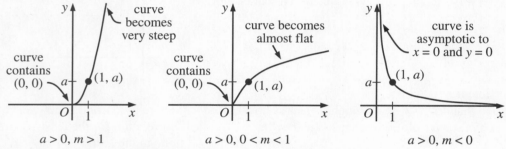

$$a > 0, m > 1 \qquad\qquad a > 0, 0 < m < 1 \qquad\qquad a > 0, m < 0$$

To find the equation for the best power curve model, find the least-squares line for the points $(\log x, \log y)$. This gives an equation of the form $\log y = b + m \log x$, which can be converted to $y = 10^b x^m$, or $y = ax^m$ $(a = 10^b)$.

## Choosing the best model

Learn to use a scientific calculator and practice on the text examples. Your work will be more accurate and less tedious.

Begin by plotting the data on a scatter plot. The arrangement of dots may suggest a model to you. Sometimes the points look like a translation of a basic model. For example, the scatter plot at the right looks like an exponential model shifted upward. In this case, work with $(x, \log (y - c))$ instead of $(x, \log y)$.

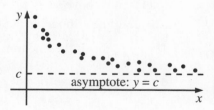

## CHECKING THE MAIN IDEAS

1. Which of the following statements are false?

   A. The correlation coefficient is the slope of the least-squares line.

   B. A negative correlation means that there is no relationship between the two variables.

   C. The least-squares line passes through more of the data points than any other line.

**2.** Find the equation of the least-squares line and the correlation coefficient for the data in the table below.

| year | 1988 | 1989 | 1990 | 1991 |
|------|------|------|------|------|
| library patrons (thousands) | 410 | 421 | 429 | 434 |

**3.** Using the results of Exercise 2, how many library patrons can be expected in 1993?

**4.** Write the equation $\log y = 42 + 6.1x$ as an exponential function.

**5.** Find the equation of the power curve that best fits the data in this table.

| age in months | 10 | 12 | 14 | 18 | 24 |
|---------------|----|----|----|----|----|
| number of words in vocabulary | 2 | 10 | 20 | 100 | 300 |

**6.** *Critical Thinking* What does the above model predict for vocabulary size at age 16 (192 months)? Is this reasonable?

## USING THE MAIN IDEAS

**Example 1** In each case, state whether you would expect a positive correlation, a negative correlation, or no correlation. Explain your reasoning.
  **a.** number of months studying French and score on a French proficiency exam
  **b.** oven temperature and roasting time for a turkey
  **c.** coffee consumption and incidence of lung cancer
  **d.** numbers on a red die and numbers on a white die when they are rolled together five times

**Solution**  **a.** We expect that the longer one studies, the better one will score on the test. Even though there are individual exceptions, in general *both variables rise* (or fall) *together*. This is positive correlation.

  **b.** As the oven temperature *rises*, the roasting time *falls*. Conversely, as the temperature *falls*, the time *rises*. This is a negative correlation.

  **c.** In the above example, oven temperature determined (caused) the roasting time. Often though, two variables are correlated only because they are both related to a *third* variable. Cigarette smokers drink more coffee *and* have a higher incidence of lung cancer than the general population, so these two variables are positively correlated. This does not mean that coffee causes lung cancer, or that lung cancer causes coffee drinking. (*Note:* Do not confuse "correlation" with "causation.")

  **d.** We expect a correlation near zero, because the rolling of two dice are independent events. However, it is possible for the results of one set of 5 rolls to be pairs of numbers. The data points would then lie on the line $y = x$ and the correlation would be 1. If the 5-roll experiment is conducted many times, the expected correlation near zero will be exhibited.

**Example 2** The table below shows the burn rate for candles of four different diameters. The candles are made of the same material.

| candle diameter (in cm) | 1 | 2 | 5 | 8 |
|---|---|---|---|---|
| length burned in 1 hr (in cm) | 96 | 26 | 4.5 | 1.5 |

Find a model relating the diameter, $x$, to the length burned, $y$.

**Solution** The scatter plot at the right suggests an asymptote at $x = 0$. This is confirmed by common sense, for if we imagine the diameter of the candle becoming larger and larger, the length burned would get smaller and smaller, but never less than zero.

The asymptote may indicate an exponential curve model of the form $y = ab^x$ with $0 < b < 1$ or a power curve model of the form $y = ax^m$ with $m < 0$. To test these models, plot both $\{(x, \log y)\}$ and $\{(\log x, \log y)\}$.

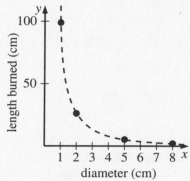

Exponential curve: $(x, \log y)$     Power curve: $(\log x, \log y)$

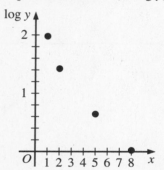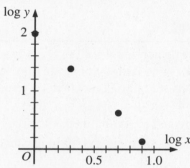

Both graphs look promising, but the second looks more linear than the first. To decide between the models, think about the burning candles. The diameter is twice the radius. The amount that burns (the volume) is length $\times \pi \times$ (radius)$^2$. So the length is

$$\frac{V}{\pi\left(\frac{1}{2}d\right)^2} = \frac{V}{\frac{\pi d^2}{4}} = \frac{4V}{\pi} \cdot d^{-2}.$$

This is a power curve of the form $y = ax^m$, $m < 0$.

To find the equation of the best power curve, determine the least-squares line that relates $\log y$ and $\log x$.

| $x$ | $y$ | $\log x$ | $\log y$ | $(\log x)(\log y)$ | $(\log x)^2$ |
|---|---|---|---|---|---|
| 1 | 98 | 0 | 1.9912 | 0 | 0 |
| 2 | 26 | 0.3010 | 1.4150 | 0.4259 | 0.0906 |
| 5 | 4.5 | 0.6990 | 0.6532 | 0.4566 | 0.4886 |
| 8 | 1.5 | 0.9031 | 0.1761 | 0.1590 | 0.8156 |
| sums | | 1.9031 | 4.2355 | 1.0415 | 1.3948 |
| means | | 0.4758 | 1.0589 | 0.2604 | 0.3487 |

To find the slope of the least-squares line, first find the variance. Recall from text page 655 that the variance of a set of numbers is the mean of the squares minus the square of the means.

---

$$s_{\log x}{}^2 = 0.3487 - (0.4758)^2 \approx 0.1223$$

↑      ↑      ↑

variance    mean of    square
of $\log x$   squares of   of mean
       $\log x$     of $\log x$

Now use the variance to find the slope of the least-squares line.

mean of products      product of means
   $(\log x)(\log y)$       of $\log x$ and $\log y$

       ↓          ↙

$$m = \frac{(0.2604) - \overset{\frown}{(0.4758)(1.0589)}}{0.1223} \approx -1.990 \text{ (See text page 684.)}$$

              ↑

       variance of $\log x$

The equation of the line has the form $\log y - \overline{\log y} = m(\log x - \overline{\log x})$.

$$\log y - 1.0589 = -1.990(\log x - 0.4758)$$

       ↑        ↑         ↑

    mean of    slope      mean of
     $\log y$             $\log x$

or $\log y = 2.006 - 1.990 \log x$.

Therefore, the power curve model is $y = 10^{2.006} x^{-1.990}$, or $y = 101.4 x^{-1.990}$, where $y$ is the length burned and $x$ is the diameter of the candle.

✔ *Check:* Substitute $x = 1, 2, 5,$ or $8$ into $y = 101.4 x^{-1.990}$. Check that $y$ is near the given value.

*Note:* If you use a calculator which finds the power curve, your answer will be somewhat different, because of rounding and differences in the method of calculating $s_x$.

## Exercises

7. Find the best exponential curve model for the data in Example 2. Plot the power curve and the exponential curve on the same axes. How do they differ?

8. **a.** Plot these data points and find the correlation coefficient $r$.

| $x$ | 1 | 3 | 4 | 6 | 10 |
|-----|---|---|---|---|----|
| $y$ | 1 | 4 | 3 | 3 | 1 |

   **b.** By eliminating a data point you can make the correlation for the remaining points much better. Eliminate the point $(1, 1)$ from the data above. Find the least-squares line for the remaining four points.

9. Find a model for the following data on salt concentration in a reservoir, measured in pounds per million gallons.

| day | 50 | 100 | 200 | 300 | 400 | 500 |
|-----|----|-----|-----|-----|-----|-----|
| concentration (pounds per million gallons) | 16 | 13.5 | 11.5 | 10.5 | 10.2 | 10.1 |

**QUICK CHECK**

*Chapter 18*

Complete these exercises before trying the Practice Test for Chapter 18. If you have difficulty with a particular problem, review the indicated section.

1. If a set of data points lie very near the line $y = 10 - 0.25x$, the correlation coefficient $r$ is closest to which of these values? *(Section 18-1)*

   **A.** 1        **B.** $-1$        **C.** $-0.25$        **D.** 0

2. Copy and complete the table below. Then find the equation of the least-squares line and the correlation coefficient for the data. *(Section 18-1)*

   | $x$ | $y$ | $xy$ | $x^2$ | $y^2$ |
   |-----|-----|------|-------|-------|
   | $-1$ | 3 | | | |
   | 3 | $-2$ | | | |
   | 4 | $-3$ | | | |
   | 5 | $-6$ | | | |

   Sums:

3. Which of these situations is *least* likely to have an exponential model? *(Section 18-2)*

   **A.** cost of painting houses of various sizes

   **B.** deaths in an animal population due to bacterial illness

   **C.** United States energy consumption over time

   **D.** value of a deposit in a savings account over time

4. Express $y$ in terms of $x$. *(Section 18-3)*

   **a.** $\log y = 0.26 - 3.8 \log x$        **b.** $\log y = 0.26 - 3.8x$

5. You take an ice cube from the freezer, which is set at 20°F, and set it in a bowl in your kitchen, where the room temperature is 70°F. You measure the temperature of the ice/water every minute. Sketch a graph of a model that is likely to fit the results of your observations. *(Section 18-4)*

**PRACTICE TEST**

*Chapter 18*

For Exercises 1–5, use the data (1, 141), (2, 130), (2, 115), and (3, 100).

1. Make a scatter plot for the data.

2. Draw a line on the scatter plot to fit the data, using just your eye.

3. Find an equation of the least-squares line.

4. Draw the least-squares line from Exercise 3 on the scatter plot.

5. Use the least-squares line to predict $y$ when $x = 5$.

**Express $y$ in terms of $x$.**

6. $\log y = -1.5x + 2.5$        7. $\ln y = 1.2x + 4.7$

8. $\log y = 3 \log x + 2.1$        9. $\ln y = 0.2 \ln x - 2$

**Fit an exponential curve to the data in the table.**

10.

| $x$ | 1 | 3 | 5 | 6 |
|-----|-----|-----|-----|-----|
| $y$ | 3.60 | 5.18 | 7.46 | 8.96 |

11.

| $x$ | 2 | 5 | 7 | 9 |
|-----|-----|-----|-----|-----|
| $y$ | 20.83 | 12.06 | 8.37 | 5.81 |

**12.** Using the curve from Exercise 10, predict $y$ when $x = 10$.

**13.** Using the curve from Exercise 11, predict $y$ when $x = 12$.

**For Exercises 14–18, use the data (0, 115.00), (1, 48.33), (2, 26.11), (6, 15.14), (7, 15.05), and (8, 15.02).**

**14.** Make a scatter plot for the data.

**15.** What seems to be the equation of the horizontal asymptote?

**16.** Write $\log(y - 15)$ as a function of $x$.

**17.** Write $y$ as a function of $x$.

**18.** Use your answer to Exercise 16 to predict $y$ when $x$ is 1.5.

**Explain how to test a set of data for a fit by the given type of curve.**

**19.** exponential curve              **20.** power curve

---

### MIXED REVIEW
*Chapters 1–18*

**Give the domain, range, and zeros of each function.**

**1.** $f(x) = \sqrt{9 - x^2}$

**2.** $g(t) = |t + 5|$

**3.** $f(x) = \dfrac{3x}{x^2 - 25}$

**4.** $h(t) = \dfrac{1}{t^2 - 10t + 25}$

**Tell whether each graph is a function. If it is, give the domain and range.**

**5.**
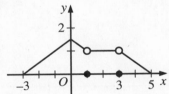

**6.**
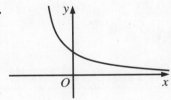

**Find the limit, if one exists, of these infinite sequences.**

**7.** $\displaystyle\lim_{n \to \infty} \frac{n^2 - 6n + 1}{3n^2 + 5n - 4}$

**8.** $\displaystyle\lim_{n \to \infty} e^{1/n}$

**9.** $\displaystyle\lim_{n \to \infty} \sqrt[n]{\pi}$

**For each infinite geometric series, find (a) the interval of convergence and (b) the sum expressed in terms of $x$.**

**10.** $1 - x^2 + x^4 - x^6 + \dots$

**11.** $1 + \dfrac{3}{x} + \dfrac{9}{x^2} + \dfrac{27}{x^3} + \dots$

**For Exercises 12–16, use the vectors $\mathbf{u} = (3, 0, -1)$ and $\mathbf{v} = (1, 1, 0)$.**

**12.** Find $\mathbf{u} \cdot \mathbf{v}$.

**13.** Find $\mathbf{u} \times \mathbf{v}$.

**14.** Find the angle between $\mathbf{u}$ and $\mathbf{v}$.

**15.** Find a unit vector perpendicular to both $\mathbf{u}$ and $\mathbf{v}$.

**16.** Find the area of a triangle having $\mathbf{u}$ and $\mathbf{v}$ as two of its sides.

**17.** Find the first five terms of the sequence defined recursively by $t_1 = 2$ and $t_n = \dfrac{1}{2}t_{n-1}^2 - 1$.

**18.** Repeat Exercise 20 using $t_1 = 4$.

---

# Limits

## Sections 19-1 and 19-2

| OVERVIEW | These sections cover limits of functions, continuity, and rational functions. Emphasis is placed on understanding, interpreting, and creating graphs. |
|---|---|

## KEY TERMS

**Limit** (pp. 717–718)

the value that a function approaches as the values of the variable approach a specific value or as they approach infinity (" $\lim_{x \to c} f(x) = L$ " means that as $x$ approaches $c$, $f(x)$ approaches $L$. " $\lim_{x \to \infty} f(x) = \infty$ " means that as $x$ gets larger (without bound), $f(x)$ becomes larger (without bound). Similar definitions are made with $-\infty$.)

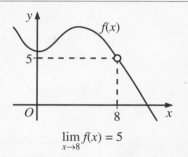

$$\lim_{x \to 8} f(x) = 5$$

**Continuous function** (p. 720)

a function $f(x)$ such that for all $c$ in the domain of $f$, as $x$ approaches $c$, $f(x)$ approaches $f(c)$ (The graph of a continuous function can be drawn without lifting the pencil.)

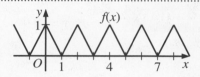

**Rational function** (p. 726)

a function of the form $y = \dfrac{p(x)}{q(x)}$, where both $p(x)$ and $q(x)$ are polynomials and $q(x) \neq 0$

$$y = \frac{4x^3 - 8x + 7}{3x^4 + 10}$$

## UNDERSTANDING THE MAIN IDEAS

### Finding limits

- To find limits like $\lim_{x \to -\infty} \log(x^2)$, ask what happens to $\log(x^2)$ as $x$ becomes very large (in size) and negative—as this happens, $x^2$ becomes very large and positive, and so the logarithm of $x^2$ becomes arbitrarily large. Therefore, $\lim_{x \to -\infty} \log(x^2) = \infty$.

- To find limits like $\lim_{x \to 2} f(x)$, look at what happens to $f(x)$ *near* 2, not *at* 2. In the figure at the right, there is no value for $f(x)$ at $x = 2$. However, a function can have a limit at a point even if the function is undefined there. In this case, there is no limit as $x$ approaches 2 because the left-hand and right-hand limits are not the same. (*Note:* The right-hand and left-hand limits do exist however, and are 3 and 0, respectively, as shown in the figure.)

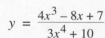
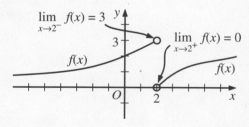

- To find the limit of a quotient $\dfrac{f(x)}{g(x)}$, look at $\dfrac{\text{limit of } f(x)}{\text{limit of } g(x)}$. If this is a number, that number is the limit. If this is $\dfrac{0}{0}$, $\dfrac{\infty}{\infty}$, or $\dfrac{\text{non-zero number}}{0}$, try to simplify $\dfrac{f(x)}{g(x)}$ using algebra. (See the list on text page 721.)

### Graphing rational functions $y = \dfrac{p(x)}{q(x)}$

- Factor the numerator and denominator and reduce to lowest terms.
- Plot zeros where the numerator is zero.
- Draw vertical asymptotes where the denominator is zero.
- Use a sign analysis to determine the sign of the function between the zeros and asymptotes.
- Set $x = 0$ to plot the $y$-intercept.
- Determine the behavior of the function as $x \to \pm\infty$ and as $f(x)$ approaches any vertical or horizontal asymptotes.
- Sketch the graph of the function.

## CHECKING THE MAIN IDEAS

1. Refer to the graph at the right. Write four limit statements, such as $\lim\limits_{x \to 4} f(x) = 0$.

2. For what values of $x$ is the function $f(x)$ above continuous?

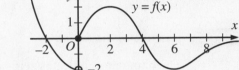

3. Find $\lim\limits_{x \to -\infty} \dfrac{\sqrt{x^4 + x^2}}{x + 7}$.

4. Find $\lim\limits_{x \to 2} \dfrac{2 - x}{3 - \sqrt{x + 7}}$.

5. The graph at the right is that of the function $y = \dfrac{k(x + a)(x + b)}{(x + c)}$. Find $a$, $b$, $c$, and $k$.

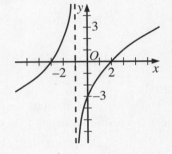

## USING THE MAIN IDEAS

**Example 1** Evaluate $\lim\limits_{x \to \infty} \sqrt{9x^2 + 12x} - 3x$.

**Solution** As $x \to \infty$, both $\sqrt{9x^2 + 12x}$ and $3x$ approach $\infty$. But what does their difference approach? "$\infty - \infty$" is meaningless.  **Caution:** Since "$\infty$" is not a number you can't subtract "$\infty - \infty$". It is not zero!

Treat $\sqrt{9x^2 + 12x} - 3x$ as the fraction $\dfrac{\sqrt{9x^2 + 12x} - 3x}{1}$

and multiply the numerator and denominator by the conjugate of

$\sqrt{9x^2 + 12x} - 3x$:

$$\frac{\sqrt{9x^2 + 12x} - 3x}{1} \cdot \frac{\sqrt{9x^2 + 12x} + 3x}{\sqrt{9x^2 + 12x} + 3x} = \frac{\left(\sqrt{9x^2 + 12x}\,\right)^2 - (3x)^2}{\sqrt{9x^2 + 12x} + 3x}$$

$$= \frac{9x^2 + 12x - 9x^2}{\sqrt{9x^2 + 12x} + 3x}$$

$$= \frac{12x}{\sqrt{9x^2 + 12x} + 3x}$$

Now take the limit as $x \to \infty$.

$$\lim_{x \to \infty} \frac{12x}{\sqrt{9x^2 + 12x} + 3x} = \lim_{x \to \infty} \frac{\frac{1}{x}(12x)}{\frac{1}{x}\left(\sqrt{9x^2 + 12x} + 3x\right)}$$

↑ Both numerator and denominator go to ∞.    ↑ Divide numerator and denominator by the highest power of $x$.

$$= \lim_{x \to \infty} \frac{12}{\sqrt{9 + \frac{12}{x}} + 3} \leftarrow \frac{1}{x}\sqrt{9x^2 + 12} = \sqrt{\frac{1}{x^2}(9x^2 + 12x)}$$

$$= \frac{12}{\sqrt{9 + 0} + 3} = \frac{12}{6} = 2$$

**Example 2** Sketch the graph of $y = \dfrac{3x^2 - 12}{x^2 + 5x + 4}$.

**Solution** First factor the function: $\dfrac{3x^2 - 12}{x^2 + 5x + 4} = \dfrac{3(x-2)(x+2)}{(x+4)(x+1)}$

The numerator indicates zeros at $x = -2$ and 2. The denominator indicates vertical asymptotes at $x = -1$ and $-4$. When $x = 0$, $y = -3$; the $y$-intercept is $-3$.

Now let $x \to \infty$ and $x \to -\infty$:

$$\lim_{x \to \infty} \frac{3x^2 - 12}{x^2 + 5x + 4} = \lim_{x \to \infty} \frac{3 - \frac{12}{x^2}}{1 + \frac{5}{x} + \frac{4}{x^2}} = \frac{3 - 0}{1 + 0 + 0} = 3$$

(The limit as $x \to -\infty$ is the same.)
This means that $y = 3$ is a horizontal asymptote.

Do a sign analysis of the function. We find that $f(x) > 0$ for $x < -4, -2 < x < -1$, and $x > 2$, and that $f(x) < 0$ for $-4 < x < -2$ and $-1 < x < 2$.

Now, put everything together. Use a calculator to find a few points $(x, y)$ that are on the graph. This serves as a check.

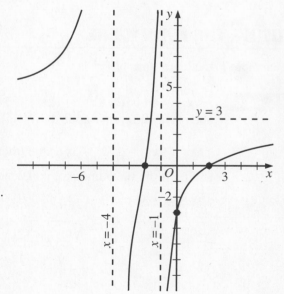

## Exercises

6. Sketch the graph of a function with $\lim_{x \to 4^-} f(x) = 0$, $\lim_{x \to 4^+} f(x) = -1$, $\lim_{x \to -\infty} f(x) = 5$, and $\lim_{x \to \infty} f(x) = -\infty$.

7. Evaluate $\lim_{x \to \pi^+} \dfrac{\sec x}{\tan x}$.

8. Evaluate $\lim_{x \to -6} \dfrac{x^2 + 12x + 36}{x^2 + 11x + 30}$.

9. Sketch the graph of $y = \dfrac{x^3 - 1}{x^2 - 1}$.

10. Sketch the graph of $y = x - \dfrac{1}{x}$.

# Series and Iterated Functions

> **OVERVIEW**
>
> In Section 19-3, areas under curves are approximated using a computer or programmable calculator. In Section 19-4, the power series of a given function is used to approximate other functions. Sections 19-5 and 19-6 introduce iterated functions and their applications.

## KEY TERMS

**EXAMPLE/ILLUSTRATION**

**Power series** (p. 733)
a function defined by an 'infinite polynomial,"
$$f(x) = a_0 + a_1 x + a_2 x^2 + a_3 x^3 + \dots$$

$$e^x = 1 + \frac{x}{1} + \frac{x^2}{2!} + \frac{x^3}{3!} + \dots + \frac{x^n}{n!} + \dots$$

**$p$-series** (p. 735)

a series of constants, $1 + \frac{1}{2^p} + \frac{1}{3^p} + \frac{1}{4^p} + \dots$
(The $p$-series with $p = 1$ is called the *harmonic series*.)

The $p$-series for $p = 2$ is $1 + \frac{1}{4} + \frac{1}{9} + \frac{1}{16} + \dots + \frac{1}{n^2} + \dots$

**Orbit of $x_0$** (p. 737)
an infinite sequence of numbers found by repeatedly applying a function $f$: $x_n = f(x_{n-1}) = f^n(x_0)$ (*Note:* $n$ indicates the number of times that $f$ has been applied to $x_0$; it does *not* indicate a power of $f$.)

Orbit of $x_0 = 12$ for iterations of the function $f(x) = \frac{1}{2}x + 4$:
$x_0 = 12$
$x_1 = 10$
$x_2 = 9$
$x_3 = 8.5$
$\vdots$

**Fixed point of a function $f$** (p. 739)
a number $x_0$ such that $f(x_0) = x_0$ (Graphically, a fixed point of $f$ occurs where the graph of $y = f(x)$ intersects the graph of $y = x$.)

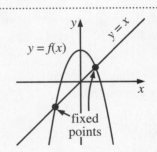

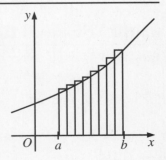

fixed points

## UNDERSTANDING THE MAIN IDEAS

### Areas under curves

The area under a curve from $x = a$ to $x = b$ can be approximated by adding the areas of rectangles with bases on the $x$-axis and tops touching the curve. Approximations are improved by using more (narrower) rectangles. Computer programs make it practical to use a great many rectangles.

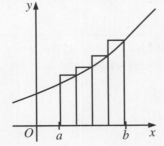

### Power series

There is a list on text page 733 of five important power series and their intervals of convergence. To find a series for $\sin \frac{1}{2}x$,

use the series for $\sin x$ $\left(x - \frac{x^3}{3!} + \frac{x^5}{5!} - \frac{x^7}{7!} + \ldots\right)$ and

substitute $\frac{1}{2}x$ for $x$. ────────────→

$$\sin \frac{1}{2}x = \frac{1}{2}x - \frac{\left(\frac{1}{2}x\right)^3}{3!} + \frac{\left(\frac{1}{2}x\right)^5}{5!} - \frac{\left(\frac{1}{2}x\right)^7}{7!} + \ldots$$

$$= \frac{1}{2}x - \frac{x^3}{2^3 3!} + \frac{x^5}{2^5 5!} - \frac{x^7}{2^7 7!} + \ldots$$

### Analyzing orbits

- A web diagram (like the one shown at the right) or programmable calculator can be used to analyze the orbit of a number $x_0$, called the *seed*.
- Fixed points can be found by solving the equation $f(x) = x$. (See Example 2 on text page 740.)

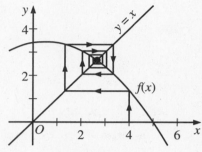

orbit of $x_0 = 4$
(approaches attracting
fixed point at $x \approx 2.65$)

### Applications of iterated functions

Iterated functions can be used to model money and population growth. The logistic function $f(x) = cx(1 - x)$ (where $x$ is the population as a fraction of capacity) models populations for which the size of the next generation is jointly proportional to the current population ($x$) and the remaining capacity ($1 - x$).

## CHECKING THE MAIN IDEAS

1. Approximate the area under the curve $y = \sqrt{x}$ from $x = 0$ to $x = 4$ using rectangles of width 0.5. (Let the first rectangle have a height of $\sqrt{0.5}$.) Use a calculator for the arithmetic but do not use a program.

2. Use a program to approximate the area in Exercise 1 using rectangles of width 0.1, 0.01, and 0.001. What appears to be the limit of these approximations?

3. Write an infinite series in expanded form for $\cos \sqrt{x}$. Give the interval of convergence.

4. Copy the graph at the right and, by analyzing orbits, decide if the fixed points 0, $\pi$, and $2\pi$ are attracting or repelling.

5. Starting with $50, every week you spend half your money and then earn $20. Use an iterated function to model your weekly balance. Does it approach a steady amount?

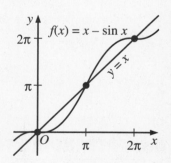

## USING THE MAIN IDEAS

**Example 1** Use the power series for $\ln (1 + x)$ to:
  a. find an infinite series whose sum is $\ln 10$.

  b. find $\lim\limits_{x \to 0} \frac{\ln (1 + x)}{x}$.

**Solution** **a.** The power series for $\ln(1 + x)$ is:

$$\ln(1 + x) = x - \frac{x^2}{2} + \frac{x^3}{3} - \frac{x^4}{4} + \dots + \frac{(-1)^{n+1}x^n}{n}, \text{ with}$$

interval of convergence $-1 < x \le 1$.

If we could set $x = 9$, then $\ln(1 + x) = \ln 10$, but 9 is *not* in the interval of convergence.

The laws of logarithms (see text page 197) will help us circumvent this difficulty. Since $\ln 10 = \ln\left(\frac{1}{10}\right)^{-1} = -\ln\left(\frac{1}{10}\right)$, the plan is to find a series for $\ln\left(\frac{1}{10}\right)$ and then change the sign.

$$\ln(0.1) = \ln(1 + (-0.9)) \leftarrow \begin{bmatrix} \text{Use } x = -0.9; \text{ it does lie in} \\ \text{the interval } -1 < x \le 1. \end{bmatrix}$$

$$= -0.9 - \frac{(-0.9)^2}{2} + \frac{(-0.9)^3}{3} - \frac{(-0.9)^4}{4} + \dots + \frac{(-1)^{n+1}(-0.9)^n}{n}$$

$$= -\frac{9}{10} - \frac{9^2}{2 \cdot 10^2} - \frac{9^3}{3 \cdot 10^3} - \frac{9^4}{4 \cdot 10^4} - \dots - \frac{9^n}{n \cdot 10^n}$$

Since $\ln 10 = -\ln 0.1$, the answer is

$$\ln 10 = \frac{9}{10} + \frac{9^2}{2 \cdot 10^2} + \frac{9^3}{3 \cdot 10^3} + \dots = \sum_{n=1}^{\infty} \frac{9^n}{n \cdot 10^n}.$$

**b.** As $x \to 0$, we know $\ln(1 + x) \to 0$, so that $\lim\limits_{x \to 0} \frac{\ln(1+x)}{x}$ takes on the indeterminate form $\frac{0}{0}$.

Clarify this by using the series

$$\ln(1 + x) = x - \frac{x^2}{2} + \frac{x^3}{3} - \frac{x^4}{4} + \dots + \frac{(-1)^{n+1}x^n}{n} + \dots.$$

Therefore, $\dfrac{\ln(1+x)}{x} = 1 - \dfrac{x}{2} + \dfrac{x^2}{3} - \dfrac{x^3}{4} + \dots + \dfrac{(-1)^{n+1}x^{n-1}}{n} + \dots$

As $x \to 0$, every term of the new series goes to zero, except for the first term, which stays at 1. Therefore $\lim\limits_{x \to 0} \dfrac{\ln(1+x)}{x} = 1$.

**Example 2** Example 2 on text page 745 uses a program to analyze the orbit of $x_0 = 0.2$ in the logistic function $f(x) = 3.1x(1 - x)$. Use a web diagram to illustrate this solution.

**Solution** The graph of logistic function $f(x) = 3.1x(1 - x) = -3.1x^2 + 3.1x$ is a parabola opening downward with zeros at $x = 0$ and $x = 1$, and with vertex at $(0.5, 0.775)$.

To find the fixed points, set $f(x) = x$.

$$-3.1x^2 + 3.1x = x$$
$$3.1x^2 - 2.1x = 0$$
$$x(3.1x - 2.1) = 0$$
$$x = 0 \text{ or } x = \frac{2.1}{3.1} \approx 0.677$$

Now plot the orbit of $x_0 = 0.2$:

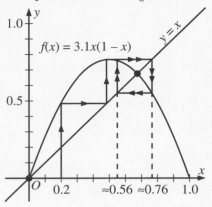

$f(x) = 3.1x(1 - x)$

$y = x$

$O$   $0.2$   $\approx 0.56$  $\approx 0.76$   $1.0$

◀▰▰▰▰ **Caution:** It is often necessary to magnify a graph to get a good picture of an orbit. Careful graphing is important!

The web settles down to traveling around a rectangle which encloses the fixed point. The $x$-values alternate between approximately 0.56 and approximately 0.76. This is the period-2 cycle described on text page 746.

## Exercises

6. ***Critical Thinking*** When you use the computer program on text page 730 to find the area under the curve $y = f(x)$ from $x = a$ to $x = b$, the area is sometimes overestimated and sometimes underestimated. Illustrate each case with a sketch. What property of the graph of $y = f(x)$ distinguishes each case?

7. Use a calculator or computer to find the sum of terms of the series for ln 10 in Example 1(a). How many terms must be added until the sum is correct in the first 2 decimal places? first 3?

8. Could the method of Example 1(b) be used to find $\lim\limits_{x \to \infty} \dfrac{\ln(1 + x)}{x}$? Explain.

9. Find the fixed points of $f(x) = x^2 + x - 2$. Describe the orbits of $x_0 = 0$, $x_0 = \dfrac{1}{2}$, and $x_0 = 2$.

10. You have a savings account paying an annual interest of 6%, into which you made an initial deposit of $10,000, and from which you withdraw $2000 each year. How long will the money in the account last?

ASK ABOUT
OUR LOW RATE
AUTO LOANS

LOAN RATES

# Chapter 19: Limits, Series, and Iterated Functions

Complete these exercises before trying the Practice Test for Chapter 19. If you have difficulty with a particular problem, review the indicated section.

1. Evaluate the given limit or state that it does not exist. *(Section 19-1)*

   a. $\lim\limits_{x \to -\infty} \dfrac{4x^5 - 8x^2 + 6}{12 - 13x^3}$

   b. $\lim\limits_{x \to 5} \dfrac{x^2 - 25}{x^2 + 4x - 45}$

2. Sketch the graph of the function $y = \dfrac{x^2 + x}{2x - 4}$. *(Section 19-2)*

3. Modify and run the program on text page 730 to approximate the area under the curve $y = 25 - x^2$ from $x = 0$ to $x = 5$. *(Section 19-3)*

4. Write an infinite series for $\text{Tan}^{-1}(1)$. How could this series be used to find an approximation of $\pi$? *(Section 19-4)*

5. Analyze all the orbits for iterations of the function $f(x) = \dfrac{1}{3}x - 4$. *(Section 19-5)*

6. Use an iterated function to solve this problem: Annual deposits of $2500 per year are placed in an account paying 5.5% interest. How long will it take to accumulate $20,000? *(Section 19-6)*

Evaluate the given limit or state that it does not exist.

1. $\lim\limits_{x \to \infty} \dfrac{2x^2 - 3x - 2}{3x^2 + x}$

2. $\lim\limits_{x \to 3} \dfrac{x^2 - 9}{x - 3}$

3. $\lim\limits_{x \to 0} \dfrac{\lfloor x \rfloor}{x}$

4. $\lim\limits_{x \to 3^+} \dfrac{2x + 5}{3 - x}$

Sketch the graph of each function.

5. $y = \dfrac{12}{x^2 - 4}$

6. $y = \dfrac{x^2 + 1}{x}$

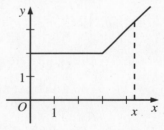

7. The area $A(x)$ of the figure at the right depends on the right boundary $x$ which can vary from 0 to 5. Sketch the graph of $A(x)$ for $0 \le x \le 5$.

Find a power series for each function.

8. $\text{Tan}^{-1}(2x)$

9. $-1 + \cos x$

10. $\sin(-x)$

11. $\cos(-x)$

**For Exercises 12 and 13, refer to your answers to Exercises 10 and 11.**

12. Explain why the sine function is an odd function.

13. Explain why the cosine function is an even function.

**In Exercises 14–16, consider the function $f(x) = 0.2x + 1$ with seed $x_0 = 2$.**

14. Find the first four terms in the orbit of $x_0$ for iterations of $f$.

15. Find any fixed points of $f$.

16. Draw a web diagram to analyze the orbit of $x_0$.

1. Find the slope of the line through $\left(-\frac{1}{2}, 2\right)$ and $(3, -1)$.

2. Write the equation of a line through $(-4, 1)$ and perpendicular to the line $3x + 5y = 1$.

3. Find the vertex of the parabola $y = \frac{1}{8}x^2 - \frac{1}{2}x - \frac{1}{2}$. Then sketch the graph of the parabola.

4. Find the points of intersection of the line $y = 2x - 2$ and the parabola $y = x^2 - 1$. Sketch the two graphs on the same set of axes.

5. Find $\lim\limits_{x \to 0} \sqrt{9 + x}$.

6. Find $\lim\limits_{x \to 0} \dfrac{\sqrt{9 + x} - 3}{x}$.

7. Use a computer or graphing calculator to find the maximum and minimum on the graph of $y = x^3 - 12x^2 + 45x - 47$ for $0 \le x \le 7$.

8. Use a computer or graphing calculator to find the approximate dimensions of the rectangle of largest area which can be inscribed under the parabola $y = 4 - x^2$ as shown in the figure at the right.

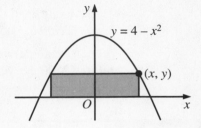

9. A cylindrical can has a volume of 2500 cm$^3$. Express the height of the can as a function of the radius. Give the domain of the function.

10. A box with a square base and no top is to be manufactured from 300 cm$^2$ of material. Express the volume of the box as a function of the length of a side of the base. Give the domain of the function.

**Sketch each pair of conic sections and determine their points of intersection.**

11. $xy = 32$
$x^2 - y^2 = 48$

12. $x^2 + y^2 - 4x + 2y = 3$
$3x - y + 1 = 0$

13. Graph $9x^2 - 16y^2 \ge 25$.

14. An object moves with constant velocity so that its position is given by $(x, y) = (0, 1) + t(-1, 6)$. Find its velocity and speed.

15. A plane is flying on a course of 290° at a speed of 510 mi/h. What are the north-south and east-west components of its velocity vector?

**Exercises 16 and 17 refer to the Venn diagram at the right, which shows the number of students who have jobs (*J*) and/or are honor students (*H*).**

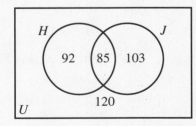

16. If a student is chosen at random, what is the probability that he or she has a job or is an honor student?

17. If a student is chosen at random, are the events "choosing a student with a job" and "choosing an honor student" independent?

18. A 26 inch (diameter) bicycle wheel revolves at 20 rpm. How fast is the bike going in ft/s?

---

# An Introduction to Calculus

| OVERVIEW | This chapter is a first look at the concepts and methods of calculus. The basic tool, the derivative, is applied to finding slopes of functions, curve sketching, solving extreme value problems, and working with velocity and acceleration. |
|---|---|

## KEY TERMS

**EXAMPLE/ILLUSTRATION**

**Slope of a curve at $P$** (p. 757)
the limit of the slopes of $\overline{PQ_i}$, as $Q_i$ approaches $P$ along the curve

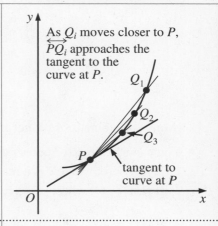

As $Q_i$ moves closer to $P$, $\overleftrightarrow{PQ_i}$ approaches the tangent to the curve at $P$.

tangent to curve at $P$

**Derivative of $f(x)$** (p. 758)
the function $y = f'(x)$ which, for each value of $x$ in the domain of $f$, gives the slope of $f(x)$ at $x$; $f'(x) = \lim_{h \to 0} \dfrac{f(x+h) - f(x)}{h}$

If $f(x) = 3x^2$, then $f'(x) = 6x$

**Local maximum (or minimum) points** (p. 765)
those points on the graph of a function which are higher (or lower) than all other nearby points on the graph

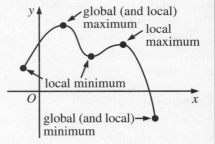

global (and local) maximum
local maximum
local minimum
global (and local) minimum

**Global maximum (or minimum) point** (p. 769)
the point which is the highest (or lowest) point on the entire graph of a function (The $y$-coordinate of this point is called the global maximum (or minimum) value of the function.)

**Instantaneous velocity** (p. 775)
the limit of the average velocity of an object as $\Delta t \to 0$:
$v(t) = \lim_{\Delta t \to 0} \dfrac{s(t + \Delta t) - s(t)}{\Delta t} = s'(t)$ (the derivative of the position function)

If $s(t) = 3t^2 + t - 1$, then $v(t) = 6t + 1$

**Instantaneous acceleration** (p. 776)
the limit of the average acceleration of an object as $\Delta t \to 0$; the derivative of the velocity function and the second derivative of the position funciton: $a(t) = v'(t) = s''(t)$

If $s(t) = 3t^2 + t - 1$, then $v(t) = 6t + 1$ and $a(t) = 6$

# UNDERSTANDING THE MAIN IDEAS

### Finding derivatives

- To find a derivative, such as the derivative of $f(x) = 3x^2$, using the limit definition, substitute into the formula and simplify algebraically:

$$\lim_{h \to 0} \frac{f(x+h) - f(x)}{h} = \lim_{h \to 0} \frac{3(x+h)^2 - 3x^2}{h}$$

$$= \lim_{h \to 0} \frac{3x^2 + 6xh + 3h^2 - 3x^2}{h}$$

$$= \lim_{h \to 0} \frac{6xh + 3h^2}{h}$$

$$= \lim_{h \to 0} \frac{h(6x + 3h)}{h}$$

$$= \lim_{h \to 0} (6x + 3h) = 6x + 0 = 6x$$

(*Note:* $h$ is going to zero, not $x$; that is why $\lim_{h \to 0} 6x = 6x$.)

- You can use these formulas rather than the limit definition:
  1. If $f(x) = c$, where $c$ is a real number, then $f'(x) = 0$.
  2. If $f(x) = x^n$, where $n$ is a nonzero real number, then $f'(x) = nx^{n-1}$.
  3. If $f(x) = cx^n$, where $c$ and $n$ are nonzero real numbers, then $f'(x) = cnx^{n-1}$.
  4. If $f(x) = p(x) + q(x)$, then $f'(x) = p'(x) + q'(x)$.

### Curve sketching

The sign of $f'(x)$ provides information about the graph of $y = f(x)$.
On intervals where $f'(x) > 0$, the graph is rising (from left to right).
On intervals where $f'(x) < 0$, the graph is falling.
At any point $c$ where $f'(c) = 0$, a local minimum or maximum occurs if $f'(x)$ changes sign at $x = c$.

### Extreme value problems

For a continuous function $f$ on an interval $a \leq x \leq b$, maximum and minimum points can occur *only* (a) where $f'(x) = 0$, (b) where $f'(x)$ is undefined, and (c) at the endpoints of the domain of $f$. To find any maximum or minimum values of the function, these are the only places you need to examine.

### Velocity and acceleration

- If the position of a moving object is given by a function such as $s(t) = t^3 - 2t^2 + 6$, the instantaneous velocity is given by the derivative of $s(t)$, $v(t) = s'(t) = 3t^2 - 4t$, and the instantaneous acceleration is given by the second derivative of $s(t)$, $a(t) = v'(t) = s''(t) = 6t - 4$.
- A positive velocity generally means motion to the right or forward or upward. Zero velocity means an instantaneous stop. With smooth motion, there will always be an instantaneous stop when the direction is reversed.

---

**184** *ADVANCED MATHEMATICS Student Resource Guide*

## CHECKING THE MAIN IDEAS

1. Find $f'(x)$ for each of these functions. Express your answers without using fractional or negative exponents.
   a. $f(x) = 12x^7 - 3x^5 + 2x - 6$
   b. $f(x) = 4\sqrt[3]{x} - 2\sqrt[4]{x} + \sqrt{7}$

2. Find the equation of the line tangent to the curve $y = \dfrac{4}{x}$ at the point $(4, 1)$.

3. Sketch a curve for which $f'(x) = 0$ at $x = -4, 1,$ and $8$; $f'(x)$ is positive for $x < -4, 1 < x < 8,$ and $x > 8$; and $f'(x)$ is negative for $-4 < x < 1$.

4. Find all extreme points (global and local) for the function $f(x) = x^3 - 6x^2 + 9x + 3$ on the interval $0 \le x \le 5$.

5. A particle travels along a number line. Its position at time $t$ is given by the function $s(t) = t^2 - 3t + 2$.
   a. Find its average velocity from $t = 0$ to $t = 4$.
   b. Find the velocity function $v(t)$.
   c. At what time does the particle reverse direction?

## USING THE MAIN IDEAS

**Example 1** A rectangular garden, $x$ feet by $y$ feet as shown in the figure at the right, will adjoin a fence and be bordered on the other three sides by a walkway 4 feet wide. The area of the garden will be 600 square feet. What dimensions of the garden will minimize the area of the walkway?

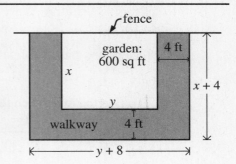

**Solution** First we need to *find a function for the quantity to be minimized*, namely, the area of the walkway.

Let $A$ = area of walkway = total area − garden area.

$A = (x + 4)(y + 8) - 600$

  ↑   ↑

width of  length of garden plus
garden plus walkways on each side
walkway

The equation $A = (x + 4)(y + 8) - 600$ has two variables!

Since the garden area $= xy = 600$, then $y = \dfrac{600}{x}$. Therefore, the walkway area is

$A(x) = (x + 4)\left(\dfrac{600}{x} + 8\right) - 600$, where $x > 0$. 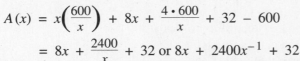 **Caution:** Always note the domain of your function.

$A(x) = x\left(\dfrac{600}{x}\right) + 8x + \dfrac{4 \cdot 600}{x} + 32 - 600$

   $= 8x + \dfrac{2400}{x} + 32$ or $8x + 2400x^{-1} + 32$

*Now* we can take the derivative: $A'(x) = 8 - 2400x^{-2}$

Set $A'(x) = 0$: $8 - 2400x^{-2} = 0$

       $x^2 = 300$

       $x = \pm10\sqrt{3}$

 *ADVANCED MATHEMATICS Student Resource Guide* **185**

Of course, $x$ is a length and cannot be negative, so eliminate $x = -10\sqrt{3}$ and check for a minimum at $x = 10\sqrt{3}$.

Do a sign analysis of $A'(x)$ to the left and right of $x = 10\sqrt{3}$.

Because the sign of $A'(x)$ changes at $x = 10\sqrt{3}$, the area of the walkway *is* a minimum when $x = 10\sqrt{3}$ and $y = \dfrac{600}{x} = \dfrac{600}{10\sqrt{3}} = 20\sqrt{3}$.

The desired dimensions of the walkway are $10\sqrt{3}$ feet by $20\sqrt{3}$ feet, or about 17 feet by 35 feet.

**Example 2** The function $h(t) = -4.9t^2 + 14t + 10$ gives the height of a cannon ball $t$ seconds after it is fired from an initial height of 10 m.
  **a.** Find the velocity function $v(t)$.
  **b.** What is the maximum height reached by the ball?
  **c.** What is the velocity of the ball when it hits the ground?

**Solution** **a.** The velocity function is the derivative of the height function:
   $v(t) = h'(t) = -9.8t + 14$
  **b.** At $t = 0$, the ball goes upward with velocity 14 m/s. Eventually, the ball reaches its maximum height and begins falling (negative velocity). *Maximum height is reached when $v(t) = 0$* (when the slope of the graph of $h(t)$ is zero).

   $v(t) = 0 \rightarrow -9.8t + 14 = 0 \rightarrow t = \dfrac{14}{9.8} = \dfrac{10}{7}$

   Therefore, the maximum height, in meters, is

   $h\left(\dfrac{10}{7}\right) = -4.9\left(\dfrac{10}{7}\right)^2 + 14\left(\dfrac{10}{7}\right) + 10 = 20$

  **c.** To find the time when the ball hits the ground, set $h(t) = 0$:
   $-4.9t^2 + 14t + 10 = 0$

   $t = \dfrac{-14 \pm \sqrt{14^2 + (4)(4.9)(10)}}{2(-4.9)} = \dfrac{-14 \pm 14\sqrt{2}}{-9.8}$

   $\approx 3.45 \text{ or } -0.59 \quad \leftarrow$ Reject $-0.59$ since the time must be positive.

   Substituting $t = 3.45$ into $v(t) = -9.8t + 14$ shows that the velocity of the ball is about $-19.8$ m/s, which means the ball is *falling* at an approximate rate of 19.8 m/s.

## Exercises

**6.** Use the derivative to graph $f(x) = 3x^4 - 4x^3$.

**7.** A package service will accept a box with a square base so long as the (height) + (width of base) $\leq 60$ inches. What is the volume of the largest such box they will accept?

**8.** A particle moves smoothly along a number line according to a position function $s(t)$. Match the statements about the particle with the statements about the graphs of $v(t)$ and $s(t)$ at the right.

  **A.** The particle is moving forward.

  **B.** The particle is changing direction.

  **C.** The particle is accelerating.

  **i.** $v(t)$ has zero slope.

  **ii.** $s(t)$ has zero slope.

  **iii.** $v(t)$ has positive slope.

  **iv.** $s(t)$ has positive slope.

# Chapter 20: An Introduction to Calculus

*Chapter 20*

Complete these exercises before trying the Practice Test for Chapter 20. If you have difficulty with a particular problem, review the indicated section.

1. Find the derivative of $f(x) = \sqrt{3}x^7 - \dfrac{6}{\sqrt[3]{x}} + \sqrt[3]{3}$. Express your answer without using fractional or negative exponents. *(Section 20-1)*

2. Graph $f(x) = x^4 - 2x^3 - 2x^2 + 1$. Identify any local or global maximum and minimum points. *(Section 20-2)*

3. Find the global extreme values of $f(x) = 1 + x^{2/3}$ on the interval $-1 \le x \le 125$. *(Section 20-3)*

4. The position function of a particle moving along a straight line is
$$s(t) = -\frac{1}{3}t^3 + 7t^2 - 40t.$$

   a. During what interval of time is the particle moving to the right?
   b. What is the maximum velocity of the particle? *(Section 20-4)*

*Chapter 20*

**Find the slope of each curve at the given point $P$.**

1. $f(x) = 2x^3 - 6x;\ P(1, -4)$     2. $f(x) = x^3 - 4x;\ P(-1, 3)$

3. $f(x) = \dfrac{2}{\sqrt{x}};\ P(4, 1)$     4. $f(x) = \dfrac{1}{x} - \dfrac{2}{x^3};\ P(2, 0.25)$

**For Exercises 5–10, consider the function $f(x) = -x^4 + 4x^3 - 10$ on the interval $-1 \le x \le 4$.**

5. Find $f'(x)$.     6. Find $x$ such that $f'(x) = 0$.

7. Identify any local maxima.     8. Find the global maximum value.

9. Find the global minimum value.   10. Sketch the graph of $y = f(x)$.

**A cylindrical can is to have a volume of $54\pi$ cubic inches.**

11. Use the formulas for the volume and total surface area of a cylinder to express the surface area as a function of the radius.

12. Find the dimensions of the can that requires the least material.

**A chair manufacturer can produce $x$ chairs each day at a total cost, in dollars, given by the function $C(x) = 50 + 25x + 0.4x^{5/2}$. Assume all the chairs can be sold and that each chair sells for $150.**

13. Express the profit for one day as a function of $x$.

14. How many chairs should be made each day to maximize the profit?

**An arrow is shot upward from 5 feet above the ground so its height in feet above the ground after $t$ seconds is given by the function $h(t) = -16t^2 + 144t + 5$.**

15. Find the velocity function $v(t)$.   16. What was the initial velocity?

17. When is the velocity zero?     18. How long is the arrow in the air?

19. What is the maximum height of the arrow?

1. A game is played with two coins. The goal is to get two heads when the coins are tossed. The coins are tossed together. If they are both heads, you win. If not, you pick up the coins that are tails and flip them again. Now if you have two heads, you win; otherwise, you lose. Draw a tree diagram depicting this game. Find the probability of winning.

2. If you play the game in Exercise 1, at the end you either have two heads, one head and one tail, or two tails. Assign dollar amounts for payoff or loss in each case so that the game is fair.

3. The equation of the standard normal distribution is $f(x) = \dfrac{1}{\sqrt{2\pi}} e^{-x^2/2}$.

   Modify the program on text page 730 to approximate the area under this curve from $x = -2$ to $x = 2$. Use rectangles of width 0.1.

4. How does your answer to Exercise 3 relate to your knowledge of the normal distribution?

**In Exercises 5 and 6, solve these equations without using a calculator.**

5. $\log_{16} 32 = x$

6. $\log x = \log 4 + 2 \log 3$

7. Find $(3 - 4i)^{12}$. Give your answer in polar form.

8. Sketch the polar graph of $r = \theta^2$ for $\theta \geq 0$.

9. Solve this matrix equation for $X$:

$$\begin{bmatrix} 1 & 3 \\ 2 & 1 \end{bmatrix} X - \begin{bmatrix} 0 & 1 \\ 1 & 4 \end{bmatrix} = \begin{bmatrix} 10 & 9 \\ 0 & 1 \end{bmatrix}$$

10. Suppose the graph of $f(x)$ has symmetry in the $y$-axis. How does $f'(x)$ compare with $f'(-x)$?

11. If the graph of $f(x)$ has symmetry in the origin, how does $f'(x)$ compare with $f'(-x)$?

12. Find the remaining roots of the polynomial equation $2x^3 - x^2 + 7x + 4 = 0$ given that $-\dfrac{1}{2}$ is a root.

13. Write a polynomial function which could be the equation of the graph at the right.

14. The figure at the right is a cross-section of a rectangular storage area to be built in an attic. What are the dimensions of the rectangle of greatest area that will fit under the roof?

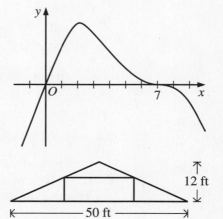

**Solve each inequality and graph the solution set.**

15. $|5x + 10| \leq 20$

16. $7 - |x - 8| \leq 5$

**Graph the solution set of each system of inequalities.**

17. $x^2 + y^2 < 144$
    $x + y > 12$

18. $y \leq 4 - x^2$
    $y \geq x^2 - 4$

19. Analyze all orbits for iterations of the function $f(x) = x^2 - 8$.

# Spanish Glossary

*inglés–español*

## A

**Absolute value of a complex number / Valor absoluto de un número complejo**   La longitud de la flecha que representa el número complejo. Si $z = a + bi$, entonces $|z| = \sqrt{a^2 + b^2}$. *(p. 403)*

**Absolute value of a number / Valor absoluto de un número**   La distancia entre el número y cero en la línea del número. *(p. 96)*

**Acceleration / Aceleración**   Medida del incremento de la velocidad de un objeto en función del tiempo. *(p. 776)*

**Additive inverse of a matrix A (denoted −A) / Inverso aditivo de una matriz A (designado como −A)**   La matriz en la cual cada elemento es el opuesto de su elemento correspondiente en A. *(p. 531)*

**Amplitude of a periodic function / Amplitud de una función periódica**   La mitad de la diferencia entre los valores máximo y mínimo de la función. *(p. 139)*

**Angle / Angulo**   La rotación alrededor de un punto de un rayo inicial hasta su rayo terminal. *(p. 257)*

**Angle of depression / Angulo de depresión**   El ángulo por el cual la línea de mira de quien observa debe deprimirse desde la horizontal hasta el punto observado. *(p. 335)*

**Angle of elevation / Angulo de elevación**   El ángulo por el cual la línea de mira de quien observa debe elevarse desde la horizontal hasta el punto observado. *(p. 332)*

**Apparent size / Tamaño aparente**   La medida de un ángulo subtendido al ojo de una persona por un objeto que se está observando. *(p. 264)*

**Argand diagram / Diagrama de Argand**   Un diagrama que representa números complejos en forma geométrica. *(p. 403)*

**Arithmetic sequence / Secuencia aritmética**   Una secuencia de números donde la diferencia de dos términos consecutivos es constante. *(p. 473)*

**Asymptote / Asíntota**   Línea que se acerca indefinidamente a una curva. *(p. 232)*

**Average / Promedio**   Un solo número, como la media, mediana y modo, que se usa para representar un conjunto entero de datos. *(p. 641)*

**Axis of symmetry / Eje de simetría**   Una gráfica tiene una línea llamada eje de simetría cuando pueden aparearse los puntos de la gráfica de manera que la línea se convierta en bisectriz perpendicular del segmento que une cada par. *(p. 133)*

## B

**Binomial coefficients / Coeficientes binomios**   Los números en el triángulo de Pascal; los coeficientes de los términos en la expansión de $(a + b)^n$. *(p. 590)*

**Box-and-whisker plot / Diagrama de líneas y bloques**   Método para exhibir la mediana, los cuartiles y los extremos de un conjunto de datos. *(p. 649)*

## C

**Circle / Círculo**   El conjunto de todos los puntos de un plano que se encuentran a una distancia fija (el *radio*) desde un punto fijo (el *centro*). *(p. 219)*

**Circular permutation / Permutación circular**   Permutación donde los elementos están dispuestos en círculo. *(p. 584)*

**Cofunctions / Cofunciones**   Seno, coseno; tangente, cotangente; secante, cosecante. *(p. 318)*

**Combination / Combinación**   Disposición de un conjunto de objetos donde el orden no es importante. *(p. 578)*

**Common logarithm of a positive real number x / Logaritmo común de un número positivo real x**   El exponente cuando el número se escribe como potencia de 10. *(p. 191)*

**Communication matrix / Matriz de comunicación**   Matriz que representa las vías mediante las cuales se pueden transmitir y recibir datos. *(p. 538)*

**Compass bearing or course / Rumbo o curso de la brújula**   El ángulo medido en el sentido de las agujas del reloj a partir de la dirección norte (con respecto a la primera ubicación) hasta la segunda dirección. *(p. 359)*

**Complement of a set A (denoted $\overline{A}$) / Complemento de un conjunto A (designado como $\overline{A}$)**   El conjunto de todos los elementos que no se encuentran en un conjunto A. *(p. 566)*

**Complex number / Número complejo**   Cualquier número de la forma $a + bi$, donde $a$ y $b$ son números reales e $i$ es la unidad imaginaria. $a$ es la *parte real* y $b$ es la *parte imaginaria* de $a + bi$. *(p. 27)*

**Complex number plane / Plano de números complejos**   El plano donde los puntos representan números complejos. *(p. 403)*

**Composite of two functions / Compuesto de dos funciones**   $(f \circ g)(x) = f(g(x))$ donde $x$ se encuentra en el dominio de $g$ y $g(x)$ se encuentra en el dominio de $f$. *(p. 126)*

**Composition / Composición**   La operación que combina dos funciones para producir el compuesto de las mismas. *(p. 126)*

**Conditional probability / Probabilidad condicional**   La probabilidad de que acontezca un suceso dado que otro suceso ha acontecido. *(p. 607)*

**Conic sections / Secciones cónicas**   Secciones transversales (círculo, elipse, hipérbola, parábola) que se producen al cortar un cono doble mediante un plano. Cuando el resultado es un solo punto, una línea o un par de líneas, la sección transversal se llama *sección cónica degenerada*. *(p. 213)*

**Conjugates / Conjugadas**   Expresiones de la forma $a\sqrt{b} + c\sqrt{d}$ y $a\sqrt{b} - c\sqrt{d}$. Los números $a + bi$ y $a - bi$ son *conjugadas complejas*. *(p. 27)*

**Constant / Constante**   Polinomio (o término de un polinomio) con grado 0. *(p. 53)*

**Continuous function / Función continua**   Una función $f(x)$ es continua en un número real $c$ si $\lim_{x \to c} f(x) = f(c)$. *(p. 720)*

**Convergent series / Serie convergente**   Serie infinita que tiene una suma. *(p. 500)*

**Coordinates / Coordenadas**   Par ordenado de números asociados a un punto en el plano. *(p. 1)*

**Correlation coefficient / Coeficiente de correlación**   Número entre $-1$ y 1, inclusive, que mide con cuánta proximidad los puntos trazados tienden a agruparse alrededor de la línea de cuadrados mínimos. *(p. 685)*

**Cosecant / Cosecante**   $\csc \theta = \frac{r}{y}, y \neq 0$, donde $P(x, y)$ es un punto en el círculo $O$ con una ecuación $x^2 + y^2 = r^2$ y $\theta$ es un ángulo en posición estándar con el rayo terminal $OP$. También, $\csc \theta = \frac{1}{\sin \theta}$, $\sin \theta \neq 0$. *(p. 282)*

**Cosine / Coseno**   $\cos \theta = \frac{x}{r}$ donde $P(x, y)$ es un punto en el círculo $O$ con una ecuación $x^2 + y^2 = r^2$ y $\theta$ es un ángulo en posición estándar con el rayo terminal $OP$. *(p. 268)*

**Cotangent / Cotangente**   $\cot \theta = \frac{x}{y}, y \neq 0$, donde $P(x, y)$ es un punto en el círculo con una ecuación $x^2 + y^2 = r^2$ y $\theta$ es un ángulo en posición estándar con el rayo terminal $OP$. También, $\cot \theta = \frac{\cos \theta}{\sin \theta}$ y $\cot \theta = \frac{1}{\tan \theta}$, $\sin \theta \neq 0$ y $\tan \theta \neq 0$. *(p. 282)*

**Coterminal angles / Angulos coterminales**   Dos ángulos en posición estándar que tienen el mismo rayo terminal. *(p. 260)*

**Cross product of two vectors / Producto cruzado de dos vectores**   Si $\mathbf{v}_1 = (a_1, b_1, c_1)$ y $\mathbf{v}_2 = (a_2, b_2, c_2)$, entonces $\mathbf{v}_1 \times \mathbf{v}_2 = \begin{vmatrix} \mathbf{i} & \mathbf{j} & \mathbf{k} \\ a_1 & b_1 & c_1 \\ a_2 & b_2 & c_2 \end{vmatrix}$. *(p. 465)*

**Cubic / Cúbico**   Polinomio (o término de un polinomio) con grado 3. *(p. 53)*

**Degree / Grado**   Unidad de medida de ángulos, $\frac{1}{360}$ de una revolución. *(p. 257)*

**Degree of a polynomial in *x* / Grado de un polinomio en *x***   La potencia de $x$ contenida en el término delantero. *(p. 53)*

**Derivative of *f(x)* / Derivada de *f(x)***   La función de inclinación $f'(x)$. *(p. 758)*

**Determinant / Determinante**   Valor asociado a una serie cuadrada de números. *(pp. 458 y 459)*

**Dimensions of a matrix / Dimensiones de una matriz**   El número de hileras y columnas de la matriz. *(p. 517)*

**Directrix of a parabola / Directriz de una parábola**   Línea fija donde todos los puntos de la parábola son equidistantes de dicha línea y un punto fijo llamado el *foco*. *(p. 238)*

**Discriminant / Discriminante**   La cantidad $b^2 - 4ac$ que aparece debajo del signo radical en la fórmula cuadrática. *(p. 31)*

**Divergent series / Serie divergente**   Serie cuyas sumas parciales se acercan al infinito o no tienen límite finito. *(p. 500)*

**Domain of a function / Dominio de una función**   El conjunto de valores para el cual se define la función. *(pp. 20, 119)*

**Dot product of two vectors (also called *scalar product*) / Producto escalar de dos vectores**   En dos dimensiones $\mathbf{v}_1 \cdot \mathbf{v}_2 = x_1 x_2 + y_1 y_2$, donde $\mathbf{v}_1 = (x_1, y_1)$ y $\mathbf{v}_2 = (x_2, y_2)$. *(p. 411)* En tres dimensiones $\mathbf{u} \cdot \mathbf{v} = x_1 x_2 + y_1 y_2 + z_1 z_2$, donde $\mathbf{u} = (x_1, y_1, z_1)$ y $\mathbf{v} = (x_2, y_2, z_2)$. *(p. 448)*

**Doubling time / Período multiplicado por dos**   El tiempo requerido para duplicar una cantidad que crece en forma exponencial. *(p. 182)*

**Dynamical system / Sistema dinámico**   Sistema que está sufriendo cambio ya sea ordenado o errático. *(p. 737)*

**E**

**e / e**   El número irracional, aproximadamente igual a $2.718\ldots$, considerado el límite a medida que $n$ se aproxima al infinito de $\left(1 + \dfrac{1}{n}\right)^n$. *(p. 186)*

**Eccentricity of a conic section / Excentricidad de una sección cónica**   La constante positiva $e$ tal como $PF:PD = e$, donde $F$ es un punto fijo (*foco*) que no se encuentra en una línea fija $d$ (*directriz*), y $PD$ es la distancia perpendicular de $P$ a $d$. La curva es una elipse si $0 < e < 1$, una parábola si $e = 1$, y una hipérbola si $e > 1$. *(p. 247)*

**Element (or entry) of a matrix / Elemento (o rubro) de una matriz**   Número de una matriz. *(p. 517)*

**Ellipse / Elipse**   (1) El conjunto de todos los puntos $P$ de un plano tal como $PF_1 + PF_2 = 2a$, donde $F_1$ y $F_2$ son los puntos fijos llamados *focos* y $a$ es una constante positiva. *(p. 226)* (2) $\dfrac{x^2}{a^2} + \dfrac{y^2}{b^2} = 1$, donde $a$ y $b$ son constantes positivas. *(p. 227)*

**Even function / Función par**   $f$ es una función par si $f(-x) = f(x)$. *(p. 137)*

**Event / Suceso**   Cualquier subconjunto de un espacio de muestra. *(p. 598)*

**Expected value / Valor anticipado**   Si una situación dada tiene una recompensa de $x_1$, con probabilidad $P(x_1)$, una recompensa $x_2$, con probabilidad $P(x_2)$, y así siguiendo, el valor anticipado será $x_1 \cdot P(x_1) + x_2 \cdot P(x_2) + \ldots + x_n \cdot P(x_n)$. *(p. 630)*

**Explicit definition of a sequence / Definición explícita de una secuencia**   Fórmula o regla que dice cómo computar cada término $t_n$ explícitamente en términos de $n$. *(p. 479)*

**Exponential equation / Ecuación exponencial**   Una ecuación que contiene una variable en el exponente. *(p. 203)*

**Exponential function with base b / Función exponencial con base b**   Cualquier función de la forma $f(x) = ab^x, a > 0, b > 0,$ y $b \neq 1$. *(p. 181)*

**Exponential growth or decay / Crecimiento o decrecimiento exponencial**   Situación donde ocurre un porcentaje fijo de crecimiento o decrecimiento en un período de tiempo fijo. *(p. 170)*

**Extreme values of a function / Valores extremos de una función**   Los valores mínimo y máximo de la función. *(p. 157)*

**F**

**Feasible solution / Solución factible**   Cualquier punto $(x, y)$ que satisface todos los requisitos del problema. El conjunto de dichos puntos se llama *región factible*. *(p. 109)*

**Fixed point / Punto fijo**   Un número $x_0$ de manera que $f(x_0) = x_0$. *(p. 739)*

**Frequency polygon / Polígono de frecuencias**   Una gráfica de líneas utilizada para exhibir un conjunto de datos. *(p. 641)*

**Frequency table / Tabla de frecuencias**   Una tabla que se usa para organizar y exhibir datos. *(p. 640)*

**Function / Función**   (1) Relación dependiente entre cantidades. *(p. 19)* (2) Correspondencia o regla que asigna a cada elemento de un conjunto $D$ (el *dominio*) exactamente un elemento del conjunto $R$ (el *margen de variación*). *(p. 119)*

**G**

**Geometric sequence / Secuencia geométrica**   Secuencia donde la razón (*razón común*) de dos términos consecutivos cualquiera es constante. *(p. 473)*

**Graph of an equation / Gráfica de una ecuación**   El conjunto de todos los puntos del plano de coordenadas correspondiente a las soluciones de una ecuación. *(p. 1)*

**H**

**Half-life / Semivida**   El tiempo que se requiere para que decremente la mitad de una cantidad que decrementa en forma exponencial. *(p. 182)*

**Histogram / Histograma**   Gráfica de barras utilizada para exhibir un conjunto de datos. *(p. 640)*

**Hyperbola / Hipérbola**   El conjunto de todos los puntos $P$ del plano tal como $|PF_1 - PF_2| = 2a$, donde $F_1$ y $F_2$ son los puntos fijos llamados *focos* y $a$ es una constante positiva. *(p. 231)*

**I**

**Identity matrix (denoted $I_{n \times n}$) / Matriz de identidad (designada como $I_{n \times n}$)**   Una matriz $n \times n$ cuyos elementos diagonales principales son 1 y los demás elementos son 0. *(p. 530)*

**Imaginary number / Número imaginario**   Un número complejo $a + bi$ donde $b \neq 0$. *(p. 27)*

**Imaginary unit $i$ / Unidad imaginaria $i$**   El número $\sqrt{-1}$ con la propiedad $i^2 = -1$. *(p. 26)*

**Inclination of a line / Inclinación de una línea**   El ángulo $\alpha$, donde $0° \leq \alpha < 180°$, que se mide desde el eje $x$ positivo hasta la línea. *(p. 296)*

**Independent events / Sucesos independientes**   Dos sucesos $A$ y $B$ son independientes si, y solamente si, el acontecimiento de $A$ no afecta la probabilidad del acontecimiento $B$. *(p. 607)*

**Infinite series / Serie infinita**   Una serie que no tiene un último término. *(p. 493)*

**Inverse functions / Funciones inversas**   Dos funciones $f$ y $g$ son funciones inversas si $g(f(x)) = x$ para todas las $x$ que se encuentran en el dominio de $f$ y $f(g(x)) = x$ para todas las $x$ que se encuentran en el dominio de $g$. $f^{-1}$ representa la inversa de $f$. *(p. 146)*

**Irrational numbers / Números irracionales**   Números reales que no son racionales, tal como $\sqrt{7}$ y $\pi$. *(p. 25)*

**Leading term of a polynomial in x / Término inicial de un polinomio en x**   El término que contiene la potencia más alta de $x$. El coeficiente del término inicial se llama *coeficiente inicial*. *(p. 53)*

**Least-squares line / Línea de cuadrados mínimos**   La línea que reduce al mínimo la suma de los cuadrados de las distancias de los puntos de datos desde la línea. *(p. 684)*

**Limit / Límite**   Ver la definición formal en las pp. 722 y 723.

**Limit of an infinite sequence / Límite de una secuencia infinita**   Un "valor objeto" al cual se acercan cada vez más los términos de una secuencia a medida que $n$ se hace cada vez más grande (es decir, a medida que $n$ "va hacia el infinito"). *(p. 493)*

**Line of reflection / Línea de reflejo**   Línea que actúa de espejo y se encuentra a mitad de camino entre un punto y su reflejo. *(p. 131)*

**Linear equation / Ecuación lineal**   Cualquier ecuación de la forma $Ax + By = C$, donde $A$ y $B$ no son ambos 0. *(p. 1)*

**Linear function / Función lineal**   Función de la forma $f(x) = mx + k$. *(p. 19)*

**Linear permutation / Permutación lineal**   Permutación donde los elementos están dispuestos en línea. *(p. 584)*

**Linear programming / Programación lineal**   Una rama de la matemática que utiliza desigualdades lineales llamadas *restricciones* para resolver ciertos problemas de toma de decisiones donde una expresión lineal deba aumentarse al máximo o reducirse al mínimo. *(p. 108)*

**Logarithm to base b of a positive number x / Logaritmo a la base b de un número positivo x**   El exponente cuando el número se escribe como potencia de $b$, donde $b > 0$ y $b \neq 1$. *(p. 193)*

**Logarithmic function with base b / Función logarítmica con base b**   La inversa de la función exponencial con base $b$. *(p. 193)*

**Markov chain / Cadena de Markov**   Secuencia de observaciones o predicciones, cada una de las cuales depende de aquélla inmediatamente precedente, y solamente de ella. *(p. 545)*

**Mathematical induction / Inducción matemática**   Método de prueba utilizado para demostrar que una afirmación es verdadera con respecto a todos los enteros positivos. *(p. 511)*

**Mathematical model / Modelo matemático**   Una o más funciones, gráficas, tablas, ecuaciones o desigualdades que describen una situación del mundo real. *(p. 20)*

**Matrix / Matriz**   Ordenamiento rectangular de números. Cada número es un *elemento* o *rubro* de la matriz. *(p. 517)*

**Mean (denoted $\overline{x}$) / Media (designada como $\overline{x}$)**   Promedio estadístico hallado dividiendo la suma de un conjunto de datos por el número de rubros de datos. *(p. 641)*

**Median / Mediana**   Promedio estadístico hallado disponiendo un conjunto de datos en orden creciente o decreciente y hallando el número medio o la media de dos números medios. *(p. 642)*

**Mode / Modo**   Promedio estadístico que consiste en el rubro de datos que acontece con mayor frecuencia. *(p. 642)*

**Multiplicative inverse of a matrix A (denoted $A^{-1}$) / Inversa multiplicativa de una matriz A (designada como $A^{-1}$)**   La matriz $A^{-1}$ tal como $A \cdot A^{-1} = A^{-1} \cdot A = I$, donde $|A| \neq 0$. *(pp. 531 y 532)*

**Mutually exclusive events / Sucesos de exclusión mutua**   Sucesos que no pueden acontecer al mismo tiempo. *(p. 599)*

**Natural logarithm of a positive number x / Logaritmo natural de un número positivo x**   $\log_e x$ ó $\ln x$. *(p. 193)*

**Normal distribution / Distribución normal**
Distribución de datos a lo largo de una curva
acampanada que alcanza su altura máxima en la media.
*(p. 662)*

**Odd function / Función impar**   *f* es una función impar
si $f(-x) = -f(x)$. *(p. 137)*

**One-to-one function / Función uno-a-uno**   Función
que tiene una inversa. Cada valor *x* corresponde
exactamente a un valor *y*, y cada valor *y* corresponde
exactamente a un valor *x*. *(p. 148)*

**Origin / Origen**   Punto de intersección de los ejes *x* e
*y*. *(p. 1)*

**Parabola / Parábola**   (1) Gráfica de una función
cuadrática. *(p. 37)* (2) El conjunto de todos los puntos *P*
en el plano que son equidistantes de un punto fijo *F*,
llamado *foco*, y una línea fija *d*, que no contenga *F*,
llamada *directriz*. *(p. 238)*

**Parametric equations / Ecuaciones paramétricas**
Ecuaciones en una variable, tal como $x = f(t)$ y
$y = g(t)$, que dan las coordenadas de los puntos de una
línea o curva en el plano; *t* se llama el *parámetro*.
*(p. 433)*

**Pascal's triangle / Triángulo de Pascal**
Ordenamiento de los coeficientes de los términos en la
expansión de $(a + b)^n$. *(p. 590)*

**Percentiles / Percentilas**   Los 99 puntos que dividen
un conjunto de datos en 100 partes iguales cuando los
datos están dispuestos en orden ascendente. *(p. 665)*

**Periodic function / Función periódica**   Una función *f*
es periódica si hay un número positivo *p*, llamado el
*período* de *f*, de manera que $f(x + p) = f(x)$ para todas
las *x* que se encuentran en el dominio de *f*. *(p. 138)*

**Permutation / Permutación**   Disposición de un
conjunto de objetos donde el orden es importante.
*(p. 578)*

**Point of symmetry / Punto de simetría**   Una gráfica
tiene un punto de simetría si es posible aparear los
puntos de la gráfica de manera que el punto es el punto
medio del segmento que une cada par. *(p. 133)*

**Polar axis / Eje polar**   El rayo de referencia,
generalmente el eje *x* no negativo, en el sistema de
coordenadas polares. *(p. 395)*

**Polar coordinates / Coordenadas polares**   El par
ordenado $(r, \theta)$ que describe la posición de un punto *P*
en el plano, *r* es la distancia entre el polo y *P*, mientras
$\theta$ es la medida de un ángulo formado por el rayo *OP* y

el eje polar. *(p. 395)*

**Polar equation / Ecuación polar**   Ecuación de una
curva dada en términos de *r* y $\theta$. *(p. 396)*

**Pole / Polo**   El origen en el sistema de coordenadas
polares. *(p. 395)*

**Polynomial in *x* / Polinomio en *x***   Expresión que
puede ser escrita en la forma $a_n x^n + a_{n-1} x^{n-1} +$
$\ldots + a_2 x^2 + a_1 x + a_0$ donde *n* es un entero no
negativo. Los números $a_n, a_{n-1}, \ldots, a_2, a_1,$ y $a_0$ se
llaman los *coeficientes* del polinomio. *(p. 53)*

**Polynomial inequalities / Desigualdades polinomias**
$P(x) < 0$ y $P(x) > 0$ donde $P(x)$ es un polinomio.
*(p. 100)*

**Population / Población**   En estadística, todo el
conjunto de los individuos u objetos bajo estudio.
*(p. 669)*

**Power series / Serie de potencia**   Una serie infinita
de la forma $a_0 + a_1 x + a_2 x^2 + \ldots + a_n x^n + \ldots$
*(p. 733)*

**Probability / Probabilidad**   (1) Un número entre 0 y 1,
inclusive, que indica la posibilidad de que acontezca un
suceso. *(p. 597)* (2) Si un espacio de muestra de un
experimento contiene resultados *n* igualmente posibles
y si el *m* de los resultados *n* corresponde a algún suceso
*A*, entonces $P(A) = \dfrac{m}{n}$. *(p. 598)*

**Pure imaginary number / Número imaginario puro**
Un número imaginario $a + bi$ para el cual $a = 0$ y
$b \neq 0$. *(p. 27)*

**Quadrantal angle / Angulo cuadrantal**   Angulo en
posición estándar con el rayo terminal a lo largo de un
eje. *(p. 259)*

**Quadrants / Cuadrantes**   Las cuatro regiones en las
cuales el plano de coordenadas está dividido por los
ejes *x* e *y*. *(p. 1)*

**Quadratic / Cuadrático**   Polinomio (o el término de
un polinomio) con grado 2. *(p. 53)*

**Quadratic equation / Ecuación cuadrática**   Cualquier
ecuación que puede ser escrita en la forma de $ax^2 +$
$bx + c = 0$, donde $a \neq 0$. *(p. 30)*

**Quartic / Cuártico**   Polinomio (o el término de un
polinomio) con grado 4. *(p. 53)*

**Quartile / Cuartil**   La mediana de la mitad inferior de
un conjunto de datos es el *cuartil inferior*; la mediana
de la mitad superior es el *cuartil superior*. *(p. 649)*

**Quintic / Quíntico**   Polinomio (o el término de un
polinomio) con grado 5. *(p. 53)*

## R

**Radian / Radián**    Una unidad de medida de ángulos. *(p. 258)*

**Range of a function / Margen de variación de una función**    (1) El conjunto de los valores resultantes de la función. *(p. 20)* (2) El conjunto de elementos al cual los elementos del dominio son asignados por la función. *(p. 119)*

**Range of a set of data / Margen de variación de un conjunto de datos**    La diferencia entre el número más grande y el número más pequeño del conjunto. *(p. 649)*

**Rational function / Función racional**    Función de la forma $y = \frac{p(x)}{q(x)}$ donde tanto $p(x)$ y $q(x)$ son polinomios y $q(x) \neq 0$. *(p. 726)*

**Rational numbers / Números racionales**    Números que constituyen la razón de dos enteros. *(p. 25)*

**Real numbers / Números reales**    El cero y todos los enteros positivos y negativos, números racionales y números irracionales. *(p. 26)* Todos los números complejos $a + bi$ donde $b = 0$. *(p. 27)*

**Rectangular coordinate system / Sistema de coordenadas rectangulares**    Sistema para localizar el punto asociado con cualquier par ordenado mediante referencia a dos líneas numéricas llamadas *ejes* que se intersecan en ángulo recto. El punto de intersección se llama el *origen*. *(p. 1)*

**Recursive definition of a sequence / Definición recursiva de una secuencia**    Condición inicial, tal como la especificación del primer término, y una fórmula o regla que dice cómo computar un término a partir de los términos anteriores. *(p. 479)*

**Reference angle / Angulo de referencia**    El ángulo positivo agudo formado por el rayo terminal de un ángulo y el eje $x$. *(p. 275)*

**Relation / Relación**    Correspondencia o regla que asigna cada elemento de un conjunto $D$ (el *dominio*) a por lo menos un elemento de un conjunto $R$ (el *margen de variación*). *(p. 121)*

**Resultant / Resultante**    Suma vectorial de dos vectores. *(p. 420)*

**Revolution / Revolución**    Movimiento circular completo. *(p. 257)*

## S

**Sample / Muestra**    Un subconjunto de la población. *(p. 669)*

**Sample space of an experiment / Espacio de muestra de un experimento**    Un juego $S$ donde cada resultado del experimento corresponde exactamente a un elemento de $S$. *(p. 598)*

**Sampling / Muestreo**    El proceso de seleccionar una muestra representativa de una población total. *(p. 669)*

**Scalar / Escalar**    En trabajos con vectores y matrices, un número real cualquiera. *(pp. 422, 518)*

**Scatter plot / Diagrama de esparcimiento**    Gráfica de pares ordenados que se utiliza para determinar si hay una relación matemática entre dos conjuntos de datos. *(p. 683)*

**Secant / Secante**    $\sec \theta = \frac{r}{x}$, $x \neq 0$, donde $P(x, y)$ es un punto en un círculo con la ecuación $x^2 + y^2 = r^2$ y $\theta$ es un ángulo en posición estándar con un rayo terminal $OP$. También $\sec \theta = \frac{1}{\cos \theta}$, $\cos \theta \neq 0$. *(p. 282)*

**Sector of a circle / Sector de un círculo**    Región limitada por un ángulo central y el arco interceptado. *(p. 263)*

**Segment of a circle / Segmento de un círculo**    Región limitada por un arco del círculo y la cuerda que conecta los puntos finales del arco. *(p. 340)*

**Sequence / Secuencia**    (1) Conjunto de números, llamados *términos*, dispuestos en un orden específico. *(p. 473)* (2) Función cuyo dominio es el conjunto de enteros positivos. *(p. 474)*

**Sequence of partial sums / Secuencia de sumas parciales**    La secuencia $S_1, S_2, S_3, \ldots$ *(p. 500)*

**Series / Serie**    Una suma indicada de los términos de una secuencia. *(p. 486)*

**Sigma notation / Notación sigma**    En una serie escrita con notación sigma, por ejemplo, $\sum_{n=1}^{10} n^2$, el término general $n^2$ se llama el sumando. La letra $n$ se llama el *índice*. El primero y el último valor de $n$ se llaman los *límites de la suma*. La letra griega $\Sigma$ (sigma) se usa para escribir una serie de forma abreviada. *(p. 506)*

**Sign graph / Gráfica de signos**    Gráfica de líneas numéricas que muestra cuándo el valor de una función polinomia es positivo, negativo o cero. *(p. 100)*

**Sine / Seno**    $\sin \theta = \frac{y}{r}$ donde $P(x, y)$ es un punto en el círculo $O$ con la ecuación $x^2 + y^2 = r^2$ y $\theta$ es un ángulo en posición estándar con el rayo terminal $OP$. *(p. 268)*

**Slope function / Función de inclinación**    $f'(x)$, la derivada de $f(x)$. *(p. 758)*

**Slope of a curve at a point $P$ / Inclinación de una curva en un punto $P$**   El límite de la inclinación de $\overline{PQ}$ a medida que $Q$ se aproxima a $P$ a lo largo de la curva, donde $Q$ es un punto en la curva cerca de $P$. *(p. 758)*

**Slope of a nonvertical line / Inclinación de una línea no vertical**   (1) Número que mide la pendiente de la línea con relación al eje $x$. *(p. 7)* (2) La tangente de la inclinación de la línea. *(p. 296)*

**Square matrix / Matriz cuadrática**   Matriz que tiene la misma cantidad de hileras y columnas. *(p. 530)*

**Standard deviation / Desviación estándar**   La raíz cuadrada positiva de la varianza. *(p. 654)*

**Standard position of an angle / Posición estándar de un ángulo**   Angulo que aparece en un plano de coordenadas con el vértice en el origen y el rayo inicial a lo largo del eje positivo $x$. *(p. 259)*

**Standard value / Valor estándar**   El número de desviaciones estándar entre un rubro de datos y la media. *(p. 656)*

**Stem-and-leaf plot / Diagrama de "tallos y hojas"**   Método para organizar y exhibir datos. *(p. 639)*

**Synthetic division / División sintética**   Método de atajo para dividir un polinomio por un divisor de la forma $x - a$. *(p. 59)*

**Synthetic substitution / Sustitución sintética**   Método para evaluar una función polinomia $P(x)$ para cualquier valor de $x$. *(p. 55)*

**Tangent / Tangente**   $\tan \theta = \dfrac{y}{x}, x \neq 0$, donde $P(x, y)$ es un punto en el círculo $O$ con la ecuación $x^2 + y^2 = r^2$ y $\theta$ es un ángulo en posición estándar con el rayo terminal $OP$. También $\tan \theta = \dfrac{\sin \theta}{\cos \theta}, \cos \theta \neq 0$. *(p. 282)*

**Tangent to a curve at a point on the curve / Tangente a una curva en un punto en la curva**   La línea que pasa a través del punto y que tiene una inclinación igual a la inclinación de la curva en ese punto. *(p. 761)*

**Transformation matrix / Matriz de transformación**   Matriz que especifica una transformación del plano. *(p. 552)*

**Transformation of the plane / Transformación del plano**   Función cuyo dominio y margen de variación son conjuntos de puntos. *(p. 551)*

**Transition matrix / Matriz de transición**   Matriz que especifica una transición de una observación a la otra. *(p. 543)*

**Transpose of a matrix / Transposición de una matriz**   La matriz obtenida intercambiando las hileras y columnas de la matriz. *(p. 517)*

**Trigonometric identity / Identidad trigonométrica**   Una ecuación que es verdadera para todos los valores de la variable para la cual se ha definido cada lado de la ecuación. *(p. 318)*

**Unit circle / Círculo unitario**   El círculo $x^2 + y^2 = 1$. *(p. 269)*

**Variance / Varianza**   Estadística que se utiliza para describir la diseminación de datos alrededor de la media. *(p. 654)*

**Vector / Vector**   (1) Cantidad descrita tanto por la dirección como por la magnitud. *(p. 419)* (2) Cantidad designada por un par ordenado *(p. 426)* o un triple ordenado *(p. 446)* de números.

**Venn diagram / Diagrama de Venn**   Diagrama utilizado para representar una relación entre conjuntos. *(p. 565)*

**Vertex of a parabola / Vértice de una parábola**   El punto donde la parábola cruza su eje de simetría. *(p. 37)*

**$x$-axis and $y$-axis / Eje $x$ y eje $y$**   Líneas perpendiculares, generalmente horizontal y vertical, respectivamente, utilizadas para armar un sistema de coordenadas. *(p. 1)*

**$x$-intercept / Intercepción $x$**   La coordenada $x$ de un punto en el cual una gráfica interseca el eje $x$. *(p. 1)*

**$y$-intercept / Intercepción $y$**   La coordenada $y$ de un punto en el cual una gráfica interseca el eje $y$. *(p. 1)*

**Zero matrix (denoted $O_{m \times n}$) / Matriz de ceros (designada como $O_{m \times n}$)**   Una matriz $m \times n$ cuyos elementos son todos ceros. *(p. 530)*

**Zero of a function / Cero de una función**   Si $f(a) = 0$, entonces $a$ es un cero de la función $f$. *(p. 19)*

**Zero of a polynomial function $P(x)$ / Cero de una función polinomia $P(x)$**   Cualquier valor de $x$ para el cual $P(x) = 0$. *(p. 53)*

# Chapter 1:
# Linear and Quadratic Functions

## Sections 1-1, 1-2, 1-3, and 1-4

**1.** $10$; $(0, 3)$; $\frac{3}{4}$  **2.**

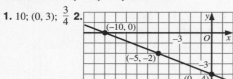

**3. a.** slope: $\frac{2}{3}$; $x$-intercept: $4.5$; $y$-intercept: $-3$  **b.** slope: $0$;

$x$-intercept: none; $y$-intercept: $5$  **4. a.** neither

**b.** perpendicular  **5. a.** $x = -4$  **b.** $y + 1 = -\frac{1}{2}(x - 3)$ or

$y - 1 = -\frac{1}{2}(x + 1)$  **c.** $\frac{y+1}{x-2} = -\frac{2}{5}$ or $2x + 5y = -1$  **6.** $-1$; $8$

**7. a.** a horizontal line with $y$-intercept $\frac{C}{B}$  **b.** a vertical line with

$x$-intercept $\frac{C}{A}$  **c.** a line with slope $-\frac{A}{B}$ passing through the origin

**8.** $(0.7, -0.3)$

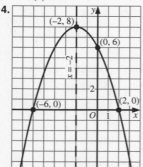

**9.** slope of $\overline{AB} = \frac{1}{3} =$
slope of $\overline{CD}$, so $\overline{AB} \parallel \overline{CD}$;
slope of $\overline{BC} \neq$ slope of
$\overline{AD}$, so $\overline{BC} \nparallel \overline{AD}$; Thus, $ABCD$
is a trapezoid.  **10.** $y = -2x + 1$
**11.** Answers will vary; a sample
answer is given. If you are given
(1) the $x$- and $y$-intercepts of the
line, (2) the slope and a point on
the line, (3) two points on the line,
or (4) an equation of the line.  **12. a.** $E(t) = 11,200 - 1500t$
**b.** 6:36 P.M.

## Sections 1-5, 1-6, 1-7, and 1-8

**1.** C  **2. a.** $-\frac{1}{2}i$  **b.** $2 + 3i$  **3.** (1) $x^2 - 2x = -5 \rightarrow x^2 - 2x + 1$

$= -4 \rightarrow (x - 1)^2 = -4 \rightarrow x - 1 = \pm 2i \rightarrow x = 1 \pm 2i$  (2) $x =$

$\frac{2 \pm \sqrt{4 - 4(1)(5)}}{2(1)} \rightarrow x = \frac{2 \pm \sqrt{-16}}{2} \rightarrow x = \frac{2 \pm 4i}{2} \rightarrow x = 1 \pm 2i$

**4.**

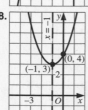

**5. a.** $A(r) = 6r - r^2$
**b.** 9 square units  **c.** domain:
$0 \le r \le 6$; range: $0 \le A \le 9$
**6.** $\frac{1}{2} - \frac{1}{2}i$  **7.** 3

**8.**

**9.** Answers will vary; a sample answer is
given. $y = 4 - 4x$ and $y = 2x^2 - 5x + 6$,
so $4 - 4x = 2x^2 - 5x + 6$; thus, $2x^2 -$
$x + 2 = 0$. The value of the discriminant is
$(-1)^2 - 4(2)(2) = -15 < 0$, so the equation
has no real roots. Therefore, there are no
points of intersection.  **10. a.** Answers will
vary; a sample answer is given. The data is
reasonable because as the price per dinner
increases, the number of dinners sold will likely decrease. The

revenue will increase if the price increase is moderate, but will
decline if the price increase is excessive.  **b.** A quadratic model; the
revenue values increase and then decrease, they neither increase
steadily nor decrease steadily.  **c.** $R(p) = -12x^2 + 372x$  **d.** $15.50$
**11.** Always; for $a$ or $b$ positive and the other negative: let $a = -4$

and $b = 9$, then $\frac{\sqrt{-4}}{\sqrt{9}} = \frac{i\sqrt{4}}{\sqrt{9}} = \frac{2i}{3}$ and $\sqrt{\frac{-4}{9}} = i\sqrt{\frac{4}{9}} = \frac{2}{3}i$

(similar results occur for $a = 4$ and $b = -9$); for $a$ and $b$ both

negative: let $a = -4$ and $b = -9$, then $\frac{\sqrt{-4}}{\sqrt{-9}} = \frac{i\sqrt{4}}{i\sqrt{9}} = \frac{2i}{3i} =$

$\frac{2i}{3i} \cdot \frac{i}{i} = \frac{2i^2}{3i^2} = \frac{-2}{-3} = \frac{2}{3}$ and $\sqrt{\frac{-4}{-9}} = \sqrt{\frac{4}{9}} = \frac{2}{3}$

## Chapter 1 Review
### Quick Check

**1.** $5\sqrt{2}$; $\left(-\frac{3}{2}, \frac{1}{2}\right)$  **2.** $(1, 1)$

**3.** 0  **4. a.** perpendicular
**b.** parallel  **c.** neither
**5.** $5x + y = -23$
**6.** $y = -\frac{1}{4}x - 3$ or

$x + 4y = -12$
**7.** $f(x) = 0.5x - 2$

**8. a.** $-3\sqrt{10}$  **b.** $9 - 5i$  **c.** $i$  **9. a.** $\frac{1 \pm i\sqrt{5}}{3}$  **b.** $2, \frac{1}{3}$  **c.** $-4 \pm \sqrt{17}$

**10. a.**

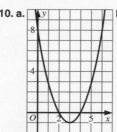

**b.**

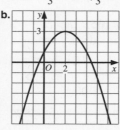

**11.** $f(x) = x^2 - x + 3$

### Practice Test

**1.** $2\sqrt{34}$  **2.** $-\frac{5}{3}$  **3.** $(-2, -1)$  **4.** $x - y = -3$  **5.** $2x - 3y = 15$

**6.** $\left(\frac{1}{2}, -3\right)$  **7.** $C(m) = 3.76m + 20$  **8.** 14 movies  **9.** $-6\sqrt{2}$

**10.** $-\frac{3}{10} + \frac{11}{10}i$  **11.** $i$  **12.** $\frac{3 \pm i\sqrt{47}}{4}$

**13.**

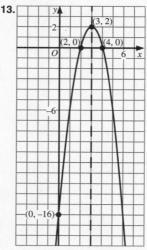

**14.**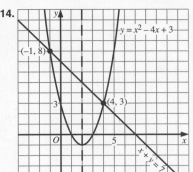

**15.** 10 seconds
**16.** 400 feet

**c.** From part (b), $x^3 + 2x^2 - 16 = (x - 2)(x^2 + 4x + 8) = 0$. If there is another root, it must be a root of $x^2 + 4x + 8 = 0$. The roots of this equation are imaginary, $-2 \pm 2i$, so these cannot be the length $x$. Therefore, $x = 2$ is the only real root of $P(x) = 16$ and 2 ft is the only length for which the volume is 16 ft$^3$.

### Sections 2-3 and 2-4

**1.** C **2.** D **3.** A **4.** B **5.** 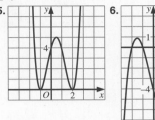 **6.**

**7.** 56.25 ft **8.** As the value of $x$ gets larger and larger, the term with the greatest absolute value is the leading term. If the coefficient of the leading term is positive, values of the function will get larger and larger as the value of $x$ increases and the graph will "point up." If the coefficient is negative, the values of the function will decrease steadily as $x$ increases and the graph will "point down." **9.** Sample answer: $y = (x + 3)(x + 1)(x - 2)(x - 4)$ **10.** Sample answer: $y = -x^2(x - 3)$ **11.** 13 cm by 6.5 cm **12.** Use the calculator to make a table of values from $x = 0$ to $x = 3$ in suitable increments, say 0.5. Since $V(x)$ increases from $x = 0$ to $x = 1$ and decreases from $x = 1.5$ to $x = 3$, the maximum must lie between 1 and 1.5. By evaluating $V(x)$ for values of $x$ in this interval, the required values of $x$ and $V(x)$ can be determined more accurately.

### Sections 2-5, 2-6, and 2-7

**1.** $-1.1, 1.4$ **2.** 2.2 **3.** $0, -3, \pm i$ **4.** A **5. a.** 5; 1
**b.**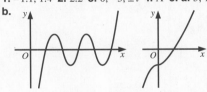

maximum         minimum

**6.** $0; -3$ **7.** $x^2 + 2x - 4 = 0$ **8.** $0, -2i$; no, the coefficients of this equation are imaginary, so the complex conjugates theorem does *not* apply. **9.** $-1, -2 \pm \sqrt{7}$ **10.** Answers will vary, a sample answer is given. The technological methods of Section 2-5 offer straightforward ways to identify and approximate most real roots; however, the location principle is a poor method if there is a double root. Also, a root that is close to another may be overlooked using technology; finally, these methods cannot be used to identify imaginary roots. Algebraic methods are useful when the equation has at least one rational root, or if the polynomial if factorable. When algebraic methods can be used, all the roots can be found exactly; however, some polynomials are not easily solved by algebraic methods.

**11.** Possible rational roots: $\pm 1, \pm 2, \pm\frac{1}{2}$; none of these satisfies the equation, so the equation has no rational roots. $P(2) = -6$ and $P(3) = 7$, so there is a real (irrational) root between $x = 2$ and $x = 3$. **12.** $b = 2, c = -2$ **13.** 0.9, 2.6

### Chapter 2 Review
#### Quick Check

**1. a.** $0, \frac{1}{2}$ **b.** $-2$ **2.** $x^3 - x^2 - x + 4; -5$ **3.** If $P(x) = x^{10} - 7x^7 - 9x^4 + 16$, then $P(2) = 2^{10} - 7(2^7) - 9(2^4) + 16 = 0$. By the factor theorem, $x - 2$ is a factor of $P(x)$.

### Mixed Review

**1.** $2\sqrt{13}$ **2.** $2x - 3y = 11$ **3.** $3x + 2y = -3$ **4.** $y = \frac{1}{9}(x + 2)^2 - 5$

**5.**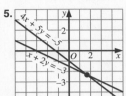

Estimates will vary; sample answer: $(2, -2.5)$; $\left(\frac{5}{3}, -\frac{7}{3}\right)$ **6.** $3x - 2y = -5$

**7.** $-9 + 3i\sqrt{2}$ **8.** $-7$

**9.**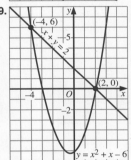

**10.** $-1.8; 3.5$ **11.** Quadratic; reasons will vary, for example, as $x$ increases by 5s, $A(x)$ increases by 175, then by 125, then by 75; for a linear model, the changes in $A(x)$ would always be the same. **12.** $A(x) = -x^2 + 50x$
**13.** 25 ft by 25 ft **14.** $\frac{-3 \pm \sqrt{29}}{10}$

## Chapter 2:
## Polynomial Functions

### Sections 2-1 and 2-2

**1. a.** no; $\pm\frac{1}{3}$ **b.** yes; 0, 4 **c.** yes; $-3, \pm\sqrt{5}$ **2. a.** > **b.** = **3.** A

**4.** quotient: $3x^2 - x + 1$; remainder: 2 **5.** Answers will vary, sample answers are given. **a.** Synthetic substitution is easier because this method requires you to find the product of just two numbers at a time, whereas direct substitution requires you to evaluate $(1 + \sqrt{3})^4$. **b.** Direct substitution is easier because $(-2)^9 - (-2)^4 + 5$ can be evaluated easily with a calculator, whereas synthetic substitution requires a ten-column array of numbers. **6. a.** $99 - 58\sqrt{3}$

**b.** $\frac{4}{x^8} - \frac{3}{x^4} - 1$ **7.** $\frac{-2 \pm \sqrt{3}}{2}$ **8.** Answers will vary, a sample answer is given. If $n$ is even, then $P(-1) = (-1)^n - 1 = 1 - 1 = 0$; thus, by the factor theorem, $x - (-1)$, or $x + 1$ is a factor of $P(x)$ when $n$ is even. However, when $n$ is odd, $P(-1) = (-1)^n - 1 = -1 - 1 = -2 \neq 0$; in this case, $x + 1$ is not a factor of $P(x)$. So $x + 1$ is not always a factor of $P(x)$. **9.** 14 **10. a.** $V(x) = x^2(x + 2) = x^3 + 2x^2$ **b.** $x^3 + 2x^2 = 16; x^3 + 2x^2 - 16 = 0$

$$\underline{2}\,|\;\;1\;\;\;2\;\;\;0\;\;-16$$
Since the remainder is 0, $x$ can represent 2 ft.
$$\;\;\;\;\;\;\;\;\;\;2\;\;\;8\;\;\;16$$
$$\overline{\;\;\;\;\;1\;\;\;4\;\;\;8\;\;\;\;\underline{|0}}$$

**4.** 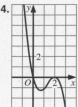  **5. a.** minimum **b.** 2  **6.** You can use the graph of $V(x)$ to approximate the maximum value or you can evaluate $V(x)$ for a suitable range of $x$-values.  **7.** (1) Graph $y = x^4 - 3x^3 + 1$ and show that the graph crosses the $x$-axis between $x = 2$ and $x = 3$, or (2) show that if $P(x) = x^4 - 3x^3 + 1$, then $P(2) = -7$ and $P(3) = 1$, so by the location principle there is a real root between $x = 2$ and $x = 3$.  **8. a.** $\pm 2, \pm i$

**b.** $\pm\sqrt{3}, -1$  **9.** $-1, -\frac{2}{3}, 2$  **10. a.** $1; -15$  **b.** $2 + i, -3$

### Practice Test

**1.** 21  **2.** $\frac{14}{27}$  **3.** $-1 + 9i$  **4.** $\frac{-1 \pm \sqrt{13}}{6}$  **5.** $8x^2 + 4x - 4; -3$  **6.** $\pm 1$,

$\pm\frac{1}{2}, \pm\frac{1}{4}, \pm\frac{1}{8}$  **7.** no  **8.**

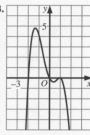

**9.** Sample answer: $y = \frac{1}{3}(x + 3)(x - 2)^2$  **10.** $2^{10} - 1$, or 1023

**11.** By the rational root theorem, the possible roots are $\pm 1$, but $f(-1) = -3$ and $f(1) = 5$. Therefore, neither $-1$ nor 1 is a root of $f(x) = x^3 + 3x + 1$, so $f$ has no rational roots. By the location principle, since $f(0) = 1$ and $f(-1) = -3$, $f$ does have an irrational root between $x = -1$ and $x = 0$.  **12. a.** 0  **b.** $-12$  **13.** $3, -1, -1 \pm i\sqrt{3}$  **14.** sum of the roots: $3 + (-1) + (-1 + 3i) + (-1 - 3i) = 0$; product of the roots: $3(-1)(-1 + 3i)(-1 - 3i) = -12$  **15.** 22

### Mixed Review

**1.** $x^3 + x + 6; -24$  **2.** $P(1) = 1 + 2 + 1 + 8 - 12 = 0$  **3.** $\pm 1$, $\pm 2, \pm 3, \pm 4, \pm 6, \pm 12; 1$ and $-3$  **4.**
**5.** $-12 + 6i$  **6.** $-3, 1, \pm 2i$
**7.** $3x - 4y = 10$  **8.** $4x + 3y = -8$  **9.** $(-0.08, -2.56)$

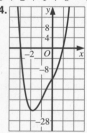

**10.** no points of intersection

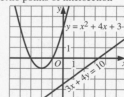

**11.** $-6i$  **12.** $-1 + 2i\sqrt{3}$
**13.** $-\frac{1}{7} + \frac{3i\sqrt{3}}{7}$  **14.** Answers will vary, sample answer : $F(t) = 1.8t + 32$; $F(t)$ gives the Fahrenheit temperature as a function of the Celsius temperature, $t$.  **15.** $-3.9$, $-0.5, 0.5$  **16.** $1, 2, -1 \pm i\sqrt{3}$
**17.** $4 \pm \sqrt{17}$  **18.** $PA = 5\sqrt{2} = PB$  **19.** $h(t) = -16t^2 + 72t + 40$; 121 ft; 5 sec  **20.** $x^3 - 2x^2 - 3x + 6 = 0$

## Chapter 3: Inequalities

### Sections 3-1 and 3-2

**1.** $x \geq -1$ 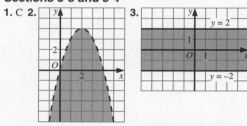 **2.** D  **3.** B  **4.** 0, 10  **5.** $x < -1$ or $-1 < x < 3$  **6.** $-4 \leq x \leq 0$ or $x \geq 2$  **7.** $x \leq -1$ or $x \geq 4$
**8.** $|x - 7| \geq 0$ is true for all real numbers; $x - 7$ represents a real number, and since the absolute value of every real number is greater than or equal to zero, $|x - 7| \geq 0$ is true for all real numbers $x$; $|x + 3| < -1$ has no solution; $x + 3$ represents a real number, and since there is no real number with a negative absolute value, $|x + 3| < -1$ has no solution.  **9.** $x \geq 5$
**10.** $x = 2, 3$  **11.** $-2 < x < \frac{8}{3}$  **12.** $-2 \leq x \leq 2$

**13.** Answers will vary. A sample answer is given. If $a \neq b$, then the graph of $P(x) = (x - a)(x - b)$ is a parabola that opens upward and has $x$-intercepts $a$ and $b$; in this case, $(x - a)(x - b) < 0$ has the solution $a < x < b$. If $a = b$, then the graph of $P(x) = (x - a)(x - b)$ is a parabola that opens upward and has its only $x$-intercept at $x = a$; in this case, the graph does not go below the $x$-axis, so $(x - a)(x - b) < 0$ has no solution.  **13.** The length of the longer side of the garden is between 5 m (the longer side must be greater than 5 m or it is not the *longer* side) and approximately 6.7 m.

### Sections 3-3 and 3-4

**1.** C  **2.** **3.**

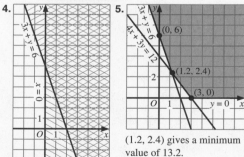

(1.2, 2.4) gives a minimum value of 13.2.

(7, 0) gives a maximum value of 21.

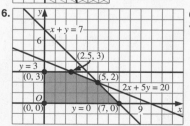

**7.** Answers will vary, a sample answer is given. Since $(a, c)$ is above $(a, b)$, the $y$-coordinate of $(a, c)$ is greater then the $y$-coordinate of $(a, b)$, that is, $c > b$; whereas, the $x$-coordinates are the same. Thus, $c > b$ and $b = ma + k$, so $c > ma + k$. That is, $(a, c)$ satisfies $y > mx + k$.

**8. a.** 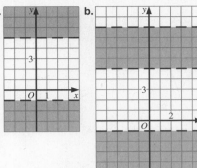 **b.**

**9.**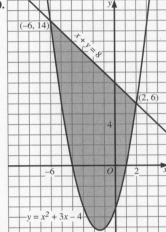

**10. a.** $P = 2.5x + 1.5y$
**b.** 100 bags of Pedigree Pellets, 700 bags of Mongrel Mash
**c.** Since $x$ and $y$ represent numbers of bags, $x$ and $y$ must be whole numbers.
**d.** No; the maximum profit would be $1463, resulting from an order of 266 bags of Pedigree Pellets and 532 bags of Mongrel Mash.

**7.** 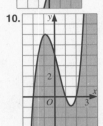 **8.**  **9.**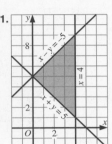

**10.** 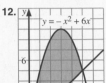 **11.**

**12.** 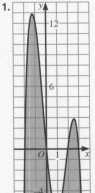 **13.** $x \geq 0, y \geq 0, x \leq 4, y \leq 3,$ $x + y \leq 6$

**14.** 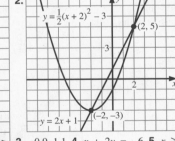 **15.** $(0, 0), (0, 4), (3, 2), (5, 0)$
**16.** 800 **17.** 1500 **18.** 1700

## Chapter 3 Review
### Quick Check

**1. a.** $x < -1$ **b.** $-1.5 \leq x \leq 1.5$ **2. a.** $x \leq -4$ or $x \geq 0$
**b.** $0 < x < 2$ or $x > 2$

**3.**  **4.**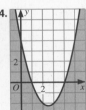

**5.** 44 **6.** $a + b \leq 200,$
$a \geq 1.5b, a \geq 0, b \geq 0$

### Practice Test

**1.** $x \leq \dfrac{5}{7}$

**2.** $-2 \leq x \leq 5$

**3.** $-5 < x < 1$

**4.** $x \leq 1$ or $x \geq 9$

**5.** $-1 < x < \dfrac{2}{3}$

**6.** $x \leq -3$ or $x = -1$ or $x \geq 2$

### Mixed Review

**1.** 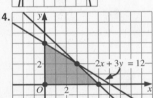 **2.**

**3.** $-0.9, 1.1$ **4.** $x + 2y = -6$ **5.** $x \geq 2.5$ or
$x \leq 1$ **6.** $\dfrac{4 \pm \sqrt{10}}{2}$ **7.** $-\dfrac{5}{3}, \dfrac{1 \pm i}{2}$

**8. a.** $l(d) = 0.25d + 0.5$ **b.** linear **9.** 10
pennies **10. a.** $(-2, 1.5)$ **b.** $AM = \dfrac{1}{2}\sqrt{61} = MB$

**11.** Sample answer:

$$\begin{array}{r|rrrrrr} 3 & 1 & 0 & -7 & -5 & -2 & -1 & -6 \\ & & 3 & 9 & 6 & 3 & 3 & 6 \\ \hline & 1 & 3 & 2 & 1 & 1 & 2 & 0 \end{array} = P(3)$$

Since $P(3) = 0, x - 3$ is a factor of $P(x)$. **12.** $0; -6$ **13.** $x = 0$
**14. a.** $-5 + 12i$ **b.** $1 - i$ **15.** $x < -3, -1 < x < 1,$ or $x > 3$
**16.** $5x + 2y = -7$

**17.**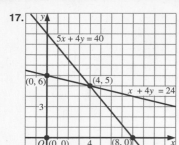

maximum: $P(4, 5) = 23$
**18.** $P(I) = 110I - 11I^2$
**19.** 0, 10 **20.** 275 watts

## Chapter 4:
## Functions

### Sections 4-1 and 4-2

**1.** yes; domain $= \{x \mid x \leq 2\}$,
range $= \{y \mid y \leq 0\}$, zero: 2
**2.** Answers will vary, a sample
answer is given. The graph of

$x = -y^2$, or $y = \pm\sqrt{-x}, x \leq 0$,
(shown at the right) fails the
vertical-line test, that is, every
negative value of $x$ has

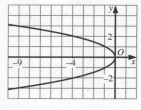

two corresponding values of $y$, not one. **3.** domain $= \{x \mid x \leq -2$ or
$x \geq 2\}$; range: $\{f(x) \mid f(x) \geq 0\}$; zeros: $\pm 2$ **4.** The graph is the same
as the graph of the line $y = x - 2$ but the graph does *not* include the
point $(-2, -4)$. The range is $\{g(x) \mid g(x) \neq -4\}$ and the only zero is 2.
**5. a.** 7; 5 **b.** $-2 \leq x \leq 2$ **c.** 11 **6.** $x^2 + 2x; -x^2 + 4$

**7.** $x^3 + 3x^2 - 4; \dfrac{1}{x-1}$ **8.** 12; $x^2 + x$ **9.** 28; $x^2 + 5x + 4$

**10.** Sample answers: function rules involving a square root symbol;
function rules with a variable in the expression in the denominator of
a fraction
**11. a.**

**b.** $\{y \mid y \leq 0\}$; 0 **c.** $y = -|x|$
**12. a.** Yes, for each value of $r$, there is
exactly one value for $A$. **b.** domain $=$
$\{r \mid r \geq 0\}$, range: $\{A \mid A \geq 1\}$
**13.** Answers will vary, a sample is given.
The addition of two real numbers

is commutative, since $x + y = y + x$ for all real numbers. However,
the composition of two functions is not always commutative.
Example 3 on text page 127 shows that $(f \circ g)(x)$ is not always equal

to $(g \circ f)(x)$. **14.** $(f \circ f)(x) = f(f(x)) = f\left(\dfrac{1}{x}\right) = \dfrac{1}{\frac{1}{x}} = x$,

provided $x \neq 0$.

### Sections 4-3, 4-4, and 4-5

**1. a.** C **b.** D **c.** F **2.** $y = |f(x)|$ **3. a.** B **b.** D **c.** A **d.** C
**4.** period $= 3$; amplitude $= 1.5$; $f(100) = 1$; $f(-100) = -2$
**5.**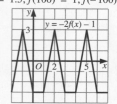

**6. a.** 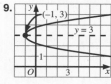 If $k > -4$, then a horizontal line through $(0, k)$
intersects the graph in two points. The function
is not one-to-one, so it has no inverse.
**b.** Let $x \geq 2$ (or let $x \leq 2$).

**c.**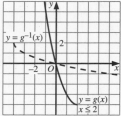

$y = g^{-1}(x)$ $y = g(x)$ $x \geq 2$
$y = g^{-1}(x)$ $y = g(x)$ $x \leq 2$

**7.** $f^{-1}(x) = x^3 - 1$ **8.** The graph has symmetry in the $x$-axis, the
$y$-axis, and the origin.

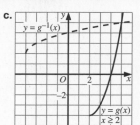

**9.**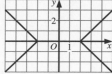

**10.** Answers will vary, a sample answer is
given. Reflect the graph of $y = |x|$ in the
$x$-axis to obtain the graph of $y = -|x|$.

Reduce each $y$-coordinate by $\dfrac{1}{2}$ to obtain

the graph of $y = -\dfrac{1}{2}|x|$. Then shift the

graph 2 units to the right and 3 units down to obtain the final graph:

$y + 3 = -\dfrac{1}{2}|x - 2|$, or $y = -\dfrac{1}{2}|x - 2| - 3$.

**11. a.** **b.** $g^{-1}(x) = 3 - \sqrt{\dfrac{x+1}{2}}$

**12. a.** $V_0 = \dfrac{273V_t}{273 + t}$

**b.** It decreases.

### Sections 4-6 and 4-7

**1.** the lengths $x$ and $z$ **2. a.** 4 **b.** 4 **c.** about 1.5 **3.** D **4.** In the first
figure, $A(x, y)$ represents a function of two variables, that is, the area
of a rectangle is a function of its dimensions $x$ and $y$. In the second
figure, $A(x, y)$ represents the location of point $A$ in the coordinate

plane. **5. a.** $h = x + 4$ **b.** $r = \sqrt{16 - x^2}$ **c.** $V(x) =$

$\dfrac{1}{3}p(16 - x^2)(x + 4)$ **6.** $0 < x < 4$ **7. a.** 840; There are 840 ways

for 4 people out of 7 to line up for a photograph. **b.** $3 \cdot p(7, 4) =$
$3 \cdot 840 = 2520; p(7, 5) = 7 \cdot 6 \cdot 5 \cdot 4 \cdot 3 = 2520 =$
$3 \cdot p(7, 4)$ **c.** Sample answers: $p(6, 6), p(6, 5)$, and $p(10, 3)$
**8.** Since $n$ and $r$ must be whole numbers with $n \geq r$, the graph of a
constant curve consists of discrete, or isolated, points in a coordinate
plane rather than a continuous curve. **9. a.** $d(x) =$

$\sqrt{1.5625x^2 - 9x + 36}$ **b.** 4.8 units (when $x = 2.88$)

**10.** $\dfrac{2048p}{81} \approx 79.4$ cubic units

### Chapter 4 Review
### Quick Check

**1.** domain $= \{$real numbers$\}$, range $= \{y \mid y \leq 6.25\}$, zeros: $-4, 1$

**2. a.** $2x - 1 - \dfrac{1}{x}$ **b.** $\dfrac{2x-1}{x}$, or $2 - \dfrac{1}{x}$ **c.** $2x^2 - x$

**d.** 0 **e.** $\dfrac{2}{x} - 1$ **f.** $\dfrac{1}{2x-1}$ **3. a.** no **b.** no **c.** yes **d.** no **4.** 12; 3; 0

**5.** $-1; f^{-1}(x) = \dfrac{1-x}{2}$ **6.** no **7. a.** $A(s, h) = 2s^2 + 4sh$

**b.** Sample answers: $(5, 2.5), (4, 4.25)$ **8.** $A(h) = \dfrac{h^4}{4}$

**Practice Test**

**1.** $\{x \mid x \neq 1, 3\}$  **2.** $\{x \mid x \leq -2 \text{ or } x \geq 2\}$
**3.** $\{y \mid y \geq -4\}; -1 \text{ and } -5$  **4.** $\{y \mid y \geq 0\}; -1 \text{ and } -5$

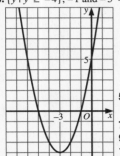

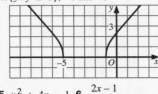

**5.** $x^2 + 4x - 1$  **6.** $\dfrac{2x - 1}{\sqrt[3]{x + 1}}$
**7.** $2\sqrt[3]{x + 1} - 1$  **8.** $x^3 - 1$
**9.** $A(x) = 6s^2$  **10.** 486 square units
**11.** 4  **12.** 1  **13.** $f$ is not one-to-one.

**14.**

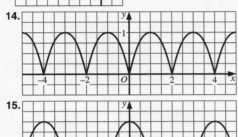

**15.**

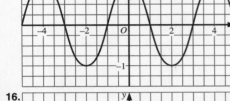

**16.**

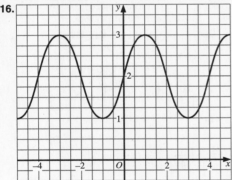

**Mixed Review**

**1.** $AB = \sqrt{(-2 - 1)^2 + (1 - 5)^2} = \sqrt{9 + 16} = \sqrt{25} = 5; BC = \sqrt{2.25 + 4} = \sqrt{6.25} = 2.5; AC = \sqrt{20.25 + 36} = \sqrt{56.25} = 7.5;$ Thus, $AB + BC = 5 + 2.5 = 7.5 = AC.$  **2.** $6x + 8y = 21$
**3.** $2x^3 + 3x^2 - 6x + 2 = 0$  **4.** $f(x) = x^2 + 4x - 1$
**5.**

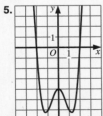

**6.** $-8 + 6i$  **7.** $x^3 - x^2 - 2x + 2; -6$  **8.** 1
**9.** $y > -\dfrac{1}{3}$  **10.** 1.6

**11. a.**

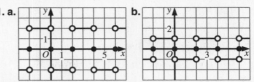

**12.** symmetry in the origin  **13.** 4; 2; 2  **14.** $x - x^2; x^3 - x^2 - x + 1; \dfrac{1}{x + 1}, x \neq \pm 1$  **15.** $A(y) = y(4 - 2y)$  **16.** 2 square units

# Chapter 5:
# Exponents and Logarithms

## Sections 5-1 and 5-2

**1.** D  **2.** A  **3.** A  **4.** $-\dfrac{x^3}{8}$  **5.** $\dfrac{1}{81x^3}$  **6.** $\dfrac{x^9}{27}$  **7.** $\dfrac{2s^2}{t} - 6t^2$  **8.** 729

**9.** $1 - x$  **10.** Answers will vary, a sample answer is given. Since $25^{1/2} = \sqrt{25} = 5$, then $-25^{-1/2} = -\dfrac{1}{5}$. However, $(-25)^{1/2}$ would be equal to $\dfrac{1}{(-25)^{1/2}}$, or $\dfrac{1}{\sqrt{-25}}$ : But $\sqrt{-25}$ is not a real number, so $(-25)^{-1/2}$ is not defined.  **11. a.** about \$15,601  **b.** about \$14,036
**12. a.** $V(t) = 0.93^t$  **b.** About \$1.24  **13.** 9  **14.** $b^{3n}$  **15.** 16%
**16.** 6.5  **17. a.** $\pm\dfrac{1}{40}$  **b.** $-3.9, -4.1$  **18.** Write each factor as a rational power of $x$ and add the exponents: $x^{1/4} \cdot x^{1/3} \cdot x^{1/6} = x^{3/12 + 4/12 + 2/12} = x^{3/4},$ or $\sqrt[4]{x^3}$.

## Sections 5-3 and 5-4

**1.** C  **2.** A  **3.** D  **4.** 2.2  **5.** $f(x) = 40(1.5)^x$  **6.** 9 years  **7.** Answers will vary, a sample answer is given. If $f(x) = ab^x$ with $a > 0$ and $b > 0$, then $f(x) > 0$ for all $x$ because $b^x > 0$ for all $x$. Because of this fact, the graph of $f(x) = ab^x$ lies entirely above the $x$-axis.
**8.** $e^3 \approx 20.1, 3^e \approx 19.8; e^3$ is greater.  **9. a.** about \$1083  **b.** about 8.3%  **10. a.** 8 cells  **b.** about 59 cells  **11. a.** $V(t) = 102 \cdot 2^{t/16}$ or $V(t) = 102(1.044)^t$  **b.** Sample answers: \$241,000, \$243,000

**12.** about 33%  **13. a.** They are identical since $g(x) = \left(\dfrac{1}{e}\right)^x = (e^{-1})^x = e^{-x} = f(x)$.  **b.** Since $f$ and $g$ are the same function and $h(x) = f(-x)$, the graph of $g$ is the reflection of the graph of $h$ in the $y$-axis.  **14. a.** yes  **b.** yes  **c.** yes  **d.** no  **15.** Answers will vary, a sample answer is given. When the money is compounded more frequently, the effective annual yield increases. For example, if it is compounded once, twice, or twelve times per year, the corresponding effective annual yields are 6%, 6.09%, and 6.17%, respectively. The maximum effective annual yield is about 6.184%, obtained by continuous compounding, so it is *not* possible to double the amount of the investment in one year if the interest rate is 6%.

## Sections 5-5, 5-6, and 5-7

**1.** C  **2.** B  **3.** D  **4.** A  **5.** $e^3 \approx 20$  **6. a.** 4  **b.** $-4$  **c.** 2.5  **d.** 0

**7.** $\dfrac{2}{3}\log M - 2\log N$  **8.** 2  **9.** Each expression is equal to 8.2; $\log 10^x = 10^{\log x} = x$ and $\ln e^x = e^{\ln x} = x$. Explanations will vary, a sample answer is given. Since $f(x) = \log x$ and $g(x) = 10^x$ are inverse functions, their composition in either order gives $x$. Similarly, $f(x) = \ln x$ and $g(x) = e^x$ are inverse functions, so their composition gives $x$.  **10.** about 5.45  **11.** 0.1  **12. a.** $-1.9084$  **b.** 2.9542  **c.** 1.4313  **13.** Answers will vary, a sample answer is given. Since 20 is between 10 and 100, $\log 20$ must be between $\log 10 = 1$ and $\log 100 = 2$. Therefore, $\log 20$ cannot be greater than 2.  **14. a.** $2e$  **b.** 3  **15.** $y = \dfrac{x^2}{10}$  **16.** in about 6.8 years

## Chapter 5 Review
### Quick Check

**1. a.** $-\frac{1}{16}$ **b.** $-12a^8$ **c.** $\frac{b^9}{a^3}$ **2.** 81¢ **3. a.** 10 **b.** 27 **4.** 3

**5.** $f(x) = 8(1.5)^x$ **6.** $P(t) = P_0 e^{rt}$; $1.05 **7.** $2^3 = 8$

**8.** $-1$ **9. a.** 1 **b.** $\frac{1}{3}$ **10.** 3.36

### Practice Test

**1.** 32 **2.** $6x^2$ **3.** $2x$ **4.** $\frac{16}{3}$ **5.** 64 **6.** 50 **7.** 3 **8.** 5 **9.** 2

**10.** $1 + \log A - 2 \log B$ **11.** $\frac{1}{\log A + \log B}$ **12.** 66.66°

**13.** 9.24 minutes **14.** $11,038.13 **15.** about 7 years

**16.** 2.48 **17.** 35.00 **18.** 0.97 **19.** $y = \frac{x^4}{27}$

**20.** 0, ln 4 ($\approx 1.3863$) **21.** 0.5, $\log_4 6$ ($\approx 1.2925$)

### Mixed Review

**1.** $0.5 \le x \le 4.5$  **2.**

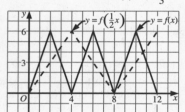

**3.** domain = {real numbers}, range = $\{y \mid y \le 1\}$, zeros: 0 and $\pm\sqrt{2}$ **4.** The graph of $x = f(y)$ is the reflection of the graph of $y = f(x)$ in the line $y = x$; no, since $f$ is not a one-to-one function. **5.** 5

**6.** $(2, -5)$, $(-1, 4)$ **7.** $\frac{-1 \pm i\sqrt{2}}{3}$

**8.** $6 + 4i$ **9.** 6 **10.** $\frac{b-a}{ab(a+b)}$

**11.** $4x + 3y = 20$ **12.** $10 + 6\sqrt{5}$

**13.** 3 and $-4$ **14.** $x$-axis, $y$-axis, origin **15.** about 34¢ **16.** 4.61

**17. a.** $\sqrt{x} + 2x - 1$ **b.** $\sqrt{2x - 1}$ **c.** $2\sqrt{x} - 1$ **18.** $A(h) = \frac{h^2\sqrt{3}}{3}$

**19.** Graphs will vary, a sample graph is shown at the right.
**20. a.** $A(r) = -r^2 + 25r$
**b.** 156.25 cm$^2$

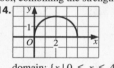

## Chapter 6:
## Analytic Geometry

### Sections 6-1, 6-2, 6-3, and 6-4

**1.** C **2.** A **3.** B **4.** $R(-2b, 2c)$; $S(-2a, 0)$ **5.** Use the midpoint formula to find $W(0, 0)$, $X(a + b, c)$, $Y(0, 2c)$, and $Z(-a - b, c)$;

$WX = \sqrt{(a + b)^2 + c^2}$, $XY = \sqrt{(a + b)^2 + (-c)^2} = \sqrt{(a + b)^2 + c^2}$, $YZ = \sqrt{(a + b)^2 + c^2}$, and

$ZW = \sqrt{(a + b)^2 + (-c)^2} = \sqrt{(a + b)^2 + c^2}$. Since $WX = XY = YZ = ZW$, quadrilateral $WXYZ$ is a rhombus.
**6.** $(x + 2)^2 + (y - 5)^2 = 4$ **7.** $(x - 4)^2 + (y - 1)^2 = 25$; center: $(4, 1)$, radius: 5

**8.** 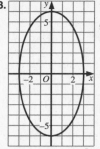 vertices: $(0, \pm 6)$, foci: $(0, \pm 3\sqrt{3})$

**9.** $\frac{x^2}{100} + \frac{y^2}{36} = 1$

**10.** 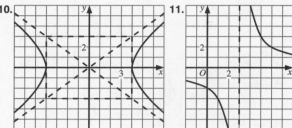 **11.**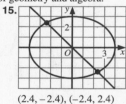

**12.** The graph of $(x - h)^2 + (y - k)^2 = r^2$ is a circle with center $(h, k)$ and radius $r$. It is obtained by shifting the graph of $x^2 + y^2 = r^2$ (a circle with radius $r$ and center at the origin) $h$ units horizontally and $k$ units vertically. Similarly, to obtain the graph of $\frac{(x - 2)^2}{25} + \frac{(y + 4)^2}{36} = 1$, shift the ellipse $\frac{x^2}{25} + \frac{y^2}{36} = 1$ to the right 2 units and 4 units down. **13.** Answers will vary, a sample answer is given. Some theorems are much easier to prove using coordinate methods than the usual (synthetic) methods. Also, coordinate geometry can provide an approach when you are not sure how to write a standard proof. Finally, coordinate geometry is a powerful tool, combining the strengths of geometry and algebra.

**14.** 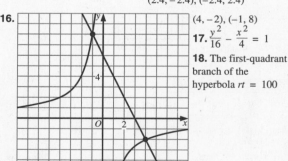 **15.**

domain: $\{x \mid 0 \le x \le 4\}$;
range: $\{y \mid 0 \le y \le 2\}$

$(2.4, -2.4)$, $(-2.4, 2.4)$

**16.**

$(4, -2)$, $(-1, 8)$

**17.** $\frac{y^2}{16} - \frac{x^2}{4} = 1$

**18.** The first-quadrant branch of the hyperbola $rt = 100$

### Sections 6-5, 6-6, and 6-7

**1.** D **2.** A **3.** C **4.** B **5. a.** vertex: $(0, 0)$; focus: $(0, 0.5)$; directrix: $y = -0.5$ **b.** vertex: $(0, 2)$; focus: $(0, 2.5)$; directrix: $y = 1.5$

**c.** vertex: $(-3, 0)$; focus: $\left(-3\frac{1}{4}, 0\right)$; directrix: $x = -2\frac{3}{4}$

**6.** $(-1, 3), (-3, -1)$ **7.** no solution **8.** Answers will vary, a sample answer is given. If the coordinates of the points of intersection are not integers, a graphical solution will generally yield approximate solutions, not exact solutions. Also, a graph may suggest that two curves intersect when they are actually close but yet do not intersect. A graph may also fail to detect two points of intersection that are close together. **9.** $\dfrac{x^2}{25} + \dfrac{y^2}{9} = 1$ **10.** hyperbola **11.** $x = \dfrac{1}{8} y^2$; a parabola containing the origin and opening to the right

**12.** $y = \dfrac{1}{4} x^2$; a parabola containing the origin and opening up

**13.** $\left( \dfrac{5\sqrt{7}}{4}, \dfrac{9}{4} \right), \left( \dfrac{5\sqrt{7}}{4}, -\dfrac{9}{4} \right), \left( -\dfrac{5\sqrt{7}}{4}, -\dfrac{9}{4} \right), \left( -\dfrac{5\sqrt{7}}{4}, \dfrac{9}{4} \right)$

**14.** Answers will vary, a sample answer is given. The graph of $4x^2 + y^2 = 0$ and the graph of $4xy = 0$ are both degenerate conics. However, the graph of $4x^2 + y = 0$ is the single point $(0, 0)$, whereas the graph of $4xy = 0$ is a pair of lines, $x = 0$ and $y = 0$. **15.** 0.2

## Chapter 6 Review
### Quick Check

**1.** $(a, a\sqrt{3})$ **2.** Show that $ZO = ZP = ZQ$. $\left( \text{Each length is } \dfrac{2a\sqrt{3}}{3}. \right)$
**3.** $(x + 3)^2 + (y - 2)^2 = 16$
**4.**
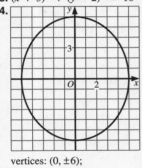
vertices: $(0, \pm 6)$;
foci: $(0, \pm\sqrt{11})$

**5.**
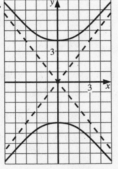
vertices: $(0, \pm 4)$, foci: $(0, \pm 5)$;
asymptotes: $y = \pm\dfrac{4}{3} x$

**6.** right; $V(0, 0)$, $F(4, 0)$, directrix: $x = -4$ **7.** $(\pm 4.8, 1.4)$
**8.** $\sqrt{(x - 4)^2 + y^2} = 2|x - 1|$; $x^2 - 8x + 16 + y^2 = 4x^2 - 8x + 4$; $3x^2 - y^2 = 12$; $\dfrac{x^2}{4} - \dfrac{y^2}{12} = 1$

### Practice Test

**1.** Answers will vary, a sample proof is given. Using the figure at the right and the midpoint formula, $P = (a, 0)$, $Q = (2a, a)$, $R = (a, 2a)$, and $S(0, a)$. By the distance formula, $PQ = QR = RS = SP = a\sqrt{2}$. By the slope formula, slope of $\overline{PQ} = 1 = $ slope of $\overline{RS}$ and slope of $\overline{QR} = -1 = $ slope of $\overline{PS}$. Since $1(-1) = -1$, each pair of adjacent sides is perpendicular to each other. Therefore, quadrilateral $PQRS$ is a square. **2.** $(x - 2)^2 + (y + 3)^2 = 25$
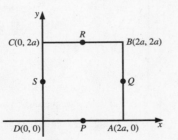
**3.** $\dfrac{x^2}{9} + \dfrac{y^2}{5} = 1$ **4.** $\dfrac{y^2}{4} - x^2 = 1$ **5.** $y + 3 = \dfrac{1}{12}(x - 2)^2$
**6.** center: $(0, 0)$; vertices: $(\pm 4, 0)$; foci: $(\pm 2\sqrt{3}, 0)$ **7.** center: $(0, 0)$; vertices: $(\pm 4, 0)$; foci: $(\pm 2\sqrt{5}, 0)$ **8.** center: $(2, -4)$; radius: 3
**9.** vertex: $(3, 1)$, focus: $\left( \dfrac{25}{8}, 1 \right)$

**10.**
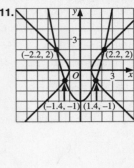

**11.**
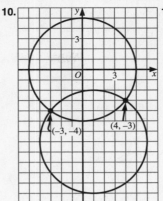

**12.** $(-3, -4), (4, -3)$ **13.** $(\pm\sqrt{5}, 2), (\pm\sqrt{2}, -1)$ **14.** hyperbola
**15.** ellipse **16.** 18.75 feet **17.** $8\dfrac{1}{3}$ feet below the vertex

### Mixed Review

**1.** 10 **2.** about 4.22 **3.** $\pm 1, \pm 2\sqrt{2}$ **4.** domain: $\{t \mid 0 \leq t \leq 6\}$; range: $\{h \mid 0 \leq h \leq 64.8\}$; zero: 6 **5.** Let $P(x) = x^3 + 2x^2 + 4$. The only possible rational roots are $\pm 1, \pm 2$, and $\pm 4$, but none of these is a root; since $P(-3) = -5$ and $P(-2) = 4$, there is an irrational root between $x = -3$ and $x = -2$; $x \approx 2.6$. **6.** $d(s) = 0.06s^2 + 0.5s + 7$; 182 ft

**7.** 4 **8.** 84.1 g **9.** $x < -2$ or $-\dfrac{1}{2} < x < 2$ **10.** $f^{-1}(x) = 2x + 10$;

$(f \circ f^{-1})(x) = f(2x + 10) = \dfrac{1}{2}(2x + 10) - 5 = x + 5 - 5 = x$;

$(f^{-1} \circ f)(x) = f^{-1}\left( \dfrac{1}{2}x - 5 \right) = 2\left( \dfrac{1}{2}x - 5 \right) + 10 =$

$x - 10 + 10 = x$ **11.** The graph of $\dfrac{y^2}{4} - x^2 = 1$ is a hyperbola with a vertical axis and vertices at $(0, \pm 2)$; it has symmetry in the $x$- and $y$-axes, and in the origin. **12.** $2x + y = 11$

**13. a.**
**b.**
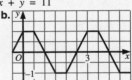

**14.** $\dfrac{x}{x + 1}$ **15.** a circle with center $(-1, 3)$ and radius 7

**16.** $-\dfrac{1}{4} < x < \dfrac{3}{4}$ **17.** $2i$; $2 - i$

## Chapter 7:
# Trigonometric Functions

### Sections 7-1 and 7-2

**1.** C **2.** C **3.** $\dfrac{10p}{9}$ **4. a.** $540°$ **b.** $-300°$ **5.** 1.5 **6.** Sample answers:

$-\dfrac{5p}{2}, \dfrac{3p}{2}$ **7.** 14 cm; 56 cm$^2$ **8.** 5.34 **9.** $232.6°$; $232°40'$ **10.** 4 cm

and 12 cm **11.** 12,700 ft

### Sections 7-3 and 7-4

**1.** A **2.** C **3.** A **4.** There are no solutions; Justifications will vary, a sample answer is given. Since $(\cos\theta, \sin\theta)$ is a point on the unit circle, the value of each coordinate is at most 1; also, the graph of $\cos\theta$ shows that there is no value of $\theta$ for which $\cos\theta = 1.5$.
**5. a.** negative **b.** negative **c.** positive **d.** negative **6. a.** $-\sin 76°$
**b.** $\cos 76°$ **c.** $\sin 40°$ **d.** $\sin 42°$ **7. a.** $-0.2957$ **b.** $0.6691$
**c.** $0.2675$ **d.** $-0.7568$ **8. a.** $-\dfrac{\sqrt{2}}{2}$ **b.** $-\dfrac{1}{2}$ **c.** $\dfrac{\sqrt{3}}{2}$ **d.** $-\dfrac{\sqrt{3}}{2}$

$\sin \theta = -\dfrac{7\sqrt{2}}{10}$; $\cos \theta = \dfrac{\sqrt{2}}{10}$ **10.** $-\dfrac{20}{29}$ **11. a.** Since $(x, y) =$ $(\cos \theta, \sin \theta)$ for every point on the unit circle, graph the line $y = -x$. The points of intesection with the unit circle can be used to find the required values of $\theta$. **b.** $135°$, $315°$ **c.** $\sin 135° = \dfrac{\sqrt{2}}{2} = -\cos 135°$; $\sin 315° = -\dfrac{\sqrt{2}}{2} = -\cos 315°$ **d.** You could graph $y = \sin \theta$ and $y = -\cos \theta$ (the reflection of $y = \cos \theta$ in the $x$-axis) for $0° \le x < 360°$ and find the points of intersection to identify the required values of $\theta$.

**12.** 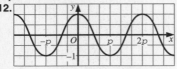 $\cos(-\theta) = \cos \theta$
**13.** about 4911 mi

### Sections 7-5 and 7-6
**1. a.** $-1.0353$ **b.** $-0.7002$ **c.** $0.4964$ **d.** $1.1049$ **2. a.** $-\cot 73°$ **b.** $-\csc 70°$ **c.** $\tan(4 - p) \approx \tan 0.86$ **d.** $\sec(2p - 6) \approx \sec 0.28$
**3. a.** $\dfrac{p}{2} + np$, where $n$ is an integer **b.** none **c.** $2np$, where $n$ is an integer **d.** $p + 2np$, where $n$ is an integer **4.** Answers will vary, a sample answer is given. Translate the graph of $y = \tan x$ to the right $\dfrac{p}{2}$ units and then reflect it in the $x$-axis. **5. a.** $53.1°$ **b.** $113.6°$
**c.** $76.0°$ **6. a.** $2.29$ **b.** $1.07$ **c.** $-0.45$ **7.** $\cos \theta = -\dfrac{11}{61}$, $\sin \theta = -\dfrac{60}{61}$, $\csc \theta = -\dfrac{61}{60}$, $\tan \theta = \dfrac{60}{11}$, and $\cot \theta = \dfrac{11}{60}$ **8.** $\tan \theta = 2$, $\sin \theta = \dfrac{2\sqrt{5}}{5}$, $\csc \theta = \dfrac{\sqrt{5}}{2}$, $\cos \theta = \dfrac{\sqrt{5}}{5}$, $\sec \theta = \sqrt{5}$ **9.** The graphs of $\tan \theta$, $\cot \theta$, $\sec \theta$, and $\csc \theta$ have asymptotes. Explanations will vary, a sample answer is given. The definitions of each of these is a fraction whose denominator cannot be zero. The asymptote for each graph occurs at a value of $\theta$ for which the denominator becomes zero.
**10. a.** $\dfrac{p}{3}$ **b.** $-\dfrac{p}{3}$ **c.** $\dfrac{5p}{6}$ **d.** $-\dfrac{p}{3}$ **11. a.** $1.09$ **b.** $\dfrac{5\sqrt{21}}{21}$ **12. a.** $1; \sqrt{2}$

**b.** Sample answer: $\text{Tan}^{-1}\dfrac{1}{\sqrt{2}}$

**c.** $35.3°$

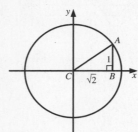

### Chapter 7 Review
**Quick Check**
**1.** $\dfrac{11p}{6}$ **2.** $540°$ **3.** $20$ cm; $100$ cm² **4.** $0; -1$ **5.** $-\dfrac{20}{29}$ **6.** $-\sin 55° \approx -0.8192$ **7.** $-\dfrac{\sqrt{3}}{2}$ **8.** $-\tan(2p - 6) \approx -0.2910$ **9.** $\sin x = \dfrac{8}{17}$, $\cos x = \dfrac{15}{17}$, $\csc x = \dfrac{17}{8}$, $\sec x = \dfrac{17}{15}$, $\cot x = \dfrac{15}{8}$ **10.** $-\dfrac{p}{3}$

### Practice Test
**1.** $\dfrac{7p}{6}$, $3.67$ **2.** $-\dfrac{5p}{9}$, $-1.75$ **3.** $300°$, $300.0°$ **4.** $417°10'$, $417.1°$
**5.** $470°$, $-250°$ **6.** $\approx 8.6$, $\approx -4.0$ **7.** $5.6$ ft **8.** $11.2$ sq ft
**9.** $3600p$ **10.** $1.4$ mi **11.** $-0.8$ **12.** $-0.75$ **13.** $\sin x = \dfrac{-2\sqrt{2}}{3}$, $\cos x = \dfrac{1}{3}$, $\tan x = -2\sqrt{2}$, $\csc x = \dfrac{-3\sqrt{2}}{4}$, $\cot x = \dfrac{-\sqrt{2}}{4}$
**14.** $\sin x = \dfrac{2\sqrt{13}}{13}$, $\cos x = -\dfrac{3\sqrt{13}}{13}$, $\tan x = -\dfrac{2}{3}$, $\sec x = \dfrac{-\sqrt{13}}{3}$, $\csc x = \dfrac{\sqrt{13}}{2}$ **15.** $-1$ **16.** $\dfrac{1}{2}$ **17.** $\dfrac{p}{6}$ **18.** $-\dfrac{p}{4}$ **19.** $-\sqrt{3}$ **20.** $-1.732$

### Mixed Review
**1.** 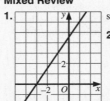 slope: $\dfrac{3}{2}$, $x$-intercept: $-3$, $y$-intercept: $4.5$
**2.** $0$; there is one real double root.
**3.**  **4.**
**5.** $-\dfrac{124}{125}$ **6.** $-3$ **7.** $\dfrac{2}{3}$ **8.** $2$ **9.** $-\dfrac{p}{4}$ **10.** $25$ **11.** $2x^3; 8x^3$
**12.** $\dfrac{2\sqrt{2}}{3}; -2\sqrt{2}$ **13.** $0.85$ **14.** $(22, -36)$ **15.** $\left(\dfrac{5}{2}, \pm\dfrac{\sqrt{11}}{2}\right)$;
$(-3, 0)$ **16.** no solution **17.** $x = \dfrac{1}{4}y^2 + 1$ **18.** $(AB)^2 + (AC)^2 = 80 + 20 = 100 = (BC)^2$, so $\triangle ABC$ is a right triangle with hypotenuse $\overline{BC}$. Also, (slope of $\overline{AB}$) • (slope of $\overline{AC}$) $= 2\left(-\dfrac{1}{2}\right) =$ $-1$, so $\overline{AB} \perp \overline{AC}$, and $\triangle ABC$ is a right triangle. **19.** $1; -0.7$ **20.** $5$
**21.** $h(a) = -0.8a + 185$; $169$ **22. a.** no **b.** yes **c.** no **d.** no

## Chapter 8:
## Trigonometric Equations and Applications

### Sections 8-1, 8-2, and 8-3
**1.** $x \approx 0.64 + 2np$ or $x \approx 5.64 + 2np$, where $n$ is an integer **2.** hyperbola; $\alpha = 9°$ **3.** C **4.** B

**5.** 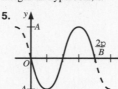 **6.** $\dfrac{p}{12}, \dfrac{p}{6}, \dfrac{7p}{12}, \dfrac{2p}{3}, \dfrac{13p}{12}, \dfrac{7p}{6},$ $\dfrac{19p}{12}, \dfrac{5p}{3}$ **7.** $7$

**8.** Answers will vary, a sample answer is given. The graph of $y = \cos x$ is translated $\dfrac{p}{6}$ units to the right and 2 units up.

The amplitude is quadrupled to 4. The minus sign in front of 4 reflects the graph about the $x$-axis. The period remains unchanged.

**9.**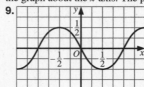

## Sections 8-4 and 8-5

**1.** B **2.** B **3.** $\cos^2 \theta$ **4.** $\cos x$ **5.** $-\sin A$ **6.** $\sec \theta$ or $\dfrac{1}{\cos \theta}$

**7.** Answers will vary, a sample answer is given. It is appropriate to divide both sides by $\cos \theta$ if $\cos \theta \neq 0$, that is, if $\theta \neq \dfrac{p}{2} + np$, where $n$ is an integer. Test the possible roots $\theta = \pm\dfrac{p}{2}$. If neither is a root, then dividing by $\cos \theta$ is appropriate. If one is a root, then it is not appropriate to divide both sides by $\cos \theta$. For example, it is appropriate to divide both sides of $\sin \theta = 2 \cos \theta$ to get $\tan \theta = 2$, but it is not appropriate to divide both sides of $\cos \theta = 2 \sin \theta \cos\theta$.
**8.** 60°, 120°, 240°, 300° **9.** 109.5°, 180°, 250.5° **10.** 0°, 180°, 270°

**11.** 0°, 63.4°, 180°, 243.4° **12.** $\dfrac{p}{4}, \dfrac{5p}{4}$ **13.** $\dfrac{p}{6}, \dfrac{5p}{6}, \dfrac{7p}{6}, \dfrac{11p}{6}$

**14.** $\dfrac{1 + \tan^2 \theta}{1 + \cot^2 \theta} = \dfrac{\sec^2 \theta}{\csc^2 \theta} = \dfrac{\sin^2 \theta}{\cos^2 \theta} = \left(\dfrac{\sin \theta}{\cos \theta}\right)^2 = \tan^2 \theta$

**15.** $\sec \theta - \cos \theta = \dfrac{1}{\cos \theta} - \cos \theta = \dfrac{1 - \cos^2 \theta}{\cos \theta} = \dfrac{\sin^2 \theta}{\cos \theta} =$
$\sin \theta \left(\dfrac{\sin \theta}{\cos \theta}\right) = \sin \theta \tan \theta$ **16.** Agree; the statement expresses in words the six cofunction relationships since $\theta$ and $90° - \theta$ are complementary angles. **17.** $\dfrac{3p}{2}$ **18.** 0.31, 2.83 **19. a.** 2 **b.** 2 = $4 \sin \theta \cos \theta$ **c.** Graph $y = \sin x \cos x - 0.5$ or square both sides of $\sin \theta \cos \theta = 0.5$ and substitute $1 - \cos^2 \theta$ for $\sin^2 \theta$. **d.** 45°

## Chapter 8 Review
### Quick Check

**1.** 113.6°, 246.4° **2.** about 68.2°
**3.** 2; $4p$

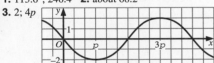

**4.** Answers will vary, a sample answer is given. Change the period from $2p$ to $\dfrac{p}{2}$ (a horizontal shrink) and change the amplitude from 1 to 3 (a vertical stretch). Then shift the resulting graph 2 units to the right and 1 unit down. **5. a.** $\sec x$ **b.** $-\cot^2 \theta$ **6.** 0°, 90°, 180°

### Practice Test

**1.** 14.0°, 194.0° **2.** 0°, 180° **3.** 30°, 70.5°, 150°, 289.5° **4.** 0.79, 2.36, 3.93, 5.50 **5.** 0, 1.05, 2.09, 3.14, 4.19, 5.24 **6.** 1.57, 3.67, 5.76

**7.** 110.6° **8.** $S(x) = 2 \sin \dfrac{p}{2}(x - 3)$ **9.** $C(x) = 2 \cos\dfrac{p}{2}x$

**10.**

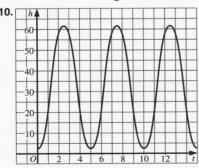

**11.** $h = 32 - 30 \cos\dfrac{2p}{5}t$ **12.** 56.3 ft

**13.**

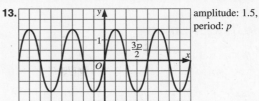

amplitude: 1.5, period: $p$

**14.**

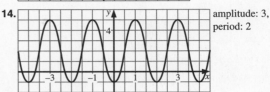

amplitude: 3, period: 2

**15.** $-\tan x$ **16.** $\cos^2 A$ **17.** $\csc^2 x$ **18.** $1 + \sin x$

### Mixed Review

**1.** $\sin x = -\dfrac{40}{41}$, $\cos x = \dfrac{9}{41}$, $\tan x = -\dfrac{40}{9}$, $\cot x = -\dfrac{9}{40}$, $\csc x = -\dfrac{41}{40}$ **2.** 500 **3.** $-1 < x < 1$ **4.** $\dfrac{1}{16}$ **5.** $\dfrac{4}{3}, \dfrac{-1 \pm \sqrt{13}}{2}$ **6.** (1, 6), (1.5, 4) **7.** $\dfrac{1}{3}$ **8.** $x \leq -2$ or $x \geq 6$ **9.** $C = (a + b, c)$; Since $AC = BD$, then $\sqrt{(a + b)^2 + c^2} = \sqrt{(a - b)^2 + (-c)^2}$ and $(a + b)^2 + c^2 = (a - b)^2 + c^2$. So $a^2 + 2ab + b^2 = a^2 - 2ab + b^2$ and $4ab = 0$; thus $a = 0$ or $b = 0$. Since $a$ is the length of $\overline{AD}$, $a \neq 0$ and therefore $b = 0$. Thus, the coordinates of $B$ are $(0, c)$ and the coordinates of $C$ are $(a + 0, c)$, or $(a, c)$. The slope of $\overline{AB}$ is undefined and the slope of $\overline{BC}$ is 0. Thus, $\overline{AB}$ is perpendicular to $\overline{BC}$ and $\angle ABC$ is a right angle. Similarly, $\angle ADC$, $\angle BCD$, and $\angle BAD$ are right angles. Therefore, parallelogram $ABCD$ is a rectangle. **10.** $f^{-1}(x) = \ln x$; zero: 1; domain: $\{x \mid x > 0\}$; range: $\{$real numbers$\}$

**11.**

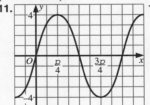

**12.**

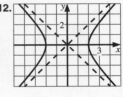

**13.** $-\dfrac{3}{4}$ **14.** 3 **15.** $\dfrac{12}{5} - \dfrac{4}{5}i$ **16.** 1 **17.** $3x - y = -5$

**18.** axis of symmetry: $x = 4$; vertex: $(4, -2)$; $x$-intercepts: 2, 6; $y$-intercept: 6 **19.** $-16$ **20.** $\sqrt{3x^2 + 1}$; $3(x - 1) + 2$, or $3x - 1$; the function $g$ has no inverse since it is not a one-to-one function.

**21.** $(x + 2)^2 + (y - 2)^2 = 45$ **22.** $0, \dfrac{p}{3}, \dfrac{5p}{3}$ **23.** sum: $\dfrac{32}{5}$, product: 4; roots: $\dfrac{2}{5}$ and $3 \pm i$; $\dfrac{2}{5} + (3 + i) + (3 - i) =$ $\dfrac{2}{5} + 6 = \dfrac{32}{5}$, $\dfrac{2}{5}(3 + i)(3 - i) = \dfrac{2}{5}(9 - i^2) = \dfrac{2}{5}(10) = 4$

**24.** symmetry in the origin; $\tan(-x) = -\tan x = -y$, so the equations $y = \tan x$ and $-y = \tan(-x)$ are equivalent. **25.** ellipse

## Chapter 9:
# Triangle Trigonometry

### Sections 9-1 and 9-2

**1.** D **2.** A **3.** cot **4.** $x = 34.3$; $z = 12.3$ **5.** $22.6°, 67.4°$

**6. a.**

**b.** $d = \sqrt{2}$ **c.** $\sin 45° = \cos 45° = \dfrac{\sqrt{2}}{2}$;

$\tan 45° = \cot 45° = 1$; $\csc 45° = \sec 45° = \sqrt{2}$ **7.** 81.6 sq units **8.** $11.5°$; $168.5°$

**9. a.** exactly one; Explanations will vary, a sample answer is given. A triangle is uniquely determined by the SAS case; the area of this unique triangle can be found using the formula $K = \frac{1}{2}ab \sin C$. **b.** In general, there are two possible angle measures: $\alpha = \mathrm{Sin}^{-1}\left(\dfrac{2K}{ab}\right)$ and $180° - \alpha$. However, if $\alpha = 90°$, then exactly one angle measure is possible.

**10.** 51.8 cm **11.** 55.5 ft **12.** $\dfrac{3\sqrt{3}}{2}$ sq units **13.** $2\sqrt{3}$ sq units

**14.** 2.60; 3.46; yes; The area of the circle is $p \bullet 1^2 = p$ sq units; the area of the circle is greater than the area of the inscribed hexagon and less than the area of the circumscribed hexagon.

### Sections 9-3, 9-4, and 9-5

**1.** B **2.** E **3.** C **4.** D **5.** $\angle C = 20°$, $a = 51.8$, $c = 35.5$ **6.** Either $\angle C = 45.8°$, $\angle A = 99.2°$, $a = 13.8$, or $\angle C = 134.2°$,

$\angle A = 10.8°$, $a = 2.61$ **7.** $\dfrac{\sin T}{t} = \dfrac{\sin R}{r}$; $\dfrac{\sin 70°}{t} = \dfrac{\sin R}{\frac{6}{5}t}$;

$\sin R = \dfrac{\frac{6}{5}t \sin 70°}{t} = \dfrac{6}{5}\sin 70° \approx 1.2(0.9397) \approx 1.1276$; Since the sine of an angle cannot be greater than 1, no such triangle is possible.

**8.** $\angle X = 27.8°$, $\angle Y = 60°$, $\angle Z = 92.2°$ **9.** $s = 20.7$, $\angle R = 36.5°$, $\angle T = 18.5°$ **10.** $27°$ **11.** about 13 nautical miles **12.** about 13.0 mi; about $122.0°$ **13.** about 2.31 km **14.** $30\sqrt{2}$ **15.** $4\sqrt{2}$ **16.** Answers will vary, a sample answer is given. An angle in standard position is measured from the positive $x$-axis; a counterclockwise rotation is positive and a clockwise rotation is negative. The course of a ship is measured clockwise from the north direction (here a clockwise rotation represents a positive angle measure). **17.** about 190 m; due east

### Chapter 9 Review
**Quick Check**

**1.** $22.6°, 67.4°$ **2.** 163 sq units **3.** Because the SSA case does not determine a unique triangle, rather, the SSA case may result in 0, 1, or 2 triangles. **4.** $\angle T = 28°$; $s \approx 46.0$; $t \approx 22.1$ **5.** SSS, SAS **6.** $128.7°, 33.1°$, and $18.2°$ **7.** $030°$

**Practice Test**

**1.** $15.9°$ **2.** 7.66 **3.** $41.4°, 41.4°, 97.2°$ **4.** 49.6 sq units **5.** 2 **6.** 1 **7.** $150.2°$ **8.** $85.5°$ **9.** 83.7 sq units **10.** 74.5 sq units **11.** You cannot be certain if the angle is acute or obtuse. **12.** $\angle A + \angle B + \angle C = 180°$, $\sin \alpha = \sin (180° - \alpha)$ **13.** about 82.7 nautical miles **14.** about $75.7°$ **15.** 200 nautical miles due east of the port **16.** 214 sq units **17.** about 11.2 sq units

**MIxed Review**

**1.** $\underline{-1\rfloor}$
| 2 | $-1$ | 13 | $-8$ | $-24$ | $-1, \dfrac{3}{2}, \pm 2i\sqrt{2}$ |
| | $-2$ | 3 | $-16$ | 24 | |
| 2 | $-3$ | 16 | $-24$ | $\lfloor 0$ | |

**2.** $V(x) = x(80 - 2x)^2$; Answers will vary, a sample answer is given. Graph $y = V(x)$ to identify the local maximum. **3.** $2 + 2i$

**4.** about 2.9 **5.** $\dfrac{9}{5}$ **6.** about 4 years **7.** $x < -3$ or $-\dfrac{1}{2} < x < \dfrac{1}{2}$

**8.** $2, -\dfrac{4}{3}$ **9.** 4.5 **10.** $(15, -6)$ **11.** $(3, 0), (2.4, 0.6)$ **12.** 2.25

**13.** $2\sqrt{61}$ cm, $2\sqrt{21}$ cm; $40\sqrt{3}$ cm$^2$ **14.** $3x + 4y = -9$ **15.** domain: {real numbers}; range: $\{y \mid y \le 4.125\}$; zeros: $\dfrac{1 \pm \sqrt{33}}{4}$ **16.** 2.47, 5.61 **17.** $f^{-1}(x) = \dfrac{7-x}{2}$; $(f \circ f^{-1})(x) = f\left(\dfrac{7-x}{2}\right) = 7 - 2\left(\dfrac{7-x}{2}\right) = 7 - (7 - x) = x$; $(f^{-1} \circ f)(x) = f^{-1}(7 - 2x)$

$= \dfrac{7 - (7 - 2x)}{2} = \dfrac{2x}{2} = x$ **18.** $\dfrac{x^2}{81} + \dfrac{y^2}{225} = 1$ **19.** $\sin \theta = -\dfrac{1}{7}$, $\cos \theta = -\dfrac{4\sqrt{3}}{7}$, $\tan \theta = \dfrac{\sqrt{3}}{12}$, $\sec \theta = -\dfrac{7\sqrt{3}}{12}$, $\cot \theta = 4\sqrt{3}$

**20.** $\sin A = \dfrac{BC}{AB} = \cos B$ **21.** $\angle R = 98°$; $RS \approx 8.30$; $ST \approx 15.5$

**22.**
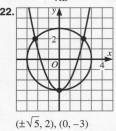
$(\pm\sqrt{5}, 2), (0, -3)$

**23.**
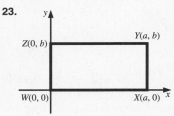
Midpoint of $\overline{WY} = \left(\dfrac{a}{2}, \dfrac{b}{2}\right) =$ midpoint of $\overline{XZ}$, so the diagonals bisect each other.

## Chapter 10:
# Trigonometric Addition Formulas

### Sections 10-1, 10-2, 10-3, and 10-4

**1.** C **2.** D **3.** A **4.** B **5.** $\cos (p - x) = \cos p \cos x + \sin p \sin x = -\cos x + 0 = -\cos x$ **6.** $\dfrac{\sqrt{2} - \sqrt{6}}{4}$ **7.** $\dfrac{7}{6}; \dfrac{9}{2}$ **8. a.** $-\sqrt{3}$ **b.** $\dfrac{\sqrt{2}}{2}$

**9.** $\cos 4x$ **10.** $\sin 35°$ **11.** $\cos 2x$ **12.** $\tan 2\alpha$, or $\dfrac{2 \tan \alpha}{1 - \tan^2 \alpha}$

**13.** $\cos \alpha$ **14.** $-2 \cos x$ **15.** Since trigonometric functions are periodic, a trigonometric equation generally has an infinite number of solutions. By limiting the domain, you get a limited number of solutions; all other solutions are coterminal with the ones found in the limited domain. **16.** $60°, 180°, 300°$ **17.** $\dfrac{7\sqrt{2}}{34}$; $-\dfrac{7\sqrt{2}}{34}$ **18.** $\dfrac{17}{81}; \dfrac{1}{3}$

**19.** Answers will vary, a sample answer is given. If $\sin \alpha = \cos \beta$ and $\alpha$ and $\beta$ are first-quadrant angles, then $\alpha = 90° - \beta$ and $\beta = 90° - \alpha$. Therefore, $\cos \alpha = \sin \beta$. Then $\sin 2\alpha = 2 \sin \alpha \cos \alpha$ and $\sin 2\beta = 2 \sin \beta \cos \beta = 2 \cos \alpha \sin \alpha = 2 \sin \alpha \cos \alpha = \cos 2\alpha$. **20.** $31.7°$

### Chapter 10 Review
**Quick Check**

**1.** $\cos 210° = -\dfrac{\sqrt{3}}{2}$ **2. a.** $\dfrac{9}{2}$ **b.** $\dfrac{7}{6}$ **3. a.** $\cos x$ **b.** $\tan 40°$ **c.** $\dfrac{1}{2}\sin 2x$

**4.** Answers will vary, a sample answer is given. Graph $y = \tan 2x$ and $y = \tan x$ on the same set of axes and find the $x$-coordinates of the points of intersection in the interval $0 \le x < 360°$; or substitute $\dfrac{2 \tan x}{1 - \tan^2 x}$ for $\tan 2x$ and solve for $x$: $0°, 60°, 120°, 180°, 240°, 300°$.

**Practice Test**

**1.** 0 **2.** $\cos x$ **3.** $-\cos 2x$ **4.** $-\sin 2x$ **5.** $-\dfrac{21}{221}$ **6.** $\dfrac{825}{2873}$ **7.** $-\dfrac{171}{140}$

**8.** $\dfrac{\sqrt{17}}{17}$ **9.** $79.7°$ **10.** $\dfrac{p}{3}, \dfrac{2p}{3}, \dfrac{4p}{3}, \dfrac{5p}{3}$ **11.** $0, \dfrac{2p}{3}, p, \dfrac{4p}{3}$ **12.** $\dfrac{7p}{6}, \dfrac{11p}{6}$

**13.** $\dfrac{p}{6}, \dfrac{5p}{6}, \dfrac{7p}{6}, \dfrac{11p}{6}$ **14.** $\cos x + \sin^2\dfrac{x}{2} = \cos x + \dfrac{1-\cos x}{2} =$

$\dfrac{2\cos x + 1 - \cos x}{2} = \dfrac{1 + \cos x}{2} = \cos^2\dfrac{x}{2}$ **15.** $\cos^4 y - \sin^4 y -$

$\cos 2y = (\cos^2 y)^2 - \sin^4 y - \cos 2y = (1 - \sin^2 y)^2 - \sin^4 y$

$- (1 - 2\sin^2 y) = 1 - 2\sin^2 y + \sin^4 y - \sin^4 y - 1 +$

$2\sin^2 y = 0$ **16.** $\dfrac{p}{6}, \dfrac{p}{2}, \dfrac{5p}{6}, \dfrac{3p}{2}$ **17.** $0 \le x < \dfrac{p}{6}, \dfrac{p}{2} < x < \dfrac{5p}{6},$

$\dfrac{3p}{2} < x \le 2p$ **18.**

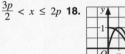
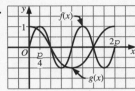

**Mixed Review**

**1.** 2.95 **2.** 0; 8 **3.**
$$\begin{array}{r} 2\underline{|\quad} \;\; -0.5 \quad 0 \quad\; 3 \quad\; 4 \\ \underline{\quad\quad -1 \;\; -2 \quad 2 \;\;} \\ -0.5 \;\; -1 \quad 1 \quad \underline{|6} \end{array}$$
Since $P(2) \ne 0$, $x - 2$ is *not* a factor of $P(x)$.

**4.** $\dfrac{5}{2}$ **5.** $\dfrac{\sqrt{3}}{3}$ **6.** $\dfrac{1}{5}$ **7.** $\dfrac{-2 \pm i\sqrt{2}}{2}$ **8.** $-4, 2 \pm \sqrt{3}$ **9.** $x < -4.3$ or

$x > -0.7$ **10.** $A(x) = x(200 - 5x)$; 2000 m² **11.** domain:

$\{x \,|\, -2 \le x \le 2\}$; range: $\{y \,|\, -2 \le y \le 0\}$; zeros: $\pm 2$; no

**12.** $x + 5y = 11$; about 169° **13.** $\csc x$ **14.** $-2 \sin x$ **15.** 45°, 225°

**16.** $g(g(x)) = g\left(-\dfrac{2}{x}\right) = -\dfrac{2}{\left(-\dfrac{2}{x}\right)} = -2\left(-\dfrac{x}{2}\right) = x$; This shows

that $g$ is its own inverse. **17. a.** symmetry in the $y$-axis **b.** symmetry

in the origin **18.** amplitude: 3, period: 4; sample points: $(-1, 5)$,

$(0, 2)$, $(1, -1)$ **19.** $AC = 19.6$; 66.6 sq units **20.** $-\dfrac{15}{8}; \dfrac{161}{289}; \dfrac{4\sqrt{17}}{17}$

**21.** $H(n) = \dfrac{1}{2}n^2 - \dfrac{1}{2}n$; 190 handshakes

# Chapter 11:
# Polar Coordinates and Complex Numbers

## Sections 11-1, 11-2, 11-3, and 11-4

**1.** C **2.** B **3.** D **4.** A **5.**

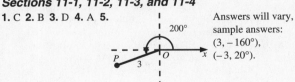

Answers will vary, sample answers: $(3, -160°)$, $(-3, 20°)$.

**6.** Answers will vary, sample answer: $\left(4\sqrt{2}, \dfrac{3p}{4}\right)$. **7.** $(\sqrt{3}, 1)$

**8.** 17 cis 298° **9.** about $1.71 + 4.70i$ **10.** 2 cis $\dfrac{3p}{4}$; $-\sqrt{2} + i\sqrt{2}$

**11.** Answers will vary, a sample answer is given. Manoj's method is

valid. Since $a = r\cos\theta$ and $b = r\sin\theta$, $\cos\theta = \dfrac{a}{r} = \dfrac{a}{\sqrt{a^2 + b^2}}$

and $\sin\theta = \dfrac{b}{r} = \dfrac{b}{\sqrt{a^2 + b^2}}$. His method specifies values for both

$\sin\theta$ and $\cos\theta$; this completely specifies $\theta$. **12. a.** $\sqrt{2}$ cis 45°

**b.** $z^{-1} = \dfrac{\sqrt{2}}{2}$ cis 315°, $z^0 = 1, z^2 = 2$ cis 90° **c.** $(1 + i)(1 + i) =$

$1 + 2i + (-1) = 2i$ and 2 cis 90° $= 2(0 + i) = 2i$; they are equal.

**13. a.** $\sqrt{3} + i, -1 + i\sqrt{3}, -\sqrt{3} - i, 1 - i\sqrt{3}$ **b.** $(\sqrt{3} + i)^4 =$

$(2 + 2i\sqrt{3})^2 = 4 + 8i\sqrt{3} + 12i^2 = -8 + 8i\sqrt{3}$; yes **14.** $(9.8, 0.4)$

**15.**

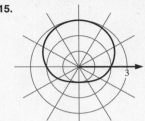

$(x^2 + y^2 - y)^2 = 4x^2 + 4y^2$

**16. a.** $34 + 14i$ **b.** $z_1 = (2\sqrt{2}, 225°)$; $z_2 \approx (13, 157.4°)$;

$z_1 z_2 \approx (26\sqrt{2}, 382.4°)$, or $(26\sqrt{2}, 22.4°) \approx 34 + 14i$

**c.**

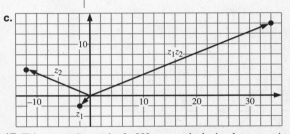

**17.** This rotates the graph of $z$ 90° counterclockwise; let $z = r$ cis $\theta$;

since $i =$ cis 90°, $iz = ($cis 90°$)(r$ cis $\theta) = r$ cis $(90° + \theta)$; thus, $iz$

has the same absolute value as $z$ but its graph is obtained by rotating

the graph of $z$ 90° counterclockwise. **18.** Answers will vary, a sample

answer is given. A radar screen usually has a circular picture of an

area with the center corresponding to the location of the radar unit.

The screen has a compass scale around the edge for direction

readings. Radar uses the distance and direction of an object to locate

it, just like the polar coordinate system does. **19. a.** $-\dfrac{1}{2} - \dfrac{1}{2}i$

**b.** $-\dfrac{1}{2} - \dfrac{1}{2}i$ **20.** cis 54°, cis 126°, cis 198°, cis 270°, cis 342°

## Chapter 11 Review
### Quick Check

**1.** $(4\sqrt{2}, -4\sqrt{2})$ **2.** Make a table of values to find values of $r$ for

known values of $\theta$ (multiples of 30° and 45°). Continue until the

points begin to repeat. Plot the points to obtain the graph. **3.** $2i =$

2 cis 270°; $-1 - i = \sqrt{2}$ cis 225°; $-2i(-1 - i) = -2 + 2i$;

(2 cis 270°)($\sqrt{2}$ cis 225°) $= 2\sqrt{2}$ cis 495° $= 2\sqrt{2}$ cis 135°, which

agrees with $-2 + 2i$. **4.** $z^4 = (\sqrt{2})^4$ cis 900°

$= 4$ cis 180° $= -4$ **5.** $4i = 4$ cis 90°;

2 cis 45° $= \sqrt{2} + i\sqrt{2}$ and 2 cis 225° $=$

$-\sqrt{2} - i\sqrt{2}$; $(\sqrt{2} + i\sqrt{2})^2 = 2 + 4i + 2i^2$

$= 4i$ and $(-\sqrt{2} - i\sqrt{2})^2 = 2 + 4i + 2i^2 = 4i$

### Practice Test

**1.** $(2, 90°)$ **2.** $(6\sqrt{2}, -45°)$ **3.** $(5, 143.1°)$ **4.** $(-2\sqrt{2}, 2\sqrt{2})$

**5.** $(-4\sqrt{3}, -4)$ **6.** $(6.43, -7.66)$

**7.** 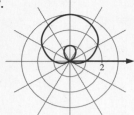  **8.**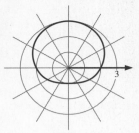

**9.**   **10.**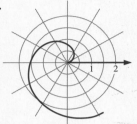

**11.** $\sqrt{2}$  **12.** $\sqrt{2}$ cis $45°$  **13.** 2 cis $240°$  **14.** $2\sqrt{2}$ cis $285°$
**15.** $-0.988 - 0.151i$  **16.** $-0.213 + 0.977i, -0.740 - 0.673i,$
$0.953 - 0.304i$
**17.**    **18.** (graph)

**19.** $z_1, z_2,$ and $z_3$ all lie on the circle $|z| = 2$ because
$|z_1| = |z_2| = |z_3| = 2$.

**Mixed Review**

**1.** {real numbers}; $\{y \mid y \geq -1\}$; 0 and 2  **2.** $(g - f)(x) = x^2 +$
$4x - 1; (f \circ g)(x) = -2x^2 - 4x + 1; (g \circ f)(x) = 4x^2 - 8x + 3$
**3.** $f^{-1}(x) = -\frac{1}{2}x + \frac{1}{2}$  **4.** $(-2 + \sqrt{5},$
$5 - 2\sqrt{5}), (-2 - \sqrt{5}, 5 + 2\sqrt{5})$
**5.** $\sqrt[3]{2}$ cis $50°, \sqrt[3]{2}$ cis $170°, \sqrt[3]{2}$ cis $290°$
**6.** tan (Cos$^{-1}$ 0) is not defined, because
tan $\frac{p}{2}$ is not defined; cos (Tan$^{-1}$ 0) $=$
cos 0 $= 1$

**7.** 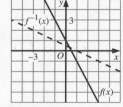  **8.** (graph)
$(-1.2, 1.6)$  $(1.2, 1.6)$
$(0, -2)$

**9. a.** about $125.1°, 30.8°,$ and $24.1°$
**b.** about $8.18$ cm$^2$  **10.** $-\sqrt{5}, -1 \pm 3i$

**11.** 3  **12.** 26  **13.** $\frac{1}{6}$  **14.** $-\cos x$  **15.** $-1024$  **16.** $\cos^2 x$  **17.** symmetry
in the origin; ellipse  **18.** $\frac{\sqrt{2-\sqrt{2}}}{2}; 1 - \sqrt{2}$  **19.** $R(n) = \frac{1}{2}n^2 + \frac{1}{2}n$
$+ 1; 8$  **20.** $-\frac{5}{13}; \frac{120}{169}; \frac{5}{13}$  **21.** 3.64  **22.** $-\frac{3}{2}; 1 \pm \sqrt{6}$  **23.** $71.6°, 135°,$
$251.6°, 315°$  **24.** $0° < \theta < 120°,$ or $180° < \theta < 240°$  **25.** $x^2 +$
$y^2 + 4x = 0,$ or $(x + 2)^2 + y^2 = 4;$ a circle with center $(-2, 0)$
and radius 2  **26.** $x - 2y = -1; (1.4, 1.2)$  **27.** Sample answer:
$y = \frac{5}{2} \cos 2\left(x - \frac{p}{4}\right)$  **28.** $A(x) = \frac{1}{4}x(8 - x); 4$ m$^2$

# Chapter 12:
# Vectors and Determinants

## Sections 12-1 and 12-2
**1.** E  **2.** F  **3.** A  **4.** C  **5.** B  **6.** D
**7.**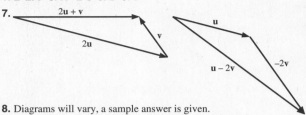

**8.** Diagrams will vary, a sample answer is given.
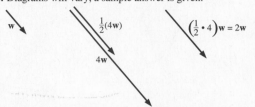

**9.** $(-4, -6); 2\sqrt{13}$  **10. a.** $(-8, -6)$  **b.** 10  **c.** $(-9.5, -15)$
**d.** $\left(-\frac{3}{5}, -\frac{4}{5}\right)$

**11.** Because $|\mathbf{v}|$ is the
magnitude or length of the
vector $\mathbf{v}$; it has no direction so it
cannot be a vector. Since $|\mathbf{v}|$
represents a length, $|\mathbf{v}|$ must be
a nonnegative real number.
**12.** 4.1 N; about $257°$  **13.** No;
$\mathbf{u} - \mathbf{v}$ is defined as $\mathbf{u} + (-\mathbf{v})$.
The vector $\mathbf{u} + (-\mathbf{v})$ is not the
same as $\mathbf{v} - \mathbf{u} = \mathbf{v} + (-\mathbf{u})$, as
the diagram at the right shows.
The vectors $\mathbf{u} - \mathbf{v}$ and $\mathbf{v} - \mathbf{u}$
have the same magnitude and
opposite directions.  **14.** $(8, -3)$
**15.** $10\sqrt{3} \approx 17.3$ knots south;
10 knots west

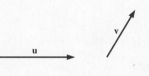

## Sections 12-3 and 12-4
**1.** A  **2.** D  **3.** Sample answer: $(x, y) = (5, 3) + t(7, 7); x = 5 + 7t,$
$y = 3 + 7t$  **4.** $(x, y) = (-6, 2) + t(4, -3); x = -6 + 4t, y =$
$2 - 3t$  **5.** $x = 3 + 2t, y = 5 - 2t; x + y = 8$  **6.** The parametric
equations $x = 3 + t$ and $y = 4 - 2t$ have the Cartesian equation
$2x + y = 10.$ When $t = -3, (x, y) = (0, 10)$ and when $t = 2,$
$(x, y) = (5, 0).$ Therefore, the curve is the first-quadrant portion of
the line $2x + y = 10,$ or a line segment with endpoints $(5, 0)$ and
$(0, 10).$  **7. a.** $8(-4) + (-6)(3) = -50; -4(8) + 3(-6) = -50;$
$\mathbf{u} \cdot \mathbf{v} = \mathbf{v} \cdot \mathbf{u}$  **b.** $\mathbf{v} \cdot \mathbf{v} = (-4)(-4) + (3)(3) = 16 + 9 = 25;$
$|\mathbf{v}|^2 = (\sqrt{(-4)^2 + 3^2})^2 = 5^2 = 25$  **c.** $(8, -6) = -2(-4, 3),$ so
$\mathbf{u} = k\mathbf{v}$ and $\mathbf{u}$ and $\mathbf{v}$ are parallel.  **8.** $126.9°$  **9.** $(x, y) = (9, 4) +$
$t(1, -4)$  **10. a.** $(5, 12); 13$  **b.** $t = -1; (-4, -13)$  **11. a.** $100.3°$
**b.** $100.3°$  **12.** 36

## Sections 12-5 and 12-6
**1.** C  **2.** C  **3.** A  **4.** B  **5.** $\left(\frac{11}{2}, -\frac{1}{2}, 8\right)$  **6. a.** $(4, -3, 12)$  **b.** 13
**c.** 7  **7.** $(x - 1)^2 + (y + 1)^2 + (z - 4)^2 = 121; (3 - 1)^2 +$
$(5 + 1)^2 + (-5 - 4)^2 = 4 + 36 + 81 = 121$  **8.** $(x, y, z) =$
$(5, -2, 4) + t(1, 3, -3); x = 5 + t, y = -2 + 3t, z = 4 - 3t$

**9.**

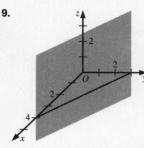

the z-axis **10.** $x + 5y - 2z = 4$
**11.** The graphs are parallel planes if $s \neq t$ since each is perpendicular to the vector $(a, b, c)$. **12.** $(-3, -4, 8)$; 11 **13.** $45.6°$ **14. a.** Sample answers: $(3, -4, -2), (2, -3, 3)$ **b.** $(x, y, z) = (-1, 0, 7) + t(-1, 1, 5)$ **15.** Answers will vary, a sample answer is given. $AB = CD = \sqrt{21}$ and $BC = AD = 7$; opposite sides are equal in length, so the quadrilateral is a parallelogram. **16. a.** $x + 4y + 3z = 13$ **b.** $6 + 4.1 + 3.1 = 13$ **c.** $PA = \sqrt{66} = PB$ **17.** $9x - 2y - 6z = -88$ **18.** Answers will vary, a sample answer is given. Show that $\mathbf{u} = k\mathbf{v}$ for some scalar $k$ or use the formula $\cos\theta = \dfrac{\mathbf{u} \cdot \mathbf{v}}{|\mathbf{u}||\mathbf{v}|}$ to show that $\cos\theta = \pm 1$. Choices and reasons will vary.

### Sections 12-7, 12-8, and 12-9

**1.** D **2.** A **3.** C **4.** Answers will vary, a sample answer is given. The value of the determinant is zero; if you expand the determinant by minors of the row containing zeros, then the determinant is equal to

$0 + 0 + 0 = 0$. **5. a.** $\begin{vmatrix} 3 & 4 & -1 \\ 2 & -2 & 5 \\ 1 & 0 & -3 \end{vmatrix} = 60;$ $\begin{vmatrix} 2 & -2 & 5 \\ 3 & 4 & -1 \\ 1 & 0 & -3 \end{vmatrix}$

$= -60;$ thus, $\begin{vmatrix} 3 & 4 & -1 \\ 2 & -2 & 5 \\ 1 & 0 & -3 \end{vmatrix} = -\begin{vmatrix} 2 & -2 & 5 \\ 3 & 4 & -1 \\ 1 & 0 & -3 \end{vmatrix}.$ **b.** If you

interchange two rows of a determinant, its value is multiplied by $-1$. **6.** $(4, -4)$ **7.** 5 **8.** $(10, -13, 4)$; $(-10, 13, -4)$ **9.** $(10, -13, 4) \cdot (3, 2, -1) = 30 - 26 - 4 = 0$, so $\mathbf{u} \times \mathbf{v}$ is perpendicular to $\mathbf{u}$.

**10.** $\frac{1}{2}\sqrt{285}$ sq units **11.** $\begin{vmatrix} a & b \\ c & d \end{vmatrix} = ad - bc;$ $\begin{vmatrix} a - kb & b \\ c - kd & d \end{vmatrix}$

$= (a - kb)d - (c - kd)b = ad - kdb - (bc - kbd) = ad - bc$ **12.** $(2b, -a)$ **13.** $(4, -3, -1)$ **14.** Answers will vary, a sample answer is given. The possible solutions are (1) a unique solution (when the three planes intersect in a single point), (2) infinitely many solutions (when the three planes intersect in a line), or (3) no solution (when the planes have no common solution; they are parallel or intersect two at a time). **15.** $y + z = 3$ **16. a.** $\begin{vmatrix} \mathbf{i} & \mathbf{j} & \mathbf{k} \\ 0 & 2 & 4 \\ 0 & 3 & 0 \end{vmatrix} =$

$(-12, 0, 0)$; $\frac{1}{2}|(-12, 0, 0)| = 6$ **b.** $\cos A = \dfrac{(0, 2, 4) \cdot (0, 3, 0)}{|(0, 2, 4)||(0, 3, 0)|}$

$= \dfrac{6}{\sqrt{20}(3)} = \dfrac{\sqrt{5}}{5}; \sin A = \sqrt{1 - \left(\dfrac{\sqrt{5}}{5}\right)^2} = \dfrac{2\sqrt{5}}{5},$

area $= \frac{1}{2}|\overrightarrow{AB}||\overrightarrow{AC}|\sin A = \frac{1}{2}(\sqrt{20})(3)\left(\dfrac{2\sqrt{5}}{5}\right) = 6$

## Chapter 12 Review
### Quick Check

**1.** Answers will vary, a sample sketch is given.

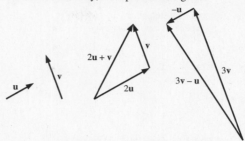

**2.** $(-1.03, 2.82)$ **3.** $(12, 5)$; 13; $x = 12t$ and $y = 4 + 5t$ **4.** $\mathbf{u} \cdot \mathbf{v} = 9(2) + (-3)(6) = 0$, so $\mathbf{u}$ and $\mathbf{v}$ are perpendicular; $\mathbf{w} = (-3, -9) = -1.5(2, 6) = -1.5\mathbf{v}$, so $\mathbf{v}$ and $\mathbf{w}$ are parallel. **5.** $135°$ **6.** Sample answers: $(2, -1, 0), (3, 0, -1), (4, 1, -2)$ **7.** $x + 2y + 3z = -4$; $x$-intercept: $-4$, $y$-intercept: $-2$, $z$-intercept: $-\frac{4}{3}$ **8. a.** $-9$ **b.** 10

**9.** $(-35, 24)$ **10.** $(1, -5, 1) \cdot (1, 0, -1) = 1 - 1 = 0; (1, -5, 1) \cdot (5, 1, 0) = 5 - 5 = 0; |(1, -5, 1)| = \sqrt{27} = 3\sqrt{3} \approx 5.20$ sq units.

### Practice Test

**1.** $\frac{5}{3}\mathbf{v}$ **2.** $\frac{8}{3}\mathbf{v}$ **3.** 13 **4.** $-7$ **5.** $\left(\frac{3}{5}, \frac{4}{5}\right)$ **6.** 18.8 knots south

**7.** 6.8 knots west **8.** $(3, 1), \sqrt{10}$ **9.** $(x, y) = (2, -3) + t(3, 1)$ **10.** 6 **11.** $(2, 2, 1)$ **12.** $\overrightarrow{AB} = (-4, 2, -4), \overrightarrow{AC} = (2, 2, -1), \overrightarrow{AB} \cdot \overrightarrow{AC} = 0$ **13.** 9 sq units **14.** $(3, 4, -2)$ **15.** Sample answer: $(1, 1, 1)$ **16.** $(-3, 0, 1)$ **17.** $(2, 1, 1)$ **18.** $(-17, -1, -10)$ **19.** 3

### Mixed Review

**1.** $\frac{1}{4} - \dfrac{\sqrt{7}}{4}i, -\frac{1}{2}$ **2.** $-\frac{1}{2}$ **3.** 7 **4.** $-512$ **5.** $-\sqrt{3} < x < -1$ or $1 < x < \sqrt{3}$ **6.** 1, 2.5 **7.** $-\frac{1}{3}, 1, \frac{3}{2}$ **8.** $\frac{3}{4}$ **9.** $\{x \mid x \neq \frac{p}{2} + np$, where $n$ is an integer$\}$ **10.** $x \approx 0.64 + 2np$ and $x \approx 2.50 + 2np$, for any integer $n$ **11.** 22.5 m; $2\frac{6}{7}$ s **12.** $111.8°$ **13.** Period; the new period is $\frac{1}{3}$ of the original period; the graph is shrunk horizontally.

**14.** 2 cis 90° $= 2i$, 2 cis 210° $= -\sqrt{3} - i$, and 2 cis 330° $= \sqrt{3} - i$ **15.** $(x, y, z) = (-3, 4, 1) + t(4, -4, -2); x + 3y - 4z = 5$ **16.** $90°$ **17.** $\dfrac{132}{493}; \dfrac{240}{161}; \dfrac{3}{7}$ **18.** $B \approx 22.3°, C \approx 137.7°, c \approx 70.8$, or $B \approx 157.7°, C \approx 2.3°, c \approx 4.2$ **19.** $2\mathbf{u} - \mathbf{v} = (-6, -2); (-6, -2) \cdot (-1, 3) = 6 - 6 = 0$, so $2\mathbf{u} - \mathbf{v}$ is perpendicular to $\mathbf{u}$. **20.** 6; 7

## Chapter 13:
# Sequences and Series

### Sections 13-1, 13-2, and 13-3

**1.** C **2.** D **3.** $0, \frac{3}{2}, \frac{8}{3}, \frac{15}{4}$; neither
**4.** Sample answer: $t_n = (-1)(-3)^n$
**5.** arithmetic; $t_n = 110 - 10n$
**6.** 10, 15, 25, 45 **7.** $t_1 = 1,$
$t_n = \frac{1}{n} \cdot t_{n-1}$ **8.** 6400
**9.** $(-1)^1 + (-1)^2 + \dots +$
$(-1)^n = \dfrac{(-1)[1 - (-1)^n]}{1 - (-1)} =$
$\dfrac{-[1 - (-1)^n]}{2} = \dfrac{(-1)^n - 1}{2}$

**10.** Answers will vary, a sample answer is given. A sequence can be thought of as a set of the ordered pairs $(1, t_1)$, $(2, t_2)$, $(3, t_3)$, and so on. **11.** 288 **12.** 121.5 **13.** $t_{n+1} = t_n + d$, where $t_1$ and $d$ are given; this equation states that each term after the first term is obtained by adding the common difference to the previous term. **14.** $V_n = 0.8V_{n-1}$; about 3 years **15.** $t_1 = a$, $t_n = a^2$, $d = 1$, and $a^2 = a + (n-1)1$, so $n = a^2 - a + 1$; $S_n = \frac{n(t_1 + t_n)}{2} = \frac{(a^2 - a + 1)(a + a^2)}{2} = \frac{a}{2}(a^2 - a + 1)(a + 1) = \frac{a(a^3 + 1)}{2}$ **16.** $t_n = 128\left(\frac{1}{2}\right)^n$; $S_0 = 0$; $S_n = S_{n-1} + 128\left(\frac{1}{2}\right)^n$ **17.** 42,925

### Sections 13-4, 13-5, 13-6, and 13-7

**1.** B **2.** A **3.** 1 **4.** 0 **5.** Sometimes; Explanations will vary, a sample answer is given. The series $8 - 4 + 2 - 1 + \ldots$ has sum $5\frac{1}{3}$ and the series $-6 + 2 - \frac{2}{3} + \ldots$ has sum $-4.5$. **6.** $10\frac{2}{3}$ **7.** 2, 1.6, 1.28 **8.** $1 + (-8) + 81 + (-1024) + 15,625$ **9.** Sample answer:

$\sum\limits_{k=1}^{\infty} 32\left(\frac{1}{4}\right)^k$ **10. a.** $\frac{1}{2} = \frac{1}{2^1}$ **b.** $\frac{1}{2} - \frac{1}{4} - \frac{1}{8} - \ldots - \frac{1}{2^k} = \frac{1}{2^k}$

**c.** $\frac{1}{2} - \frac{1}{4} - \frac{1}{8} - \ldots - \frac{1}{2^k} - \frac{1}{2^{k+1}} = \frac{1}{2^k} - \frac{1}{2^{k+1}} = \frac{1}{2^k} \cdot \frac{2}{2} - \frac{1}{2^{k+1}} = \frac{2}{2^{k+1}} - \frac{1}{2^{k+1}} = \frac{1}{2^{k+1}}$; Thus, the statement is true for $n = k + 1$. **11.** 1 **12.** does not exist **13.** $-0.4$ **14.** $-\frac{1}{2} < x < \frac{1}{2}$; $\frac{1}{1 + 2x}$

**15.** Answers will vary, a sample answer is given. A repeating decimal can be written as an infinite geometric series with the repeating digits determining the first term and with a negative power of ten as the common ratio. The formula $S = \frac{t_1}{1 - r}$ can be used to express the decimal as a fraction. **16.** 4 m **17.** $-100$ **18.** When $n = 1$, $(1 \cdot 2) = 2$ and $\frac{1(1+1)(1+2)}{3} = 2$, so the statement is true. Assume that $(1 \cdot 2) + (2 \cdot 3) + (3 \cdot 4) + \ldots + k(k+1) = \frac{k(k+1)(k+2)}{3}$. Then $(1 \cdot 2) + (2 \cdot 3) + (3 \cdot 4) + \ldots + k(k+1) + (k+1)(k+2) = \frac{k(k+1)(k+2)}{3} + (k+1)(k+2) = \frac{(k^3 + 3k^2 + 2k) + (3k^2 + 9k + 6)}{3} = \frac{k^3 + 6k^2 + 11k + 6}{3} = \frac{(k+1)(k+2)(k+3)}{3} = \frac{(k+1)[(k+1)+1][(k+1)+2]}{3}$. Thus, the statement is true for $n = k + 1$. Therefore, the statement is true for all positive integers.

### Chapter 13 Review
#### Quick Check

**1.** $\sqrt{2}$, 2, $2\sqrt{2}$, 4; geometric **2.** arithmetic; $t_n = 2n + 1$ **3.** 2, 4, 16, 256 **4.** 840 **5.** 1111.11 **6. a.** $\infty$ **b.** does not exist **7.** $\frac{8}{7}$ **8.** Sample answer: $\sum\limits_{k=0}^{\infty} \left(\frac{1}{8}\right)^k$ **9.** First, show that the statement is true for $n = 1$, that is, show that $4 = 2(1)(1 + 1)$. Then, assume that the statement is true when $n = k$, where $k$ is a positive integer. Finally, use this assumption to prove that the statement must be true if $n = k + 1$.

**1.** geometric; $t_n = 24\left(\frac{1}{2}\right)^{n-1}$ **2.** neither; $t_n = n^2 + 4$ **3.** arithmetic; $t_n = 88 - 4n$ **4.** $t_1 = 3$, $t_n = 2 \cdot t_{n-1}$ **5.** $t_1 = 3$, $t_2 = 4$, $t_n = t_{n-1} + t_{n-2}$ **6.** 145 **7.** 5120 **8.** 60,700 **9.** 47.988 **10.** 1 **11.** does not exist **12.** $1 < x < 3$ **13.** $\frac{1}{x - 1}$

**14.** $\sum\limits_{n=0}^{\infty} (-1)^n(x - 2)^n$ **15.** 0.000009 **16.** For $n = 1$, $1 = \frac{1[3(1) - 1]}{2} = \frac{2}{2} = 1$. Assume $1 + 4 + 7 + \ldots + (3k - 2) = \frac{k(3k - 1)}{2}$; then for $n = k + 1$, $1 + 4 + 7 + \ldots + (3k - 2) + [3(k + 1) - 2] = \frac{k(3k - 1)}{2} + [3(k + 1) - 2] = \frac{3k^2 - k}{2} + 3k + 1 = \frac{3k^2 - k + 2(3k + 1)}{2} = \frac{3k^2 + 5k + 2}{2} = \frac{(k + 1)[3(k + 1) - 1]}{2}$. Thus, the statement is true for $n = k + 1$.

**Mixed Review**

**1.** $x + 2y = 17$; $x = 5 - 4t$ and $y = 6 + 2t$ **2.** $-\frac{37}{12}$, $\frac{12\sqrt{3} + 35}{74}$, $\frac{1081}{1369}$, $-\frac{\sqrt{37}}{37}$ **3.** $-1$ **4.** $\frac{2\sqrt{2}}{3}$ **5.** $-512i$ **6.** 6 **7.** 12.5 **8.** $-1$ **9.** $\tan(p + \theta) = \frac{\tan p + \tan \theta}{1 - \tan p \tan \theta} = \frac{0 + \tan \theta}{1 - 0(\tan \theta)} = \frac{\tan \theta}{1} = \tan \theta$; this shows that $y = \tan \theta$ is a periodic function with fundamental period $p$. **10.** $3375 **11.** $(-10.5, -22.5)$ **12.** $(10, 40)$, $(2, 0)$ **13.** $(2.5, 6)$, $(2.5, -6)$, $(-2.5, 6)$, $(-2.5, -6)$ **14.** about 134.8° and 45.2° **15.** $\pm\frac{1}{2}$, $-2 \pm i\sqrt{2}$ **16.** $-\frac{3}{2} < x < -1$ or $x > \frac{3}{2}$ **17.** 0.16 **18.** about 2.86 **19.** $\left(\frac{1}{2}x - 6\right)^2 = \frac{1}{4}x^2 - 6x + 36$; $\frac{1}{2}x^2 - 6$ **20.** 26.6°, 90°, 206.6°, 270° **21. a.** 800 **b.** $2n^2$ **22.** about 16.6; about 240.5 sq. units **23.** $-28$; $(-40, -60, 10)$; about 111.0° **24.** symmetry in the line $y = x$ and in the origin; hyperbola **25.** 1 **26.** If $n = 1$, then $2 = 2 \cdot 1^2$, so the statement is true for $n = 1$. Suppose the statement is true for $n = k$. Then $2 + 6 + 10 + \ldots + (4k - 2) = 2k^2$. Then $2 + 6 + 10 + \ldots + (4k - 2) + [4(k + 1) - 2] = 2k^2 + [4(k + 1) - 2] = 2k^2 + 4k + 2 = 2(k^2 + 2k + 1) = 2(k + 1)^2$. Therefore, the statement is true for $n = k + 1$.

### Chapter 14:
### Matrices

#### Sections 14-1, 14-2, and 14-3

**1.** $B + C$; $\begin{bmatrix} 0 & -3 \\ 5 & 3 \\ 1 & -2 \end{bmatrix}$ **2.** $AB$; $\begin{bmatrix} 4 & 0 \\ -4 & 6 \end{bmatrix}$ **3.** $\begin{bmatrix} 5 & 7 \\ 0 & -3 \\ 1 & -4 \end{bmatrix}$; $\begin{bmatrix} 4 & 7 \\ -5 & -1 \\ 2 & -4 \end{bmatrix}$ **4.** $\begin{bmatrix} 2 & 0 \\ 10 & -4 \\ -2 & 0 \end{bmatrix}$; $\begin{bmatrix} 3 & 3 \\ 10 & -9 \\ -4 & 2 \end{bmatrix}$ **5.** $-D = \begin{bmatrix} -3 & 2 \\ 1 & -2 \end{bmatrix}$; $D^{-1} = \begin{bmatrix} \frac{1}{2} & \frac{1}{2} \\ \frac{1}{4} & \frac{3}{4} \end{bmatrix}$; $D^2 = \begin{bmatrix} 11 & -10 \\ -5 & 6 \end{bmatrix}$

**6.** $x = -20$, $y = 30$ **7.** A matrix has no inverse if it is not a square matrix or if its determinant is equal to zero. **8.** $A + B = \begin{bmatrix} 150 & 2 & 75 \end{bmatrix}$; $A - B = \begin{bmatrix} 70 & 2 & -45 \end{bmatrix}$;

Answers will vary, a sample answer is given. The matrix $A + B$ gives the total number of calories, grams of fat, and milligrams of sodium in a serving of granola and skim milk; the matrix $A - B$ is a way to compare the number of calories, fat content, and sodium content in the cereal and in the milk. Since the first two elements of matrix $A - B$ are positive and the third element is negative, the cereal has a higher calorie count and a lower sodium content than the milk.

**9. a.** no **b.** no **10.** $\begin{bmatrix} 3 & -1 \\ -\frac{4}{3} & 2 \end{bmatrix}$ **11.** $\begin{bmatrix} 8 & 7.5 \\ -3 & -2.5 \end{bmatrix}$

### Sections 14-4, 14-5, and 14-6

**1.** B **2.** A **3.** C **4.**
$\begin{array}{c} \\ A \\ B \\ C \\ D \end{array}\begin{array}{cccc} A & B & C & D \\ \end{array}$
$\begin{array}{c} A \\ B \\ C \\ D \end{array}\begin{bmatrix} 0 & 0 & 1 & 0 \\ 1 & 0 & 1 & 0 \\ 1 & 0 & 0 & 0 \\ 1 & 0 & 1 & 0 \end{bmatrix}$ **5. a.**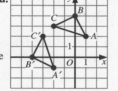

**b.** $\begin{array}{c} A \\ B \\ C \\ D \end{array}\begin{bmatrix} 0 & 1 & 1 & 1 \\ 1 & 0 & 0 & 0 \\ 1 & 1 & 0 & 1 \\ 1 & 0 & 0 & 0 \end{bmatrix}$ **c.** $\begin{bmatrix} 3 & 2 & 1 & 2 \\ 1 & 1 & 1 & 1 \\ 3 & 2 & 1 & 2 \\ 1 & 1 & 1 & 1 \end{bmatrix}$ A to A; A—B—A,

A—C—A, A—D—A; A to B: A—C—B, A—B; A to C: A—C;

A to D: A—D, A—C—D **6. a.** $T = \begin{bmatrix} 0.85 & 0.15 \\ 0.10 & 0.90 \end{bmatrix}$

**b.** $[\ 50,000 \quad 10,000\ ]$ **c.** 43,500 acres; 38,625 acres

**d.** $[\ 24,000 \quad 36,000\ ]\begin{bmatrix} 0.85 & 0.15 \\ 0.10 & 0.90 \end{bmatrix} = [\ 24,000 \quad 36,000\ ]$

**7.** Since $T = \begin{bmatrix} 2 & 3 \\ 4 & 6 \end{bmatrix}$ and $\begin{vmatrix} 2 & 3 \\ 4 & 6 \end{vmatrix} = 0$, $T$ maps every point

onto a line through the origin. Since $(1, 0)$ is mapped to $(2, 4)$ and $(0, 1)$ is mapped to $(3, 6)$, the line contains $(2, 4)$ and $(3, 6)$. The equation of the line is $y = 2x$. **8. a.**
**b.** reflection in the line $y = -x$
**c.** $\begin{bmatrix} 0 & -1 \\ -1 & 0 \end{bmatrix}$; $-1$ **d.** The

areas are equal; the orientations are opposite. **9.** There is no communication possible between

the points in the network. **10.** $M + M^2 + \ldots + M^n$ **11. a.** All the market shares remain constant, that is, no one changes brands.
**b.** Product $A$ has the total market share, whereas products $B$ and $C$ have none, that is, everyone buys product $A$. **12.** $[\ 0.4 \quad 0.6\ ]$

**13. a.** $\begin{bmatrix} 5 & 8 \\ 3 & -12 \end{bmatrix}$ ; $\begin{bmatrix} 3 & 6 \\ 9 & -10 \end{bmatrix}$ **b.** Part (a) shows that $S \circ T$

maps $(1, 0)$ to $(5, 3)$, $(0, 1)$ to $(8, -12)$, and $(x, y)$ to $(5x + 8y, 3x - 12y)$. $T \circ S$ maps $(1, 0)$ to $(3, 9)$, $(0, 1)$ to $(6, -10)$, and $(x, y)$ to $(3x + 6y, 9x - 10y)$. Since $S \circ T$ maps $P$ to a different point than does $T \circ S$, $(S \circ T)(P) \neq (T \circ S)(P)$. **14.** 6 units

to the right and 2 units up; $\begin{bmatrix} x \\ y \end{bmatrix} + \begin{bmatrix} 6 \\ 2 \end{bmatrix} = \begin{bmatrix} x' \\ y' \end{bmatrix}$.

### Chapter 14 Review
#### Quick Check

**1.** Because $A$ is a $2 \times 3$ matrix and $B$ is a $3 \times 2$ matrix;
$\begin{bmatrix} -13 & 1 \\ 0 & -7 \\ 13 & 2 \end{bmatrix}$ **2.** $\begin{bmatrix} -13 & -7 \\ -21 & 3 \end{bmatrix}$ ; $\begin{bmatrix} 13 & 2 & 35 \\ -9 & -3 & 0 \\ -1 & 1 & -20 \end{bmatrix}$ ; no

**3.** $\begin{bmatrix} -1 & 0 & -5 \\ 3 & 1 & 0 \end{bmatrix} + \begin{bmatrix} 1 & 0 & 5 \\ -3 & -1 & 0 \end{bmatrix} = \begin{bmatrix} 0 & 0 & 0 \\ 0 & 0 & 0 \end{bmatrix}$

**4. a.**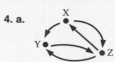
**b.** Calculate $M^2$; each element of $M^2$ gives the number of two-step communication paths between the specified points. **c.** 3 (X—Y—Z,

X—Z—X, X—Z—Y) **5. a.** 65% **b.** $[\ 0.6 \quad 0.4\ ]\begin{bmatrix} 0.8 & 0.2 \\ 0.3 & 0.7 \end{bmatrix}$

$= [\ 0.6 \quad 0.4\ ]$; over a long period of time, the percents will stabilize with 60% buying a national brand and 40% buying the store brand. **6. a.** $(x, -2y)$; $(x, y)$ **b.** $\dfrac{\text{area of } \triangle A'B'C'}{\text{area of } \triangle ABC} = \dfrac{2}{1}$

#### Practice Test

**1.** $\begin{bmatrix} -6 & 15 \\ 0 & 5 \end{bmatrix}$ **2.** not possible **3.** $\begin{bmatrix} 18 & -27 \\ 0 & 0 \\ 22 & -33 \end{bmatrix}$

**4.** $\begin{bmatrix} 1 & 2 & 3 \\ 7 & -4 & 5 \end{bmatrix}$ **5.** $\begin{bmatrix} \frac{1}{14} & \frac{3}{14} \\ \frac{2}{7} & -\frac{1}{7} \end{bmatrix}$ **6.** does not exist

**7.** $2x + 3y = -6$ and $4x - y = 16$ **8.** $\begin{bmatrix} 3 \\ -4 \end{bmatrix}$

**9.** $\begin{array}{c} A \\ B \\ C \\ D \end{array}\begin{array}{cccc} A & B & C & D \\ \end{array}\begin{bmatrix} 0 & 1 & 0 & 1 \\ 1 & 0 & 1 & 1 \\ 1 & 0 & 0 & 1 \\ 0 & 0 & 1 & 0 \end{bmatrix}$ ; $\begin{bmatrix} 1 & 0 & 2 & 1 \\ 1 & 1 & 1 & 2 \\ 0 & 1 & 1 & 1 \\ 1 & 0 & 0 & 1 \end{bmatrix}$ **10.** D

**11.** $\begin{array}{c} \\ \text{UW} \\ \text{Other} \end{array}\begin{array}{cc} \text{UW} & \text{Other} \\ \end{array}\begin{bmatrix} 0.95 & 0.05 \\ 0.10 & 0.90 \end{bmatrix}$ **12.** $\begin{array}{cc} \text{UW} & \text{Other} \\ \end{array}[\ 0.30 \quad 0.70\ ]$

**13.** 40.17% **14.** $\begin{bmatrix} 2 & 1 \\ 1 & 0 \end{bmatrix}$ **15.** $(0, 0), (6, 0), (16,8)$

#### Mixed Review

**1.** $(-28,-40)$ **2. a.** 1050 **b.** 5,242,875 **3.** about 109.5°; $3\sqrt{2}$
**4.** $-32,768 + 32,768i\sqrt{3}$; 2 cis 165° $\approx -1.93 + 0.52i$ and

2 cis 345° $\approx 1.93 - 0.52i$ **5.** $\begin{array}{c} A \\ B \\ C \\ D \end{array}\begin{array}{cccc} A & B & C & D \\ \end{array}\begin{bmatrix} 1 & 3 & 2 & 2 \\ 0 & 1 & 1 & 1 \\ 1 & 2 & 1 & 1 \\ 1 & 2 & 2 & 1 \end{bmatrix}$ **6.** $-1.6$

**7.** 0.75 **8.** 3.75 **9.** $-\frac{1}{2}$ ; $x + 2y = 0$; $T = \begin{bmatrix} 6 & -2 \\ -3 & 1 \end{bmatrix}$

**10.** about 466.0 knots; about 220.6° **11.** $\frac{13}{85}$ ; $\frac{7}{25}$ ; $\frac{\sqrt{17}}{17}$ **12.** 0; does

not exist **13.** 2, 5, 10, 17, 26, 37; $t_n = n^2 + 1$

### Chapter 15:
# Combinatorics

#### Sections 15-1 and 15-2

**1.** 9 employees **2.** 15 **3.** 8 **4.** Exercise 2 involved using the multiplication principle because the problem involved a two-step choice. Exercise 3 involved using the addition principle because choosing the ice cream and choosing the cookie are mutually

exclusive events. **5.** $\frac{5}{6}$ **6.**

**7.** C **8.** 2205 **9.** 1140

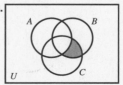

#### Sections 15-3, 15-4, and 15-5

**1.** 60 **2.** 24 **3.** 3,268,760 **4.** 715 **5.** 60 **6.** 12,600 **7.** 56 **8.** 282,240

**9.** 5511 **10.** $128x^7 - 448x^4 + 672x - 560x^{-2} + 280x^{-5} - 84x^{-8} + 14x^{-11} - x^{-14}$

## Chapter 15 Review
### Quick Check

**1. a.** those who voted Republican in both races **b.** those who did not vote Republican in either race **c.** those who did not vote Republican in both races **d.** those who voted for the Republican congressional candidate but not for the Republican presidential candidate **2.** 700 **3.** 48 **4.** $_{10}P_6 = 151,200$; $_{10}C_6 = 210$; $_{10}P_6$ is the number of arrangements of six different items from a set of 10 items when order is important; $_{10}C_6$ is the number of different ways to pick 6 items from a set of 10 items when order is not important. **5.** 5148 **6.** 180 **7.** $_{15}C_6 \cdot 3^6 = 3,648,645$

### Practice Test

**1.** 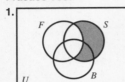  **2.**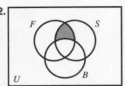

**3.** $\overline{F} \cap S$: seniors not minoring in French; $F \cap S \cap \overline{B}$: senior French minors not majoring in Biology

**4.**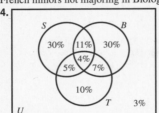

**5.** 70% **6.** 720 **7.** 30,240
**8.** 56 **9.** 95,040 **10.** 720
**11.** 420 **12.** 720 **13.** 15! $\approx$
$1.31 \times 10^{12}$ **14.** 12,441,600
**15.** $-3240a^7$ **16.** $405a^8$

### Mixed Review

**1.** 64 **2.** $P'(0, 0)$, $Q'(13, -6)$; $R'(3, 0)$ **3.** $T = \begin{bmatrix} 3 & 1 \\ -2 & 0 \end{bmatrix}$;

$|T| = 2$; The transformation $T$ preserves the orientation of the

triangle. **4.** $T^{-1} = \begin{bmatrix} 0 & -\frac{1}{2} \\ 1 & \frac{3}{2} \end{bmatrix}$; $T^{-1}: (x, y) \rightarrow \left( -\frac{1}{2}y, x + \frac{3}{2}y \right)$

**5.** 627 **6.** $-\frac{1}{2}$ **7.** does not exist

**8.**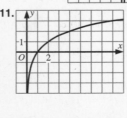

**9.**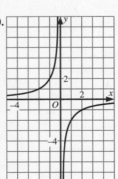

**10.** 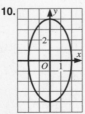 **11.**

**12.** 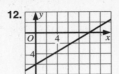 **13.**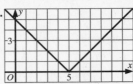

**14.** 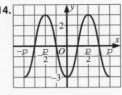 **15.**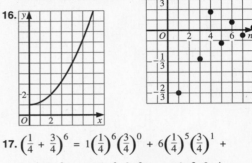

**16.**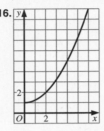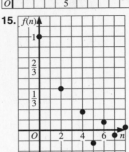

**17.** $\left( \frac{1}{4} + \frac{3}{4} \right)^6 = 1\left(\frac{1}{4}\right)^6\left(\frac{3}{4}\right)^0 + 6\left(\frac{1}{4}\right)^5\left(\frac{3}{4}\right)^1 +$

$15\left(\frac{1}{4}\right)^4\left(\frac{3}{4}\right)^2 + 20\left(\frac{1}{4}\right)^3\left(\frac{3}{4}\right)^3 + 15\left(\frac{1}{4}\right)^2\left(\frac{3}{4}\right)^4 +$

$6\left(\frac{1}{4}\right)^1\left(\frac{3}{4}\right)^5 + 1\left(\frac{1}{4}\right)^0\left(\frac{3}{4}\right)^6 = \frac{1}{4096} + \frac{18}{4096} +$

$\frac{135}{4096} + \frac{540}{4096} + \frac{1215}{4096} + \frac{1458}{4096} + \frac{729}{4096} = \frac{4096}{4096} = 1$;

$\left( \frac{1}{4} + \frac{3}{4} \right)^6 = 1^6 = 1$

## Chapter 16:
## Probability

### Sections 16-1, 16-2, 16-3, and 16-4

**1.** {1H, 1T, 2H, 2T, 3H, 3T, 4H, 4T, 5H, 5T, 6H, 6T}; There are 12 elements in the sample space. **2.** $\frac{1}{12}$ **3.** $\frac{1}{4}$ **4.** $\frac{1}{2}$ **5.** $\frac{7}{12}$ **6.** $\frac{1}{18}$ **7.** $\frac{1}{2}$

**8.** The probability that the building was demolished given that there was an earthquake; the probability that there was an earthquake given that the building was demolished; the probability that the building was demolished given that there was not an earthquake; the probability that there was not an earthquake given that the building was not demolished **9.** False; since the fifty states do not all have the same population. **10.** A **11.** $\approx 0.0004$ **12.** Keep track of the forecaster's predictions and the actual weather over a long period of time.

### Sections 16-5 and 16-6

**1.** $a = 0.4$, $b = 0.2$, $c = 0.2$ **2.** 0.3 **3.** 0.62 **4.** A: 50¢, B: −50¢
**5.** Player A; Answers will vary, a sample answer is given. Player A could pay player B 50¢ before each round. **6.** about 69% **7.** $\frac{3}{4}$ **8.** $\frac{2}{5}$

**9.** The marbles are not equally likely to be drawn. Each marble in bowl #1 has a probability of $\frac{1}{10}$ for being drawn, while each marble in bowl #2 has only a probability of $\frac{1}{20}$ for being drawn. **10.** The firm must feel that their probability of winning the case is less than 20%.
**11.** 0.7 dresses

## Chapter 16 Review
### Quick Check

**1.** {SP, SA, SI, SN, PA, PI, PN, AI, AN, IN}; $\frac{2}{5}$ **2.** $\frac{1}{4}$

**3.** $\frac{14C_7}{2^{14}} \approx 0.209$ **4.** $\frac{36}{52C_5} \approx 0.00001$ (about 0) **5.** $\frac{5}{9}$ **6.** $3500

### Practice Test

**1.** Answers will vary, a sample answer is given. Selecting a red card and selecting a spade from a well-shuffled deck of cards. **2.** Answers will vary, a sample answer is given. Selecting a king from a well-shuffled deck of cards and rolling a 3 on a die. **3.** $\frac{2}{17}$ **4.** $\frac{11}{1105}$

**5.** $\frac{1201}{5525}$ **6.** $\frac{352}{425}$ **7.** $\frac{5}{12}$ **8.** $\frac{1}{6}$ **9.** $\frac{2}{9}$ **10.** $\frac{1}{4}$ **11.** 0.42 **12.** 0.88 **13.** 0.215

**14.** 0.383 **15.** $\frac{75}{442}$ **16.** $\frac{389}{442}$ **17.**

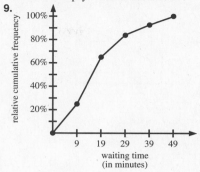

**18.** $\frac{62}{105}$ **19.** $\frac{10}{31}$ **20.** $16

**21.** $16

### Mixed Review

**1. a.**

| $r$ | 0 | 1 | 2 | 3 | 4 | 5 | 6 |
|-----|-----|-----|-----|-----|-----|-----|-----|
| $p(r)$ | 0.05 | 0.19 | 0.31 | 0.28 | 0.14 | 0.04 | 0.00 |

**b.**

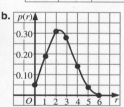

**2. a.**

| $r$ | 0 | 1 | 2 | 3 | 4 | 5 | 6 |
|-----|-----|-----|-----|-----|-----|-----|-----|
| $p(r)$ | 0.02 | 0.09 | 0.23 | 0.31 | 0.23 | 0.09 | 0.02 |

**b.**

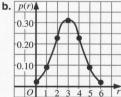

**3.** Answers will vary, a sample answer is given. The graph in Exercise 1 is skewed to the left, while the graph in Exercise 2 is symmetrical about $r = 3$. The maximum and minimum values of the graphs are about the same.

**4.** $\frac{693}{80} = 8.6625$ **5.** $x = \frac{3p}{2} + 2np$, for any integer $n$ **6.** $x = \frac{p}{6} + 2np$ or $x = \frac{5p}{6} + 2np$, for any integer $n$ **7.** $x = \frac{2p}{3} + 2np$ or $x = \frac{4p}{3} + 2np$, for any integer $n$

**8.** $x \approx 72.9° + 180n$, for any integer $n$ **9.** does not exist **10.** 40 **11.** 0 **12.** $e^{0.06}$ **13.** $x \approx 25.1$ **14.** $\angle C = 124.0°, \angle B = 23.0°$

## Chapter 17:
# Statistics

### Sections 17-1, 17-2, 17-3, and 17-4

**1.** 9; 9; 12 **2.** 9; 3 **3.** The second class has the higher mean and standard deviation. **4.** approximately $z = 0.7$ and $z = -0.7$ **5.** 0.26% **6.** Mode; Reasons will vary, a sample answer is given. Finding the mean or median call number would be meaningless. **7.** Median; Reasons will vary, a sample answer is given. The mean would be skewed by the highly-paid executives, and the mode would not reflect their pay at all. **8.** about 10%

**9.**

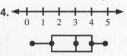

### Sections 17-5 and 17-6

**1.** Answers will vary, a sample answer is given. Those who answered the phone on Thursday night were home to watch the convention, and those who watched were most likely to be Republicans. **2.** Answers will vary, a sample answer is given. There may be students who do their homework only on Thursdays but not on other days, as well as students who might forget to do homework on a Thursday while doing their homework on all other days. **3.** 55% **4.** $0.026 < p < 0.134$ **5.** No; no; Answers will vary, a sample answer is given. You could sample more than 180 calls. This will decrease the standard deviation, $s$. **6.** Answers will vary. Check students' answers. **7.** $0.68 < p < 0.92$ **8.** C **9.** Answers will vary. Check students' answers. **10.** Answers will vary. Check students' answers.

### Chapter 17 Review
#### Quick Check

**1.**

| 4 | 0 3 3 6 8 9 |
|---|---|
| 3 | 1 2 3 7 9 |
| 2 | 0 5 6 7 |
| 1 | 0 2 |
| 0 | 3 4 6 |

3|2 represents a circumference of 3.2 m.

**2.** 2.87 m; 3.15 m
**3.**

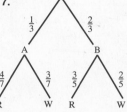

**4.**

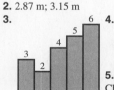

**5.** 2.17; 1.47 **6.** 0.62% **7.** Answers will vary. Check students' answers. **8.** $0 < p < 0.115$

#### Practice Test

**1.**

| 6 | 2 8 |
|---|---|
| 7 | 2 3 3 4 7 |
| 8 | 3 4 5 |
| 9 | 1 6 8 |

**2.**

**3.** 79.7 **4.** 77 **5.** 73 **6.** 36 **7.** Measures of central tendency describe the middle of the data while measures of variability describe how the data is spread out. **8.** Standard values are used to compare data from different normal distributions by converting the data to the same scale.

**9.**

| Earnings | Frequency |
|---|---|
| $20,000 | 11 |
| $40,000 | 3 |
| $200,000 | 1 |

**10.** $36,000 **11.** 1,984,000,000 **12.** $44,542 **13.** 2.28% **14.** 13.59% **15.** 84 **16.** about 0.046

**17.** $0.208 < p < 0.392$ **18.** $0.163 < p < 0.437$ **19.** $\approx 61\%$

**1.** $y - 4 = -\frac{1}{3}(x - 17)$ or $x + 3y = 29$  **2.** $8\frac{2}{3}$  **3.** $y = 3x$

**4.** 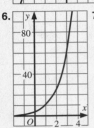 **5.**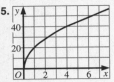

**6.** 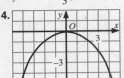 **7.**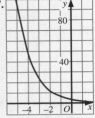

**8.** $y = 10^{1.3}(10^{3.4})^x$ or $y = 19.95(2511.89)^x$  **9.** $y = 4x^7$
**10.** $\log y = \log 4 + x \log 2.4$ or $\log y = 0.38x + 0.60$
**11.** $\log y = \log 2 - \log 3 - 4 \log x$ or $\log y = -4 \log x - 0.18$

**12.** $f(x) = \begin{cases} -x, & 0 \le x < 2 \\ 2x - 6, & 2 \le x < 4 \\ 2, & 4 \le x < 6 \\ -2x + 14, & 6 \le x < 7 \\ 0, & 7 \le x \le 8 \end{cases}$ **13.**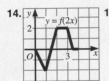

**14.** 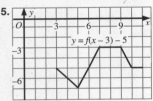 **15.** 

## Chapter 18:
## Curve Fitting and Models

### Sections 18-1, 18-2, 18-3, and 18-4

**1.** A, B, C  **2.** For $x =$ years and $y =$ patrons (in thousands), $y = -15,493 + 8x$; $r = 0.986$  **3.** 451,000  **4.** $y = 10^{42}(10^{6.1})^x$
**5.** Answers will vary due to the points chosen, a sample answer is given. For $x =$ months and $y =$ number of words, $y = 0.0000064x^{5.64}$  **6.** Answers will vary according to the equation found in Exercise 5; for $y = 0.0000064x^{5.64}$, the predicted vocabulary at age 16 (192 months) is 48,305,758 words, which is unreasonable.

**7.**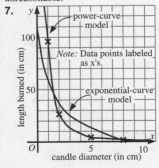

**8. a.** 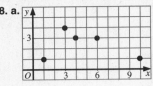 $r = -0.25$  **b.** By eliminating $(1, 1)$, the value of $r$ becomes $-0.96$. The least-squares line is then $y = 5 - 0.39x$.  **9.** For $x =$ day and $y =$ concentration (in pounds per million gallons), $y = 10 + 9(0.99)^x$

## Chapter 18 Review
### Quick Check

**1.** B  **2.**

| $x$ | $y$ | $xy$ | $x^2$ | $y^2$ |
|---|---|---|---|---|
| $-1$ | $3$ | $-3$ | $1$ | $9$ |
| $3$ | $-2$ | $-6$ | $9$ | $4$ |
| $4$ | $-3$ | $-12$ | $16$ | $9$ |
| $5$ | $-6$ | $-30$ | $25$ | $36$ |
| Sums: | $11$ | $-8$ | $-51$ | $51$ | $58$ |

$y = 1.84 - 1.4x$  **3.** A  **4. a.** $y = 1.82x^{-3.8}$  **b.** $y = 1.82(10^{-3.8x})$
**5.**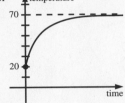

### Practice Test

**1–2.** 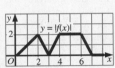 **3.** $y = -20.5x + 162.5$

**4.**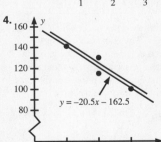

**5.** 60  **6.** $y = 316.2(10^{-1.5x})$ or $y = 316.2(0.032)^x$
**7.** $y = 109.9e^{1.2x}$ or $y = 109.9(3.320)^x$  **8.** $y = 125.9x^3$
**9.** $y = e^{-2x}x^{0.2}$ or $y = 0.135x^{0.2}$  **10.** $y = 3.0(1.2)^x$
**11.** $y = 30.0(0.834)^x$  **12.** 18.6
**13.** 3.4

**14.** 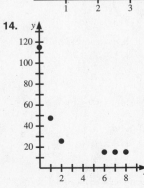 **15.** $y = 15$  **16.** $\log (y - 15) = -0.47x + 2$  **17.** $y = 15 + 100(0.34)^x$  **18.** 34.82  **19.** See if the set of points $(x, \log y)$ can be approximated by a linear function.
**20.** See if the set of points $(\log x, \log y)$ can be approximated by a linear function.

**1.** domain: $\{x \mid -3 \le x \le 3\}$; range: $\{y \mid 0 \le y \le 3\}$; zeros: $\pm 3$
**2.** domain: {real numbers}; range: $\{g(t) \mid g(t) \ge 0\}$; zeros: $-5$
**3.** domain: $\{x \mid x \ne \pm 5\}$; range: {real numbers}; zeros: 0  **4.** domain: $\{t \mid t \ne 5\}$; range: $\{h(t) \mid h(t) \ge 0\}$; zeros: none  **5.** yes; domain: $\{x \mid -3 \le x \le 5\}$; range: $\{y \mid 0 \le y \le 1.5\}$  **6.** yes; domain: {real numbers}; range: $\{y \mid y > 0\}$  **7.** $\frac{1}{3}$  **8.** 1  **9.** 1  **10. a.** $-1 < x < 1$

**b.** $\frac{1}{1 + x^2}$  **11. a.** $|x| > 3$  **b.** $\frac{x}{x-3}$  **12.** 3  **13.** $\mathbf{i} - \mathbf{j} + 3\mathbf{k}$ or $(1, -1, 3)$

**14.** $47.9°$  **15.** $\left( \frac{\sqrt{11}}{11}, -\frac{\sqrt{11}}{11}, \frac{3\sqrt{11}}{11} \right)$  **16.** $\frac{\sqrt{11}}{2}$ square units

**17.** $2, 1, -\frac{1}{2}, -\frac{7}{8}, -\frac{79}{128}$  **18.** 4, 7, 23.5, 275.125, 37,845.8828125

# Chapter 19:
# Limits, Series, and Iterated Functions

## Sections 19-1 and 19-2

**1.** $\lim\limits_{x \to -\infty} f(x) = \infty$; $\lim\limits_{x \to 0^-} f(x) = -2$; $\lim\limits_{x \to 0^+} f(x) = 0$; $\lim\limits_{x \to \infty} f(x) = 0$  **2.** $x \ne 0$  **3.** $-\infty$  **4.** 6  **5.** $a = -2, b = 3, c = 1$, $k = \frac{1}{2}$ or $a = 3, b = -2, c = 1, k = \frac{1}{2}$

**6.** Graphs will vary, a sample graph is given. 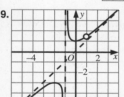  **7.** $-\infty$  **8.** 0

**9.** 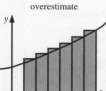  **10.**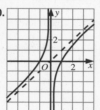

## Sections 19-3, 19-4, 19-5, and 19-6

**1.** 5.765  **2.** 5.427, 5.343, 5.334; $5.\overline{3}$  **3.** $1 - \frac{x}{2!} + \frac{x^2}{4!} - \frac{x^3}{6!} + \dots$, the series converges for all $x \ge 0$.  **4.** 0: attracting; $p, 2p$: repelling
**5.** $f(x) = \frac{1}{2}x + 20, x_0 = 50$; yes; \$40

**6.**
overestimate          underestimate
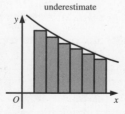

The areas of increasing functions will be overestimated; the areas of decreasing functions will be underestimated.  **7.** 41 terms; 53 terms
**8.** No; the interval of convergence of the power series for $\ln (1 + x)$ is $-1 < x \le 1$.  **9.** fixed points: $\pm\sqrt{2}$; The orbit of $x_0 = 0$ is 0, $-2$, 0, $-2$, ... which is a period 2 orbit; the orbit of $x_0 = \frac{1}{2}$ is the period 3 orbit: ..., 0.80, $-0.55$, $-2.25$, 0.80, $-0.55$, $-2.25$, ...; the orbit of $x_0 = 2$ is 2, 4, 18, 340, ... , which goes to infinity.  **10.** 7 years

## Chapter 19 Review
### Quick Check

**1. a.** $-\infty$  **b.** $\frac{5}{7}$  **2.**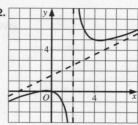

**3.** 83.3 square units  **4.** $\text{Tan}^{-1} 1 = 1 - \frac{1}{3} + \frac{1}{5} - \frac{1}{7} + \dots$; Since $\text{Tan}^{-1} 1 = \frac{p}{4}$, to approximate $p$ you could add the terms of this series and multiply the sum by 4.  **5.** All orbits approach the fixed point $x = -6$.  **6.** 6 years

### Practice Test

**1.** $\frac{2}{3}$  **2.** 6  **3.** does not exist  **4.** $-\infty$

**5.** 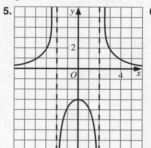  **6.**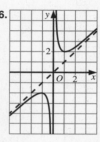

**7.**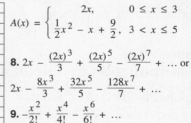
$$A(x) = \begin{cases} 2x, & 0 \le x \le 3 \\ \frac{1}{2}x^2 - x + \frac{9}{2}, & 3 < x \le 5 \end{cases}$$

**8.** $2x - \frac{(2x)^3}{3} + \frac{(2x)^5}{5} - \frac{(2x)^7}{7} + \dots$ or $2x - \frac{8x^3}{3} + \frac{32x^5}{5} - \frac{128x^7}{7} + \dots$
**9.** $-\frac{x^2}{2!} + \frac{x^4}{4!} - \frac{x^6}{6!} + \dots$

**10.** $-x + \frac{x^3}{3!} - \frac{x^5}{5!} + \dots$  **11.** $1 - \frac{x^2}{2!} + \frac{x^4}{4!} - \frac{x^6}{6!} + \dots$

**12.** From Exercise 10, the power series for sin $(-x)$ is the opposite of the power series for sin $x$, that is, sin $(-x) = -\sin x$. Thus, since $f(-x) = -f(x)$, sin $x$ is an odd function.  **13.** From Exercise 11, the power series for cos $(-x)$ is the same as the power series for cos $x$, that is, cos $(-x) = \cos x$. Thus, since $f(-x) = f(x)$, cos $x$ is an even function.  **14.** 2, 1.4, 1.28, 1.256  **15.** 1.25
**16.**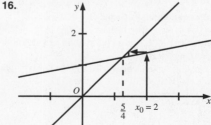

**1.** $-\dfrac{6}{7}$ **2.** $-5x + 3y = 23$ or $5x - 3y = -23$

**3.**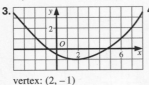
vertex: $(2, -1)$

**4.**
$y = x^2 - 1$  $(1, 0)$  $y = 2x - 2$

**5.** 3 **6.** $\dfrac{1}{6}$ **7.** maximum: $(7, 23)$; minimum: $(0, -47)$ **8.** 2.30 units by

2.67 units **9.** $h = \dfrac{2500}{pr^2}$; domain: $\{r \mid r > 0\}$ **10.** $V = \dfrac{1}{4}s(300 - s^2)$;

domain: $\{s \mid 0 < s < 10\sqrt{3}\}$

**11.**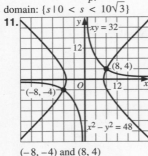
$xy = 32$  $(8, 4)$  $(-8, -4)$  $x^2 - y^2 = 48$
$(-8, -4)$ and $(8, 4)$

**12.**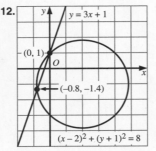
$y = 3x + 1$  $(0, 1)$  $(-0.8, -1.4)$  $(x - 2)^2 + (y + 1)^2 = 8$
$(0, 1)$ and $(-0.8, -1.4)$

**13.**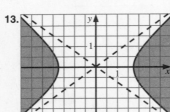

**14.** velocity: $(-1, 6)$;
speed: $\sqrt{37}$ **15.** 174 mi/h N,
479 mi/h W **16.** $\dfrac{7}{10}$ **17.** no
**18.** 2.27 ft/s

# Chapter 20:
# An Introduction to Calculus

## Sections 20-1, 20-2, 20-3, and 20-4

**1. a.** $84x^6 - 15x^4 + 2$ **b.** $\dfrac{4}{3\sqrt[3]{x^2}} - \dfrac{1}{2\sqrt[4]{x^3}}$ **2.** $y - 1 = -\dfrac{1}{4}(x - 4)$

or $x + 4y = 8$ **3.** Graphs will vary, a sample graph is given.

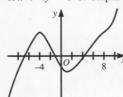

**4.** $(0, 3)$ and $(3, 3)$ are global minima;
$(3, 3)$ is a local minimum; $(1, 7)$ is a
local maximum; $(5, 23)$ is the global
maximum. **5. a.** 1 **b.** $v(t) =$
$2t - 3$ **c.** $t = \dfrac{3}{2}$

**6.** $f'(x) = 12x^3 - 12x^2 = 12x^2(x - 1)$, so $f'(x) = 0$ when $x = 0$
and $x = 1$; thus, the critical points are $(0, 0)$ and $(1, -1)$. Since
$f'(x) < 0$ for $x < 0$ and $0 < x < 1$, the graph of $y = f(x)$ is
decreasing for $x < 0$ and $0 < x < 1$. Since $f'(x) > 0$ for $x > 1$, the
graph of $y = f(x)$ is increasing for $x > 1$.

**7.** 32,000 cubic inches (or 18.5 cubic feet)
**8.** A: iv, B: ii, C: iii

---

# Chapter 20 Review
## Quick Check

**1.** $7\sqrt{3}x^6 + \dfrac{2}{x\sqrt[3]{x}}$ **2.**
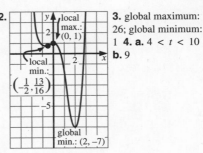
local max.: $(0, 1)$
local min.: $\left(-\dfrac{1}{2}, \dfrac{13}{16}\right)$
global min.: $(2, -7)$

**3.** global maximum:
26; global minimum:
1 **4. a.** $4 < t < 10$
**b.** 9

## Practice Test

**1.** 0 **2.** $-1$ **3.** $-\dfrac{1}{8}$ **4.** $\dfrac{1}{8}$ **5.** $-4x^3 + 12x^2$ **6.** 0, 3 **7.** $(3, 17)$
**8.** 17 **9.** $-15$

**10.**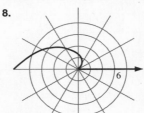

**11.** $A(r) = 2pr^2 + \dfrac{108p}{r}$

**12.** $r = 3$ inches, $h = 6$ inches
**13.** $P(x) = -50 + 125x - 0.4x^{5/2}$
**14.** 25 chairs **15.** $v(t) = -32t + 144$
**16.** 144 ft/s **17.** $t = 4.5$ s **18.** about 9.0 s
**19.** 329 ft

## Mixed Review

**1.**
$\dfrac{1}{4}$  $\dfrac{1}{2}$  $\dfrac{1}{4}$   probability of winning: $\dfrac{9}{16}$
2H  1H, 1T  2T
$\dfrac{1}{2}$  $\dfrac{1}{2}$  $\dfrac{1}{4}$  $\dfrac{1}{2}$  $\dfrac{1}{4}$
1H, 1T  2H  2T  1H, 1T  2H

**2.** Answers will vary, a sample answer is given. Heads-Heads wins
\$2, Heads-Tails loses \$2, and Tails-Tails loses \$6. **3.** about 0.954
square units **4.** The area within two standard deviations, $2s$, of the
mean is about 95% of the total area under the curve. **5.** $x = 1.25$
**6.** $x = 36$ **7.** $5^{12}(\cos 82° + i \sin 82°)$

**8.**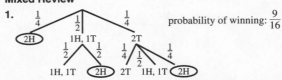

**9.** $X = \begin{bmatrix} -1.4 & 1 \\ 3.8 & 3 \end{bmatrix}$

**10.** $f'(x) = -f'(-x)$
**11.** $f'(x) = f'(-x)$ **12.** $\dfrac{1 \pm i\sqrt{15}}{2}$
**13.** Answers will vary, sample
answer: $f(x) = -x(x - 7)^3$
**14.** 25 ft by 6 ft

**15.** $\{x \mid -6 \le x \le 2\}$;
(number line from $-6$ to $2$)

**16.** $\{x \mid x \le 6$ or $x \ge 10\}$;
(number line from 4 to 12)

**17.**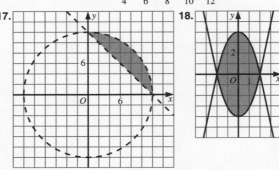

**18.**

**19.** The fixed points $x = \dfrac{1 \pm \sqrt{33}}{2}$ are repellers; all orbits go to
infinity.

---

$(2i)^2 + 5(2i) + 6$

$-4 + 10i + 6$